# Communications in Computer and Information Science

**2472**

Series Editors

Gang Li , *School of Information Technology, Deakin University, Burwood, VIC, Australia*
Joaquim Filipe , *Polytechnic Institute of Setúbal, Setúbal, Portugal*
Zhiwei Xu, *Chinese Academy of Sciences, Beijing, China*

### Rationale

The CCIS series is devoted to the publication of proceedings of computer science conferences. Its aim is to efficiently disseminate original research results in informatics in printed and electronic form. While the focus is on publication of peer-reviewed full papers presenting mature work, inclusion of reviewed short papers reporting on work in progress is welcome, too. Besides globally relevant meetings with internationally representative program committees guaranteeing a strict peer-reviewing and paper selection process, conferences run by societies or of high regional or national relevance are also considered for publication.

### Topics

The topical scope of CCIS spans the entire spectrum of informatics ranging from foundational topics in the theory of computing to information and communications science and technology and a broad variety of interdisciplinary application fields.

### Information for Volume Editors and Authors

Publication in CCIS is free of charge. No royalties are paid, however, we offer registered conference participants temporary free access to the online version of the conference proceedings on SpringerLink (http://link.springer.com) by means of an http referrer from the conference website and/or a number of complimentary printed copies, as specified in the official acceptance email of the event.

CCIS proceedings can be published in time for distribution at conferences or as post-proceedings, and delivered in the form of printed books and/or electronically as USBs and/or e-content licenses for accessing proceedings at SpringerLink. Furthermore, CCIS proceedings are included in the CCIS electronic book series hosted in the SpringerLink digital library at http://link.springer.com/bookseries/7899. Conferences publishing in CCIS are allowed to use Online Conference Service (OCS) for managing the whole proceedings lifecycle (from submission and reviewing to preparing for publication) free of charge.

### Publication process

The language of publication is exclusively English. Authors publishing in CCIS have to sign the Springer CCIS copyright transfer form, however, they are free to use their material published in CCIS for substantially changed, more elaborate subsequent publications elsewhere. For the preparation of the camera-ready papers/files, authors have to strictly adhere to the Springer CCIS Authors' Instructions and are strongly encouraged to use the CCIS LaTeX style files or templates.

### Abstracting/Indexing

CCIS is abstracted/indexed in DBLP, Google Scholar, EI-Compendex, Mathematical Reviews, SCImago, Scopus. CCIS volumes are also submitted for the inclusion in ISI Proceedings.

### How to start

To start the evaluation of your proposal for inclusion in the CCIS series, please send an e-mail to ccis@springer.com

Alexander Dudin · Anatoly Nazarov ·
Alexander Moiseev
Editors

# Information Technologies and Mathematical Modelling

Queueing Theory and Related Fields

23rd International Conference, ITMM 2024
Karshi, Uzbekistan, October 20–26, 2024
Revised Selected Papers

*Editors*
Alexander Dudin 
Belarusian State University
Minsk, Belarus

Anatoly Nazarov 
Tomsk State University
Tomsk, Russia

Alexander Moiseev 
Tomsk State University
Tomsk, Russia

ISSN 1865-0929　　　　　　　ISSN 1865-0937 (electronic)
Communications in Computer and Information Science
ISBN 978-3-031-88306-4　　　ISBN 978-3-031-88307-1 (eBook)
https://doi.org/10.1007/978-3-031-88307-1

© The Editor(s) (if applicable) and The Author(s), under exclusive license
to Springer Nature Switzerland AG 2025

This work is subject to copyright. All rights are solely and exclusively licensed by the Publisher, whether the whole or part of the material is concerned, specifically the rights of translation, reprinting, reuse of illustrations, recitation, broadcasting, reproduction on microfilms or in any other physical way, and transmission or information storage and retrieval, electronic adaptation, computer software, or by similar or dissimilar methodology now known or hereafter developed.
The use of general descriptive names, registered names, trademarks, service marks, etc. in this publication does not imply, even in the absence of a specific statement, that such names are exempt from the relevant protective laws and regulations and therefore free for general use.
The publisher, the authors and the editors are safe to assume that the advice and information in this book are believed to be true and accurate at the date of publication. Neither the publisher nor the authors or the editors give a warranty, expressed or implied, with respect to the material contained herein or for any errors or omissions that may have been made. The publisher remains neutral with regard to jurisdictional claims in published maps and institutional affiliations.

This Springer imprint is published by the registered company Springer Nature Switzerland AG
The registered company address is: Gewerbestrasse 11, 6330 Cham, Switzerland

If disposing of this product, please recycle the paper.

# Preface

The series of scientific conferences Information Technologies and Mathematical Modelling (ITMM) was started in 2002. In 2012, the series acquired an international status, and selected revised papers have been published in *Communications in Computer and Information Science* since 2014. The conference series was named after Alexander Terpugov, one of the first organizers of the conference, an outstanding scientist of the Tomsk State University and a leader of the famous Siberian school on applied probability, queueing theory, and applications.

Traditionally, the conferences have about ten sections in various fields of mathematical modelling and information technologies. Throughout the years, the sections on probabilistic methods and models, queueing theory, and communication networks have been the most popular ones at the conference. These sections gather many scientists from different countries. Many foreign participants come to this Siberian conference every year because of our warm welcome and serious scientific discussions. In 2024, the conference was held in the Republic of Uzbekistan - the city of Karshi. The conference was organized by the National Research Tomsk State University of Russia, Karshi State University of Uzbekistan, Peoples' Friendship University of Russia (RUDN University), Trapeznikov Institute of Control Sciences of the Russian Academy of Sciences, Lobachevsky State University of Nizhni Novgorod, Siberian Federal University of Russia, Baku Engineering University of Azerbaijan, and Shahrisabz State Pedagogical Institute of Uzbekistan.

This volume presents selected papers from the 23th ITMM conference. The conference received 136 submissions, from which 26 were selected to be published in the current collection. The papers passed single-blind peer review and each of them had at least three reviewers.

The papers are devoted to new results in queueing theory and its applications, and also related areas of probabilistic analysis. Its target audience includes specialists in probabilistic theory, random processes, and mathematical modeling as well as engineers engaged in logical and technical design and operational management of data processing systems, communication, and computer networks.

October 2024

Alexander Dudin
Anatoly Nazarov
Alexander Moiseev

# Organization

## International Program Committee Chairs

| | |
|---|---|
| Alexander Dudin | Belarusian State University, Belarus |
| Anatoly Nazarov | Tomsk State University, Russia |
| Shavkat Ayupov | Uzbekistan Academy of Sciences, Uzbekistan |
| Svetlana Moiseeva | Tomsk State University, Russia |
| Azam Imomov | Karshi State University, Uzbekistan |

## International Program Committee

| | |
|---|---|
| Abdurahim Abdushukurov | Romanovsky Institute of Mathematics of the Academy of Sciences of the Republic of Uzbekistan, Uzbekistan |
| Ivan Atencia | University of Málaga, Spain |
| Abdulla Azamov | Romanovsky Institute of Mathematics of the Academy of Sciences of the Republic of Uzbekistan, Uzbekistan |
| Nikolay Borisov | Lobachevsky State University of Nizhni Novgorod, Russia |
| Rui Dinis | Universidade Nova de Lisboa, Portugal |
| Dmitry Efrosinin | Johannes Kepler University Linz, Austria |
| Mais Farhadov | Institute of Control Sciences, Russian Academy of Sciences, Russia |
| Shakir Formanov | Romanovsky Institute of Mathematics of the Academy of Sciences of the Republic of Uzbekistan, Uzbekistan |
| Yulia Gaydamaka | Peoples' Friendship University of Russia (RUDN University), Russia |
| Erol Gelenbe | Institute of Theoretical and Applied Informatics, Polish Academy of Sciences, Poland |
| Tareq Hamadneh | Al-Zaytoonah University, Jordan |
| Yakubzhan Husanbaev | Romanovsky Institute of Mathematics of the Academy of Sciences of the Republic of Uzbekistan, Uzbekistan |
| Uygun Jamilov | Romanovsky Institute of Mathematics of the Academy of Sciences of the Republic of Uzbekistan, Uzbekistan |

| | |
|---|---|
| Bara Kim | Korea University, South Korea |
| Che Soong Kim | Sangji University, South Korea |
| Udo Krieger | Universität Bamberg, Germany |
| B. Krishna Kumar | Anna University, India |
| Achyutha Krishnamoorthy | Cochin University of Science and Technology, India |
| Quan-Lin Li | Yanshan University, China |
| Yury Malinkovsky | Francisk Skorina Gomel State University, Belarus |
| Natalia Markovich | Institute of Control Sciences, Russian Academy of Sciences, Russia |
| Agassi Melikov | National Aviation Academy of Azerbaijan, Azerbaijan |
| Sherzod Mirakhmedov | Romanovsky Institute of Mathematics of the Academy of Sciences of the Republic of Uzbekistan, Uzbekistan |
| Alexander Moiseev | Tomsk State University, Russia |
| Evsey Morozov | Institute of Applied Mathematical Research, Karelian Research Centre of Russian Academy of Sciences, Russia |
| Rein Nobel | Vrije Universiteit Amsterdam, The Netherlands |
| Hamzagha Orujov | Baku Engineering University, Azerbaijan |
| Michele Pagano | Pisa University, Italy |
| Svetlana Paul | Tomsk State University, Russia |
| Tuan Phung-Duc | University of Tsukuba, Japan |
| Gul'nara Raimova | Romanovsky Institute of Mathematics of the Academy of Sciences of the Republic of Uzbekistan, Uzbekistan |
| Jacques Resing | Eindhoven University of Technology, The Netherlands |
| Svetlana Rozhkova | Tomsk Polytechnical University, Russia |
| Alexander Rumyantsev | Institute of Applied Mathematical Research, Karelian Research Centre of Russian Academy of Sciences, Russia |
| Vladimir Rykov | Gubkin Russian State University of Oil and Gas, Russia |
| Konstantin Samouylov | Peoples' Friendship University of Russia (RUDN University), Russia |
| Daria Semenova | Siberian Federal University, Russia |
| Ahmadzhan Soleev | Samarkand State University, Uzbekistan |
| Olimzhon Sharipov | Romanovsky Institute of Mathematics of the Academy of Sciences of the Republic of Uzbekistan, Uzbekistan |
| Stanislav Shidlovskiy | Tomsk State University, Russia |

| | |
|---|---|
| Sergey Suschenko | Tomsk State University, Russia |
| János Sztrik | University of Debrecen, Hungary |
| Henk Tijms | Vrije Universiteit Amsterdam, The Netherlands |
| Gurami Tsitsiashvili | Institute of Applied Mathematics, Far Eastern Branch of Russian Academy of Sciences, Russia |
| Vladimir Vishnevsky | Institute of Control Sciences, Russian Academy of Sciences, Russia |
| Anton Voitishek | Institute of Computational Mathematics and Mathematical Geophysics, Siberian Branch, Russian Academy of Sciences, Russia |
| Alexander Zamyatin | Tomsk State University, Russia |
| Andrey Zorin | Lobachevsky State University of Nizhni Novgorod, Russia |
| Amdjed Zraiqat | Al-Zaytoonah University, Jordan |

## Local Organizing Committee

Ekaterina Fedorova (Chair)
Azam Imomov (Co-chair)
Ivan Lapatin (Co-chair)
Ilkhomjon Bekpulatov
Aleksandr Bulavchuk
Elena Danilyuk
Yana Izmailova
Irina Kochetkova
Danil Korolev
Vladimir Kulikov
Andrey Larionov
Ekaterina Lisovskaya
Olga Lizyura
Alexander Moiseev
Svetlana Moiseeva
Anna Morozova
Misliddin Murtazaev
Zuhriddin Nazarov
Ekaterina Nevenchenko
Ekaterina Pankratova
Svetlana Paul
Vyacheslav Romanov
Daria Semenova
Eduard Sopin
Radmir Salimzyanov
Daria Salimzyanova
Dmitry Shashev
Maria Shklennik
Alexey Shkurkin
Aleksandr Soldatenko
Konstantin Voytikov
Lyubov Zadiranova
Andrey Zorine

# Contents

On the Application of a Queuing Network to Simulate Container
Transshipment in a Sea Container Terminal .............................. 1
   *Alexander Kazakov, Anna Lempert, Giang Vu, and Maxim Zharkov*

On a Limiting Structure of Discrete-Time Stochastic Branching Systems
with the Eventually Awaiting Death Moment ............................. 17
   *Azam A. Imomov and Yorqinoy Ibrohimova*

A Power-Law Adjustable Coefficient Dynamic Pricing Model Considering
Shortages and Leftovers ................................................ 29
   *Anna Kitaeva and Yu Cao*

Inferences of Modularity for Graphs Evolved by the Clustering Attachment
Model .................................................................. 44
   *Natalia Markovich and Maksim Ryzhov*

Queueing Model with Correlated Arrival Process, Simultaneous Service
of a Finite Number of Customers, and Pre-processing, Post-processing
and Co-processing of Customers ......................................... 54
   *Alexander Dudin and Olga Dudina*

Reliability Analysis of a k-out-of-n System with External Service
and Non-preemptive Priority Under N-Policy with Multiple Server
Vacations .............................................................. 70
   *Binumon Joseph and K. P. Jose*

On Recursive Marginal and MAP Inference in State Observation Models ...... 85
   *Branislav Rudić, Valentin Sturm, and Dmitry Efrosinin*

Asymptotic Analysis of Sojourn Time in Retrial Queueing System
with Non-persistent Customers and Feedback ............................. 98
   *Ekaterina Fedorova, Anatoly Nazarov, and Daria Nikolaeva*

Implementation of a Convolution Algorithm to the Evaluation of Stationary
Characteristics of Resource Loss System with Resource-Dependent
Service Times .......................................................... 108
   *Artem Nazarin and Eduard Sopin*

Diffusion Approximation for the MAP/GI/1 Retrial Queue with Two-Way
Communication .................................................... 119
   *Anatoly Nazarov and Olga Lizyura*

Investigation of M/G/1//N System with Impatient Customers, Unreliable
Primary and a Backup Server ........................................ 134
   *Ádám Tóth and János Sztrik*

G-Network with Rewards as a Cluster System Model .................... 146
   *Tatiana Rusilko and Dmitry Salnikov*

On Application of Karamata Slowly Varying Functions in the Theory
of Noncritical Markov Branching Systems ............................. 158
   *Azam A. Imomov and Zuhriddin A. Nazarov*

Asymptotic Analysis of a Multiserver Retrial Queue with Disasters ... 171
   *Natalya Meloshnikova and Ekaterina Fedorova*

Analysis of a Batch Arrival Queue with Power Saving Mode ............ 181
   *Yuta Sakai and Tuan Phung-Duc*

Lumping and Numerical Analysis for Multi-Server Job Model ........... 193
   *Sergey Astafiev*

Unloading Time Martingale Relations in a Cyclic Queueing System
in Random Environment ............................................... 208
   *Andrei V. Zorine*

On Estimation of Some Functional of the Distribution Function
for Dependent Incomplete Observations in Queuing Theory ............. 219
   *Rustamjon S. Muradov and Nurlan T. Dushatov*

Analysis of a Queueing System Providing Service to Regular and Ad Hoc
Clients ............................................................. 230
   *Sergei Dudin, Alexander Dudin, and Olga Dudina*

Modeling and Analysis of Cyclic Control of Periodic Conflict Flows .. 244
   *V. L. Tsodikov and Andrei V. Zorine*

Statistical Testing for Long-Range Dependence in the Workload
of a Single-Server Queue ............................................ 260
   *V. Igolkin and A. Rumyantsev*

Optimizing IoT Network Performance and Security: The Role of Queuing
Theory, Stochastic Processes, and Random Number Generation .............. 273
   *Mirkhon Muhammadovich Nurullaev*

Algorithmic Approach to Study Queueing-Inventory Systems
with Queue-Dependent Hybrid Replenishment Policy ...................... 288
   *Serife Ozkar and Agassi Melikov*

A Stochastic Approach for Optimizing a Discrete Time (s, S) Perishable
Inventory System with Modified N-Policy ................................ 303
   *K. P. Jose, Jijo Joy, and M. P. Anilkumar*

Algorithm for Calculating the Stationary Probability Distribution
of a System $M_2|1|(N_1, N_2)$ with Priorities .............................. 316
   *Natalia Haustova, Svetlana Moiseeva, Ekaterina Pakulova,*
   *and Oybek Khurramov*

Profit Optimization of a Perishable Inventory System with Retrial
of Customers and Unreliable Server ..................................... 327
   *Bobina J. Mattam and K. P. Jose*

**Author Index** ......................................................... 343

# On the Application of a Queuing Network to Simulate Container Transshipment in a Sea Container Terminal

Alexander Kazakov[1], Anna Lempert[1], Giang Vu[2], and Maxim Zharkov[1(✉)]

[1] Matrosov Institute for System Dynamics and Control Theory of Siberian Branch of Russian Academy of Sciences (IDSTU SB RAS), Irkutsk, Russia
zharkm@mail.ru
[2] Irkutsk National Research Technical University, Irkutsk, Russia
http://idstu.irk.ru, https://www.istu.edu/

**Abstract.** The article concerns mathematical modeling and simulation of container handling in a sea container terminal based on queueing theory. The model has the form of a queueing network (QN), and various request flows are applied to describe the arrival of transport from sea and land. QN nodes describe the operation of the terminal structural elements, so they differ in the types of queuing systems used, queue length, number of servicing channels, and their performance. To describe two container movement routes, we use different types of requests with individual route matrices. The object of the study is the largest sea container terminal in the northern part of Vietnam. The QN describing its operation includes 15 nodes, two types of requests, and two incoming flows, one of which is batch. Based on the results of the simulation, the current level of terminal load and its maximum throughput are determined. In addition, we assess the operability of the object when organizing a new route followed by ultra-large container vessels.

**Keywords:** mathematical model · queuing theory · simulation · sea transport · container shipping

## 1 Introduction

Due to the expansion and complexity of the structure of computer and telecommunication systems, queuing theory (QT), as one of the most commonly used research methods in this field, has seen significant development [1]. New types of models have been developed to describe such systems. In particular, queuing systems (QSs) with correlated request flows, QSs operating in a random environment, and retrial QSs appeared [2]. At the same time, similar models have proved applicable to the description of technical systems in other areas, such as the service sector [3], production [4], and inventory management [5]. In the field of transportation, QSs are applied to describe and analyze logistics operations at transport hubs [6] and railway stations [7]. In port logistics, this mathematical tool is considered one of the most suitable for modeling the operation of berths in container terminals [8,9].

Sea container terminals provide the transshipment of containers between sea and land transport, their temporary storage to ensure transport connection, and conduct customs operations. The efficiency of these systems significantly impacts international trade volume and cargo delivery speed since up to 80% of all container shipping occurs by sea (see, https://www.mordorintelligence.com/industry-reports/global-container-shipping-market). The typical structure of container terminals includes three subsystems and two routes between them. Therefore, to consider these properties during modeling, researchers typically utilize queuing networks (QNs), which consist of a finite number of interconnected QSs [1].

Pioneering works on applying QNs in port logistics include [10] and [11]. These studies simulate the transshipment of containers from warehouses to ships and optimize the operation of gantry cranes in the warehouses. There are numerous articles on port systems simulation; we will address only the most relevant ones. Book [12] focuses on the efficiency of various warehouse configurations, while paper [13] discusses the process of transferring containers from ships to rail transport. Articles [14,15] are devoted to applying non-stationary QNs to account for daily fluctuations in the intensity of transport arrivals at the port. Articles [16,17] deal with models of container transportation between a berth and a warehouse on internal trucks and [18] – throughout the entire territory of the sea terminal. The models take into account that the number of such trucks is limited and assume that some requests (containers) come from an external source, while others circulate within the system.

It should be noted that all the works mentioned model specific subsystems of container terminals or consider only one direction of container movement. In the context of the substantial growth in container transportation (see, https://www.mordorintelligence.com/industry-reports/global-container-shipping-market), it is necessary to comprehensively model the operation of sea terminals, encompassing all subsystems, various types of transport, and directions of material flows within the system. This research addresses this pressing issue.

We have experience in utilizing the queueing theory to model systems aggregating different types of transport, whose operation is affected by random factors. In [19], we propose an approach to studying the operation of ground transport systems, and based on it, we construct models of the railway network sections [20,21]. The closest work to the current study is [22], which focuses on modeling a marine coal terminal. Coal arrives there by train and is exported in large batches by ship. The terminal's railway station also services passenger and transit freight trains of different lengths. We employ a QN with group servicing in nodes and a complex request flow consisting of several batch subflows to account for these features.

In this study, we apply the developed methodology for scenario analysis and forecasting a sea container terminal operation. It differs from the coal terminal in that containers arrive from land and sea and depart the system also in both directions. While modeling, these points lead to appearing two independent flows of requests with different parameters. Besides, the work of the berth and the ground cargo yard is described by pairs of QSs that correspond to the processes of unloading and loading of transport vehicles.

The article is organized as follows. Section 2 describes the object of the study. Section 3 presents a structural model of a sea container terminal in the form of a QN, which is then identified for the selected object. In Sect. 4, we conduct a computational experiment to assess the current throughput of the system and forecast its operation in the future.

## 2 Subject Description

This section focuses on the object of study. First, we outline the general characteristics of the system, followed by a description of the container flows arriving at the terminal and its structural elements.

A sea container terminal is a facility that specializes in the temporary storage, handling, and transshipment of containers. Its main tasks are to provide efficient loading and unloading and to adjust discrepancies between the receipt and departure of containers. A typical structure of such a facility includes three main subsystems: a cargo yard for land transport, a container yard, and a berth with quay cranes. Containers move in two directions: vessel – container yard – land transport and back.

This study considers Haiphong International Container Terminal (HICT), a deep-water seaport with a typical structure. It is located in the northern economic region of Vietnam and serves 17 regular international routes. The primary directions are the USA (7), China (5), and India (3). Regarding the number of containers handled per year, HICT ranks first in northern Vietnam and fourth among all systems in the country. Figure 1 shows the HICT location and its general appearance.

### 2.1 Container Flows

The terminal handles loaded and empty 20-foot and 40-foot containers, including refrigerated, which arrive by road and sea transport. The typical unit of measurement for their quantity is TEU (twenty-foot equivalent unit), which corresponds to one 20-foot container. Table 1 presents statistical data on the number of containers processed in 2020–2023 and the forecast for 2024, obtained from open sources (see. https://hict.net.vn/thu-vien/Pages/thu-vien.aspx).

On average, HICT receives 10.2 carriers weekly, with a capacity ranging from 950 to 13000 TEUs. The arriving containers have an average size of 1526.1 TEUs, while the shipped ones are 1600.4 TEUs. Containers also arrive on external trucks with a daily intensity of 1163.5 trucks. In the first five months of 2024, 337260

**Fig. 1.** Haiphong International Container Terminal

**Table 1.** Number of handled containers

| Year | 2020 | 2021 | 2022 | 2023 | 2024 (forecast) |
|---|---|---|---|---|---|
| TEUs (thousand) | 662 | 697 | 1181 | 1273 | 1642 |
| Increase | - | 5% | 70% | 7.8% | 29% |

TEUs (48.8% of the total) arrived from the sea, of which 39% were empty, and 353693 TEUs (51.1%) came from land, of which 8% were empty. These two transport flows are independent and are processed in separate subsystems.

### 2.2 Terminal Structural Elements

There are 6 main structural elements in HICT, which are shown in Fig. 2.

The modeling aims to evaluate the terminal throughput capacity, focusing solely on elements directly involved in container transshipment.

The berth area (No. 1) includes two berths, each with four quay cranes (No. 2). The average loading/unloading speed of each is 30 containers per hour.

The container yard (warehouse No. 3) includes a section for loaded containers with a capacity of 34,560 TEUs and a section for empty ones with a capacity of 9,000 TEUs. The yard employs 24 gantry cranes (eRTG) and four reach stackers, each capable of handling an average of 50 containers per hour.

The track gate (No. 4) has 9 entry and 5 exit lanes, each serving an average of 80 trucks per hour. There is also a site out of the gate that can accommodate up to 48 trucks.

Thirty internal trucks move containers between the berths and the warehouse. The average travel time is five minutes, and the capacity is two TEUs. External trucks transport containers between the gate and the warehouse. Based on satellite imagery analysis, we found that up to 52 cars can be on the territory simultaneously. Their average travel time from the gate to the warehouse is 4 min.

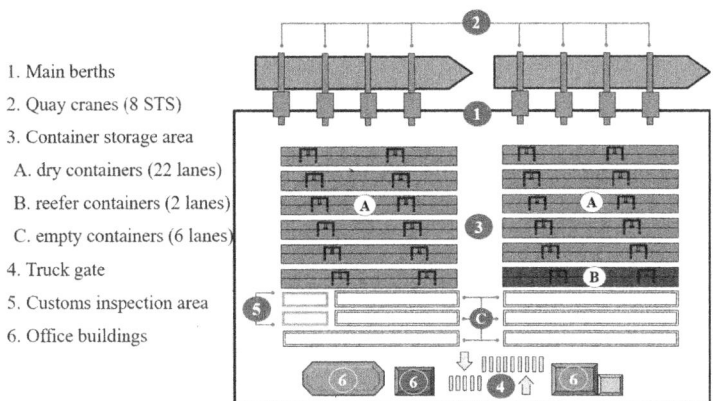

**Fig. 2.** Functional diagram

It is also necessary to distinguish the anchorage located 34 km away in the South China Sea, as the vessels wait there for permission to moor. The travel time from the anchorage to the HICT, including mooring, is 2.3 h on average.

Summarizing above, HICT has three core areas: the berth area, the container yard and the gate, and the anchorage. Containers are delivered by two modes of transport, which are independent and serviced in different subsystems.

## 3 Mathematical Modeling

This section provides the structure of a mathematical model of a sea container terminal. Besides, we conduct its parametric identification for the terminal under consideration.

The model has the form of an open queuing network, which consists of $S$ nodes [1]. Node $y$ is a queueing system that includes $1 \leq n_y < \infty$ homogeneous channels and $0 \leq m_y < \infty$ spots in the queue. The external source is considered an additional (dummy) node. The route of request movement between nodes is given by the route matrix $P$ having dimensions $(S+1) \times (S+1)$.

In a sea container terminal, unlike a coal terminal, the container flow comes both from land and sea and leaves the system after being handled in these two directions. Containers within the terminal move along two oppositely directed routes through the same structural elements. To take these features into account during modeling, we introduce two independent request flows with different parameters. Besides, each subsystem is described by two nodes of various types, including a QS operating in a random environment.

Next, the mathematical model of HICT is constructed in two stages. The first stage characterizes the incoming container flows. The second stage describes the process of their maintenance in subsystems and the movement of individual containers within the system.

**Incoming Flows.** Up to 70% of containers arriving at HICT are 40-foot containers. At the same time, trucks carry either two 20-foot containers or a 40-foot one. We further assume that one request for service is a 40-foot container. Thus, an external truck is considered one request, and a carrier vessel is a group of requests.

To describe the arrival of external trucks and carriers, we apply homogeneous or stationary Poisson point process [17,18], since these vehicles run independently of each other (orderliness and complete independence), and over long periods, the unevenness of the flows can be neglected (stationarity). The container flow generated by land transport is modeled by the stationary Poisson process $M_1$, and the one arising by sea transport is modeled by the group Poisson process $M_2$. Each flow consists of two types of requests. The first one corresponds to a loaded container, and the second type corresponds to an empty one. Based on the results of statistical data analysis (see https://hict.net.vn/thu-vien/Pages/thu-vien.aspx), it was found that the sizes of container groups obey the discrete uniform law $U(a_{\min}, a_{\max})$. The probability of arrival of a group of $i$ requests is

$$p_i = \frac{1}{a_{\max} - a_{\min} + 1},$$

where $a_{\min}$ and $a_{\max}$ correspond to the smallest and largest observed batches of containers.

Table 2 presents the parameters of incoming request flows. In the table, $\lambda_i$ is the intensity of arriving groups of requests per hour, $V_i$ is the distribution law of their sizes, $i = 1, 2$; $p_j$ is the probability that an individual request in a group has type $j$, $j = 1, 2$.

**Table 2.** Incoming Container Flow Models

| $i$ | Flow | Vehicle | $\lambda_i$ | $V_i$ | $p_1$ | $p_2$ |
|---|---|---|---|---|---|---|
| 1 | $M_1$ | External truck | 48,48 | 1 | 0,92 | 0,08 |
| 2 | $M_2$ | Carrier | 0,06 | U(475; 1050) | 0,61 | 0,39 |

**System Operation.** The operation of HICT elements is modeled by 15 nodes of different types. Nodes 0 and 14 are fictitious; they play the role of sources of incoming flows $M_1$ and $M_2$, respectively.

The operation of the gates, container storage, and running of internal and external trucks inside the terminal is modeled as follows. Each element is described by two multi-channel QSs with limited waiting or losses, where the channels are the gate lanes, cranes, or the trucks themselves.

When modeling the berth area, we assume that one (first) berth is intended only for carrier loading and the second one is for unloading. The operation of the first berth is modeled by a multi-channel QS. The channels are quay cranes, and the number of spots in the queue corresponds to the maximum size of arriving batches of containers.

The process of loading containers at the second berth has two peculiarities. Firstly, only the specific number of containers to be loaded is transferred from the container storage to the berth. Secondly, the vessel does not depart until all the assigned containers have been loaded. The queue corresponds to the size of the batch of containers being loaded, so the number of spots changes at the moment when the service of group $X_i$ begins and becomes equal to the size of the next group $X_{i+1}$, which will be accepted for service. The channel displays the running of a loaded vessel along the route to the anchorage, and the carrier itself is considered a queue. Such a system belongs to the class of QSs operating in a random environment [23]. Therefore, service cannot begin until the loading is completed, in other words, until the number of requests in the queue reaches $X_i$. Because of this, the channel is periodically idle, and the actual service intensity becomes less than the specified one (0.06). Preliminary calculations show that the intensity of handling needs to be increased to 0.069 (by 15%) to eliminate this difference.

The anchorage operation is described by a single node, where the size of the groups served corresponds to the maximum carrier capacity, and the number of channels equals the average number of vessels that arrive in the system per week.

When requests arrive in batches, there are different rules, or disciplines, for handling them. The most common is complete rejection discipline: if there is not enough space for even one request from a group, the entire group is lost. If there is a queue in the node, the most natural for transport systems is FIFO (First In, First Out) discipline. If there are no free spots left in the node, and requests from preceding nodes continue to arrive, their channels are temporary blocked.

Table 3 presents the QS nodes in Kendall's notation. Here, $T_y$ is a distribution law of service time, in hours; $X_y$ is a distribution law of the size of the serviced

Table 3. Models of HICT structural elements

| Node | Element | Model | $T_y$ | $X_y$ |
|---|---|---|---|---|
| 1 | Arrival gate | $M_1/M/9/48$ | exp(80) | 1 |
| 2 | Departure gate | $*/M/5/0$ | exp(80) | 1 |
| 3, 4 | External trucks | $*/M/26/0$ | exp(15) | 1 |
| 5 | Yard cranes and reach stackers | $*/M/14/0$ | exp(50) | 1 |
| 6 | Warehouse for empty containers | $*/M/4/4500$ | exp(50) | 1 |
| 7 | Warehouse for laden containers | $*/M/10/17280$ | exp(50) | 1 |
| 8, 9 | Internal trucks | $*/M/15/0$ | exp(12) | 1 |
| 10 | Quay cranes | $*/M/4/0$ | exp(30) | 1 |
| 11 | Berth for unloading ships | $*/M/4/1050$ | exp(30) | 1 |
| 12 | Berth for loading ships | $*/M^X/1/X$ | exp(0.069) | U(475; 1125) |
| 13 | Anchorage | $M_2/G^X/10/0$ | N(2.3; 0.17) | 1050 |

group in the channel of node $y$; $\exp(\lambda)$ is an exponential distribution law with probability density function

$$f_\lambda(t) = \begin{cases} \lambda e^{-\lambda t}, & t \geq 0, \\ 0, & t < 0, \end{cases} \quad (1)$$

$\lambda$ is the intensity; $N(\mu; \sigma)$ – the truncated normal distribution with the probability density function

$$f_{\mu,\sigma}(t) = \begin{cases} 0, & \text{if } t \leq t_a, \\ \dfrac{1}{\sigma} \dfrac{\phi\left(\frac{t-\mu}{\sigma}\right)}{\Phi\left(\frac{t_b-\mu}{\sigma}\right) - \Phi\left(\frac{t_a-\mu}{\sigma}\right)}, & \text{if } t_a < t < t_b, \\ 0, & \text{if } x \geq t_b, \end{cases} \quad (2)$$

$\mu$ is a mathematical expectation, $\sigma$ is a standard deviation of a normal distribution, $t_a = \mu - 3\sigma$, $t_b = \mu + 3\sigma$, $\phi(x) = \frac{1}{\sqrt{2\pi}} \exp\left(-\frac{1}{2}x^2\right)$, $\Phi(x) = \frac{1}{2}\left(1 + \text{erf}(x/\sqrt{2})\right)$.

Figure 3 shows a diagram of the queuing network in the form of a directed graph. The routing matrices $P_1$ and $P_2$ are sparse. Therefore, non-zero transition probabilities are given as graph weights. The elements of $P_1$ are highlighted in blue, and $P_2$ in red.

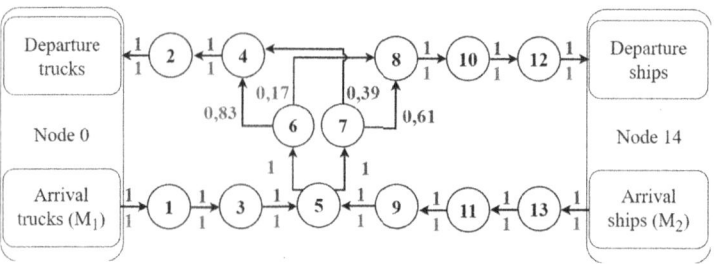

**Fig. 3.** QN diagram

Thus, the mathematical model describing the operation of HICT is a QN with two incoming flows, 15 nodes, and two types of requests. The first incoming flow is the stationary Poisson process. It simulates the arrival of containers on external trucks. The second flow is a batch Poisson flow describing the arrival of containers on vessels. QN nodes differ in type, number of channels, and capacity. Two nodes are fictitious. One is a multi-channel QS with losses and group servicing. It describes the operation of the anchorage. Three nodes are multi-channel QSs with a finite queue. They correspond to the berth for unloaded vessels, the container storage, and the entrance truck gates. Seven nodes are multi-channel QSs with losses. These nodes reflect the operation of quay cranes, the exit gates, and the running of internal and external trucks inside the system. The last node is a QS with a variable number of queue spots. It simulates the loading of the vessel and its departure from the system.

**The purpose of the modeling** is to obtain the performance indicators of the QN, such as the loss probability $P_L$; the average sojourn time in the QN – $T_S$ and separately $\bar{t}_y$ in node $y$; the duration of channel blocking – $\bar{b}_y$; the average number $V_y$ of requests arriving at node $y$, the average number $\bar{K}_y$ of operating channels, and the average length $L_y$ of the queue in node $y$. Next, we evaluate the quality of HICT operation and forecast its behaviour taking into account changes in the volume of traffic flows in the future.

It is possible to find performance indicators analytically only for the simplest types of QNs (see [1]). Currently, the generally accepted practice for determining these indicators is simulation [9,17,18], which helps to construct illustrative models of complex system operations with any level of detail.

## 4 Computational Experiment

In this section, we numerically study the QN obtained using a simulation model. It is based on the discrete-event modeling approach and Monte Carlo methods, and is implemented as software in Object Pascal [22]. The software implemented allows one to determine the performance indicators of a QN. It can have up to 100 nodes, the same number of batch request flows, and up to 10 types of requests with individual route matrices. Besides, modeling of QN nodes with varying queue lengths is acceptable. Three computational experiments are performed using the software.

- The purpose of the first experiment is to analyze the current HICT load.
- The second experiment focuses on the maximum number of containers that the system is able to handle yearly. Two of the most significant cases are considered. In the first case, the size of container lots on carriers increases. In the second case, the intensity of carrier arrivals increases.
- The third experiment investigates the behavior of the HICT when ultra-large container vessels arrive.

Each experiment includes ten program launches. The virtual simulation time for each launch is 365 days. The QN is assumed empty at the initial moment. The average results of ten launches are presented below in tables and figures.

Additionally, we simulated the case when there were $m_6/2$ and $m_7/2$ requests in the queue of Nodes 6 and 7 (the warehouse) at the initial moment. It turned out that the obtained performance indicators differ insignificantly from those for the empty QN.

### 4.1 Experiment 1

Table 4 show the average results of the numerical study. In Table 5 and further, $K_y = \bar{K}_y/n_y$ is a channel occupancy rate; $P_L$, $V_y$, $L_y$, $\bar{t}_y$, and $\bar{b}_y$ mean the same as in paragraph "the purpose of the modeling".

**Table 4.** The results of Experiment 1

| Arrival | $M_1$ | $M_2$ | Loss | $M_1$ | $M_2$ |
|---|---|---|---|---|---|
| Group | 414817 | 521 | Requests | 1164.5 | 0 |
| Requests | 414817 | 397087 | $P_L$ | 0.0014 | |
| | $V_y$ | $K_y$ | $L_y$ | $\bar{l}_y$ (h.) | $\bar{b}_y$ (h.) |
| Node 1 | 413595.0 | 0.07 | 0.16 | 0.02 | 32.21 |
| Node 2 | 400713.0 | 0.12 | - | 0.01 | 0 |
| Node 3 | 413594.8 | 0.13 | - | 0.07 | 33.61 |
| Node 4 | 400714.3 | 0.12 | - | 0.07 | 3.87 |
| Node 5 | 815576.5 | 0.14 | - | 0.02 | 34.87 |
| Node 6 | 193223.7 | 0.55 | 157.68 | 7.24 | 3826.12 |
| Node 7 | 622350.8 | 0.82 | 4225.52 | 59.23 | 5879.47 |
| Node 8 | 412224.5 | 0.84 | - | 0.27 | 5077.92 |
| Node 9 | 401991.3 | 0.26 | - | 0.08 | 29.80 |
| Node 10 | 412209.5 | 0.85 | - | 0.07 | 3992.95 |
| Node 11 | 402285.0 | 0.40 | 190.37 | 4.17 | 74.96 |
| Node 12 | 412205.5 | 0.82 | 586.35 | 26.48 | 0 |
| Node 13 | 402725.7 | 0.02 | - | 3.76 | 68.80 |

The verification of the simulation results is performed as follows. We compare the following samples:

- the calculated number of arrived containers $V_1$ with the volume of container exports, and $V_{13}$ – with the volume of container import in 2024;
- the proportions of containers arriving by sea $V_{13}/(V_1 + V_{13})$ and by land $V_1/(V_1 + V_{13})$ – with statistical data for different modes of transport;
- the ratio $V_6/V_7$ with the ratio of empty and loaded containers.

The largest relative error is 1.57%, and the average one is 0.86%. Such an error is acceptable for solving applied problems in the field of transport.

Let us interpret the simulation results in terms of the object considered and analyze them.

1. On average, 1164.5 (0.14%) of requests from flow $M_1$ are rejected at Node 1 (the gate). At the same time, it has a low load since $K_1 = 0.07 \ll 1$. Therefore, requests are rejected due to the blocking of Node 5, caused by an overflow of Nodes 6 and 7 (the warehouse).
2. The average queue length at Node 6 is $L_6 = 167.98$, and at Node 7 is $L_7 = 3881.79$, which corresponds to 3.7% and 22.5% of the queue capacity of these nodes. Consequently, the warehouse overflow occurs only during peak loads, which happen relatively rarely.
3. The average duration of channel blocking at Node 13 is $\bar{t}_{13} - 2.3 = 1.43$ h, which can be interpreted as the average waiting time at the anchorage.

Thus, HICT effectively copes with the load. In particular, the average waiting time at the anchorage is 1.43 h, and the average storage time of containers in the storage is relatively small – $\bar{t}_6 = 7.24$ and $\bar{t}_7 = 59.23$ h. The number of lost containers from external trucks is insignificant since up to 1500 containers can be kept on the territory of the terminal outside the warehouse.

### 4.2 Experiment 2

**Case 1.** Compared with Experiment 1, we increase the following model parameters by 10%, 20%, 30%, and 40%.

1. the intensity of $M_1$ flow;
2. the average size of the request groups of $M_2$ flow;
3. the average size of the serviced group of requests at Node 12.

The change in the size of the serviced groups in Node 12 occurs due to the fact that the delivery and export of containers in the vast majority of cases is carried out by the same carriers. Table 5 and in Fig. 4 present the simulation results. $V_L$ is the number of rejected requests from the $M_1$ flow.

**Table 5.** The results of Experiment 2, case 1

| No | $M_1 + M_2$ | $V_L$ | $P_L$ | $T_S$ | $L_6$ | $L_7$ |
|---|---|---|---|---|---|---|
| 1. 10% | 895554.5 | 1934.8 | 0.0022 | 69.4 | 169.2 | 5027.5 |
| 2. 20% | 969333.0 | 4483.8 | 0.0046 | 74.6 | 188.0 | 5968.9 |
| 3. 30% | 1056243.2 | 10638.2 | 0.0101 | 82.1 | 184.9 | 7239.7 |
| 4. 40% | 1147448.9 | 22548.6 | 0.0197 | 88.4 | 182.2 | 8278.0 |

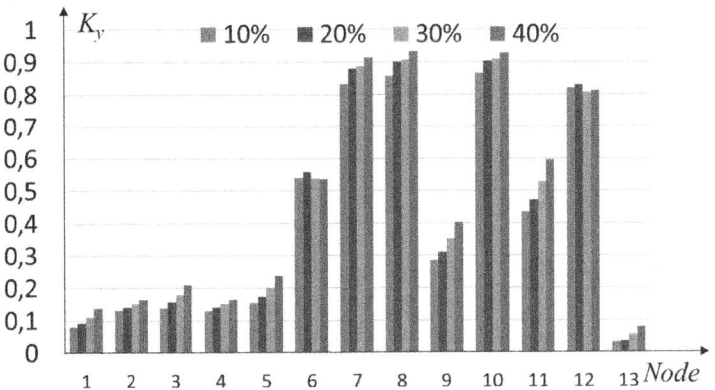

**Fig. 4.** Experiment 2, case 1. The channel occupancy rate ($K_y$)

**Case 2.** Compared with Experiment 1, we increased by 10%, 20%, 30%, and 40% the intensity of servicing for both flows and at Node 12. The increase in the intensity of service is explained by the growing number of ships arriving and, consequently, departing from HICT per year. Table 6 and in Fig. 5 present the simulation results.

Table 6. The results of Experiment 2, case 2

| No | $M_1 + M_2$ | $V_L$ | $P_L$ | $T_S$ | $L_6$ | $L_7$ |
|---|---|---|---|---|---|---|
| 1. 10% | 905970.6 | 1959.4 | 0.0022 | 67.9 | 156.1 | 5127.0 |
| 2. 20% | 981807.7 | 4859.7 | 0.0049 | 73.5 | 154.7 | 6249.4 |
| 3. 30% | 1054240.4 | 8532.2 | 0.0081 | 82.1 | 150.3 | 7771.9 |
| 4. 40% | 1143965.7 | 22573.0 | 0.0197 | 89.9 | 128.2 | 9263.7 |

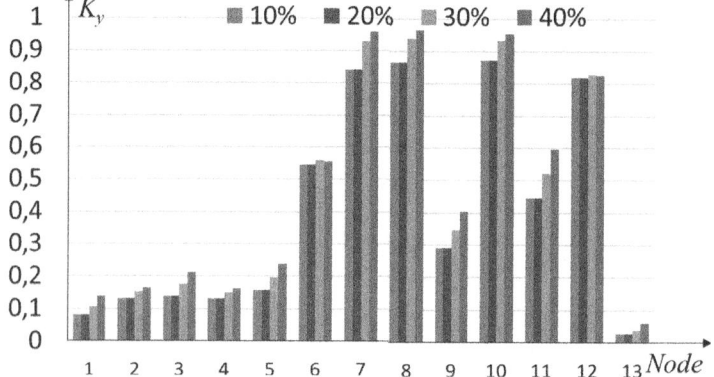

Fig. 5. Experiment 2, case 2. The channel occupancy rate ($K_y$)

Let us interpret the simulation results of Experiment 2. Tables 5 and 6 show that when the volume of container flows increases by 40% (line 4), the system in both cases stops coping with the load. The average number of rejected containers per year for two experiments is 22560.8 (1.97%), which is critical. With an increase in load by 30%, on average 9585.2 (0.91%) containers will not be serviced. According to experts, up to 8–9 thousand containers can be stored outside and handled during periods of low system load. As a result, the volume of transshipped containers can be increased to 1.96–2 million TEUs yearly, 20% more than the volume of incoming containers expected in 2024. However, this requires a clear and smooth operation of HICT, including the timely arrival and departure of carriers.

Let us pay attention to Figs. 4 and 5. The highest channel occupancy rates are observed at Nodes 7 (the warehouse), 8 (inner trucks), and 10 (quay cranes). Moreover, Nodes 7 and 8 have the longest blocking duration, which is taken into account when calculating $K_y$. Therefore, the obtained values of $K_7$ and $K_8$ follow from the high load of Node 10. Hence, the quay cranes should be considered the limiting element. An increase in the number of containers by 30% and 40% leads to $K_8 > 0.9$, which indicates a high load on internal trucks. Their number will be insufficient with further growth of container flows in the future.

### 4.3 Experiment 3

Let us now consider HICT operation if it receives 12 and 15 ultra-large container vessels (ULCVs) yearly (see https://hict.net.vn/en/about/Pages/history.aspx). We assume that such a vessel carries an average of 5000 containers [18].

Compared with Experiment 1, we change the model as follows.

1. We add a new request flow $M_3$, which describes the arrival of ULCVs. The size of request groups has the distribution U(4500;5500), the intensity of arrival $\lambda_a = 0.0014$ and $\lambda_b = 0.0017$ request groups per hour.
2. The intensity of incoming requests from flow $M_1$ is increased by $5000\lambda_a = 7$ and $5000\lambda_b = 8.5$ per hour to maintain the ratio of the volumes of requests coming from different sources.
3. The size of the service group at Node 13 and the queue capacity in Node 11 are increased to 5500.
4. At Node 12, the average size of the serviced batch of requests is increased by 12% and corresponds to the average size of the arriving batches of requests from flows $M_2$ and $M_3$. The service intensity is also increased by $\lambda_a$ and $\lambda_b$.

These changes are a result of the overall number of carriers leaving HICT increasing due to new routes for ULCVs, which can carry more containers than the others. Table 7 presents the simulation results.

**Table 7.** The results of Experiment 3

| No | $M_1 + M_2 + M_3$ | $V_L$ | $P_L$ | $T_S$ | $L_6$ | $L_7$ |
|---|---|---|---|---|---|---|
| 1. $\lambda_a$ | 920561.7 | 3818.0 | 0.0041 | 77.4 | 172.5 | 5816.6 |
| 2. $\lambda_b$ | 958748.6 | 11455.9 | 0.0119 | 85.7 | 176.1 | 6782.9 |

The results show that with the arrival of 12 ULCVs per year (line 1 in Table 7), on average 3818 (0.41%) containers are not accepted for servicing. This value does not significantly impact the system operation due to the possibility of keeping containers directly on the territory of the sea terminal and processing them when the system resources are freed. A further increase in the intensity of arrival of such vessels is heavy to maintain since HICT can only operate effectively without peak loads. In particular, with the arrival of 15 ULCVs per year, the warehouse should be loaded by no more than 40–50% immediately before the arrival of such a vessel. This can be achieved by forced waiting at the anchorage, which leads to financial losses due to the downtime of ships.

Summing up, it should be noted that HICT currently has a capacity reserve that allows handling up to 2 million TEUs per year, as well as timely servicing of up to 12 ULCVs per year upon the introduction of the new route. Further growth in container handling volumes will require upgrading the system. It is recommended to increase the number of quay cranes from 8 to 10, and the number of internal trucks up to at least 50 units to ensure uninterrupted operation of these cranes.

## 5 Conclusion

The article focuses on mathematical modeling and simulation of the operation of sea container terminals. It introduces a mathematical model for container transshipment in these systems, which is based on a specific type of queueing network. The model incorporates a node representing a QS operating in a random environment and allows different request types with individual route matrices. As a result, the model describes the operating parameters of various structural elements of the system, takes into account the arrival and departure of large batches of containers on carriers, including their different sizes, and also allows considering two oppositely directed routes of requests that move through the same terminal subsystems.

The constructed model has been identified for the international container terminal HICT, one of the largest in the Asia-Pacific region, which has a typical configuration. Based on the results of the simulation, the current and maximum permissible system loads have been estimated. Besides, we have forecasted its operation if it is included in the routes of ultra-large container vessels. As a result, we have formulated reasonable recommendations to increase the fleet of transshipment equipment.

Further research in this direction can be aimed at developing the proposed approach, in particular, by taking into account the non-stationarity of transport entering the system. It also seems interesting to adapt and apply the presented tools for modeling other types of sea terminals: oil, grain, and universal ports.

This research was supported by the Russian Science Foundation (project No. 24-21-00264).

## References

1. Bolch, G., Greiner, S., de Meer, H., Trivedi, K.S.: Queueing Networks and Markov Chains: Modeling and Performance Evaluation with Computer Science Applications. Wiley, New York (2006). https://doi.org/10.1002/0471791571
2. Dudin, A., Klimenok, V., Vishnevsky, V.: The Theory of Queuing Systems with Correlated Flows. Springer, Cham (2019)
3. Nazarov, A., Moiseev, A., Moiseeva, S.: Mathematical model of call center in the form of multi-server queueing system. Mathematics **9**(22), 2877 (2021). https://doi.org/10.3390/math9222877

4. Chiacchio, F., Oliveri, L., Khodayee, S.M., D'Urso, D.: Performance analysis of a repairable production line using a hybrid dependability queueing model based on Monte Carlo simulation. Appl. Sci. **13**(1), 271 (2023). https://doi.org/10.3390/app13010271
5. Seyedhoseini, S.M., Rashid, R., Kamalpour, I., Zangeneh, E.: Application of queuing theory in inventory systems with substitution flexibility. J. Ind. Eng. Int. **11**, 37–44 (2015). https://doi.org/10.1007/s40092-015-0099-5
6. Kazakov, A., Vu, G., Zharkov, M.: A stochastic model of a passenger transport hub operation based on queueing networks. In: Dudin, A., Nazarov, A. (eds.) ITMM 2023. CCIS, vol. 2163, pp. 48–62. Springer, Cham (2024). https://doi.org/10.1007/978-3-031-65385-8_4
7. Bychkov, I., Kazakov, A., Lempert, A., Zharkov, M.: Modeling of railway stations based on queuing networks. Appl. Sci. **11**, 2425 (2021). https://doi.org/10.3390/app11052425
8. Edmond, E.D., Maggs, R.P.: How useful are queue models in port investment decisions for container berths? J. Oper. Res. Soc. **29**, 741–750 (1978)
9. Kozan, E.: Comparison of analytical and simulation planning models of seaport container terminals. Transp. Plan. Technol. **20**(3), 235–248 (1997). https://doi.org/10.1080/03081069708717591
10. Canonaco, P., Legato, P., Mazza, R.M., Musmanno, R.: A queuing network model for the management of berth crane operations. Comput. Oper. Res. **35**(8), 2432–2446 (2008). https://doi.org/10.1016/j.cor.2006.12.001
11. Legato, P., Canonaco, P., Mazza, R.M.: Yard Crane management by simulation and optimisation. Maritime Econ. Logist. **11**(1), 36–57 (2009). https://doi.org/10.1057/mel.2008.23
12. Roy, D., de Koster, R.: Optimal stack layout configurations at automated container terminals using queuing network models. In: Bose, J.W. (eds.) Handbook of Terminal Planning. Operations Research/Computer Science Interfaces Series. Springer, Cham (2020). https://doi.org/10.1007/978-3-030-39990-0_19
13. Karla, B., Svjetlana, H., Mirano, H.: Capacity utilization of the container terminal as multiphase service system. Eur. Transp./Trasporti Europei. **86**, 1–15 (2022). https://doi.org/10.48295/ET.2022.86.4
14. Chen, G., Govindan, K., Golias, M.M.: Reducing truck emissions at container terminals in a low carbon economy: proposal of a queueing-based bi-objective model for optimizing truck arrival pattern. Transp. Res. Part E: Logist. Transp. Rev. **55**, 3–22 (2013). https://doi.org/10.1016/j.tre.2013.03.008
15. Zhang, X., Zeng, Q., Chen, W.: Optimization model for truck appointment in container terminals. Procedia. Soc. Behav. Sci. **96**, 1938–1947 (2013). https://doi.org/10.1016/j.sbspro.2013.08.219
16. Legato, P., Mazza, R.M.: Queueing analysis for operations modeling in port logistics. Maritime Bus. Rev. **5**(1), 67–83 (2020). https://doi.org/10.1108/MABR-09-2019-0035
17. Legato, P., Mazza, R.M.: Queueing networks for supporting container storage and retrieval. Maritime Bus. Rev. **8**(4), 301–317 (2023). https://doi.org/10.1108/MABR-01-2023-0009
18. Roy, D., van Ommeren, J.-K., de Koster, R., Gharehgozli, A.: Modeling landside container terminal queues: exact analysis and approximations. Transp. Res. Part B: Methodol. **162**, 73–102 (2022). https://doi.org/10.1016/j.trb.2022.05.012
19. Bychkov, I.V., Kazakov, A.L., Lempert, A.A., Bukharov, D.S., Stolbov, A.B.: An intelligent management system for the development of a regional transport logistics

infrastructure. Autom. Remote Control **77**(2), 332–343 (2016). https://doi.org/10.1134/S0005117916020090
20. Kazakov, A., Lempert, A., Zharkov, M.: Modeling a section of a single-track railway network based on queuing networks. In: Dudin, A., Nazarov, A. (eds.) ITMM 2022. CCIS, vol. 1803, pp. 266–277. Springer, Cham (2023). https://doi.org/10.1007/978-3-031-32990-6_4
21. Kazakov, A., Lempert, A., Zharkov, M.: An approach to railway network sections modeling based on queuing networks. J. Rail Transp. Plan. Manag. **27**, 100404 (2023). https://doi.org/10.1016/j.jrtpm.2023.100404
22. Kazakov, A., Lempert, A., Zharkov, M.: Modeling of a coal transshipment complex based on a queuing network. Appl. Sci. **14**, 6970 (2024). https://doi.org/10.3390/app14166970
23. Polin, E.P., Moiseeva, S.P., Moiseev, A.N.: Heterogeneous queueing system with Markov renewal arrivals and service times dependent on states of arrival process. Discret. Continuous Models Appl. Comput. Sci. **31**(2), 105–119 (2023). https://doi.org/10.22363/2658-4670-2023-31-2-105-119

# On a Limiting Structure of Discrete-Time Stochastic Branching Systems with the Eventually Awaiting Death Moment

Azam A. Imomov(✉) and Yorqinoy Ibrohimova

Karshi State University, Karshi 180100, Uzbekistan
imomov_azam@mail.ru

**Abstract.** We examine the population growth model called the Galton-Watson Branching system. This forms a discrete-time Markov chain with state space in the set of non-negative integers. The system under consideration is a special case of a more general stochastic system called the Bellman-Harris branching system, which has a stepwise distribution function of particle lifetimes. We are interested in the limits on the probability of population distribution along the trajectories awaiting its death moment.

**Keywords:** Branching system · Markov chain · generating function · transition probabilities · extinction time · invariant distribution

## 1 Introduction and Backgrounds

The stochastic branching system is one of the important and descriptive mathematical models describing population growth. These systems have been instrumental in describing systems ranging from population dynamics to data structures. A discrete-time stochastic branching system evolves in generations, with each individual producing offspring independently, according to a fixed probability distribution. The origin of mathematical models of the theory of branching systems is due to the prospect of interest in estimating the survival probability of family trees of individuals. The simplest branching model was initiated by the English statisticians Henry Watson and Francis Galton in the second half of the 19th century. Studying statistics on the reproductive rates of English peers, they have offered a mathematical model that describes population family growth. Their model is now called the Galton-Watson Branching (GWB) system. By present time numerous branching schemes have been developed depending on a context of considering problems, which are necessary to be solved; see [1–3, 7, 16, 19–21] and [30].

The modern framework of the theory of branching systems is heavily influenced by the theory of Karamata regularly varying functions. Zolotarev [33] was one of the first to demonstrate the potential of Karamata functions in studying the properties of branching systems in continuous-time cases. The main advantage of regularly varying functions lies in their specific properties, which allow for the investigation of the asymptotic behavior of the trajectories in systems where the law of particle reproduction exhibits unknown variations. Subsequently, the authors of works [11–14, 23–25, 28] and [29] successfully applied the properties of Karamata functions, thereby revealing deeper insights into the branching behavior of various models.

In contrast to classical models like the GWB system, this study focuses on systems where extinction is guaranteed, either due to subcritical reproduction rates or imposed constraints. Understanding the limiting structure of such systems provides valuable theoretical insights and practical implications, particularly in fields such as genetics, epidemiology, and ecological modeling.

Consider the set of natural numbers $\mathbb{N}$ and $\mathbb{N}_0 = \{0\} \cup \mathbb{N}$. Let $Z(n)$ be a population size at the moment $n$ in the GWB system. An evolution of the individuals' population occurs according to the branching rates law $\{p_k, k \in \mathbb{N}_0\}$. It means, that each individual lives a unit length-lifetime and then produces $k \in \mathbb{N}_0$ descendants with probability $p_k$. Newborn individuals undergo the same process of evolving under the same scheme. In the paper, we consider the system $\{Z(n)\}$ in the Schröder case, i.e.

$$p_0 > 0, \; p_1 > 0 \quad \text{and} \quad p_0 + p_1 < 1.$$

The stochastic system $\{Z(n), n \in \mathbb{N}_0\}$ is a reducible, homogeneous-discrete-time Markov chain with a state space consisting of two classes: $\mathcal{S}_0 = \{0\} \cup \mathcal{S}$, where $\{0\}$ is absorbing state, and $\mathcal{S} \subset \mathbb{N}$ is the class of possible essential communicating states.

The system under consideration is a special case of a more general stochastic system called the Bellman-Harris branching system, which has a stepwise distribution function of particle lifetimes; see [4–6, 17, 18, 31, 32].

Let

$$P_{ij}(n) := \mathbb{P}\left\{Z(n+k) = j \mid Z(k) = i\right\} \qquad \text{for any} \quad k \in \mathbb{N}_0$$

be the $n$-step transition probabilities of the chain. Putting $\mathsf{p}_k(n) := P_{1k}(n)$, and using the Kolmogorov-Chapman equation, we observe that the probability-generating function (GF)

$$\sum_{j \in \mathcal{S}_0} P_{ij}(n) s^j = \left[f(n; s)\right]^i, \tag{1.1}$$

where the GF $f(n; s) = \sum_{k \in \mathcal{S}_0} \mathsf{p}_k(n) s^k$ is $n$-fold iteration of the offspring GF

$$f(s) := \sum_{k \in \mathcal{S}_0} p_k s^k.$$

Throughout the paper symbols $\mathbb{P}_i\{\cdot\}$ and $\mathbb{E}_i[\cdot]$ always stand for the probability and mean functions provided that the system is initiated by $i$ individual-founder, i.e. $\mathbb{P}\{Z(0) = i\} = 1$. Therewith we will write $\mathbb{P}\{\cdot\}$ and $\mathbb{E}[\cdot]$ instead of $\mathbb{P}_1\{\cdot\}$ and $\mathbb{E}_1[\cdot]$ respectively.

We assume that
$$m := \sum_{k \in \mathcal{S}} k p_k < \infty.$$

Then $m = f'(1-)$ is the mean per-capita offspring number, which regulates the classification of $\mathcal{S}$. In fact, it follows from relation (1.1), that

$$\sum_{j \in \mathcal{S}} j P_{ij}(n) s^{j-1} = i f^{i-1}(n;s) \frac{\partial f(n;s)}{\partial s}. \quad (1.2)$$

Taking $s = 1$ in (1.2), we have $\mathbb{E}_i[Z(n)] = i\mathbb{E}[Z(n)]$, where

$$\mathbb{E}[Z(n)] = m^n.$$

This means that the mathematical expectation of $Z(n)$ asymptotically behaves differently depending on the value of the parameter $m$. So, the state space of the chain $\{Z(n)\}$ is classified as subcritical, critical and supercritical if $m < 1$, $m = 1$ and $m > 1$ respectively. Evidently that $f(n;0) = \mathsf{p}_0(n)$ is a vanishing probability of the system at the moment $n$ initiated by a single founder. This probability approaches monotonously to a finite limit $q$ as $n \to \infty$, which is called the extinction probability of the system, i.e. $\lim_{n\to\infty} \mathsf{p}_0(n) = q$. This probability is the smallest nonnegative root of the fixed-point equation $f(s) = s$ on the domain of $\{s : s \in [0,1]\}$, therewith $q = 1$ if $m \leq 1$ and $q < 1$ if $m > 1$. Moreover $f(n;s) \to q$ as $n \to \infty$ uniformly in $s \in [0,r]$ for any $r < 1$; see [2, Ch.I, §§1–5]. Secondly, it can be found an asymptotic expansion of the function $\partial f(n;s)/\partial s$ as $n \to \infty$. Let

$$\mathcal{U}_q[0,1) = \Big\{[0,q) \cup (q,1)\Big\}$$

be a unit interval with a punctured point $q$. It was proved in [8] that

$$\frac{\partial f(n;s)}{\partial s} = \frac{\mathcal{A}_q^2(s)}{(q-s)^2} \beta^n \big(1 + r_n(s)\big), \quad (1.3)$$

provided that $m \neq 1$ and $f''(1-) < \infty$ for $m < 1$, where $\beta = f'(q)$, and

$$\frac{1}{\mathcal{A}_q(s)} = \frac{1}{q-s} + \gamma_q, \quad (1.4)$$

at that

$$\gamma_q = \frac{f''(q)}{2\beta(1-\beta)},$$

and $r_n(s) \to 0$ as $n \to \infty$ uniformly in $s \in \mathcal{U}_q[0,1)$. This equation provides essential information about the asymptotic behavior of noncritical GWB systems. To confirm this, we note that as $s = 0$ in (1.3) it immediately yields the following limiting representation for the local probabilities:

$$\frac{\mathsf{p}_1(n)}{\beta^n} = \frac{1}{q^2} \mathcal{K}_q^2 \big(1 + o(1)\big) \quad \text{as} \quad n \to \infty, \quad (1.5)$$

where $\mathcal{K}_q$ is called the extended Kolmogorov constant and

$$\mathcal{K}_q = \mathcal{A}_q(0) = \frac{q}{1+q\gamma_q}; \qquad (1.6)$$

see [8] and [9].

Let the random variable $\mathcal{H}$ be the time of extinction of the direct line of descendants of one individual. Then it can be written as

$$\mathcal{H} = \min\{n \in \mathbb{N} : Z(n) = 0\}$$

provided that the system is initiated by a single founder individual. The event $\{n < \mathcal{H} < k\}$ means that the system survives at moment $n$, but at the same time expects its death in the next $k$ steps. The classical extinction theorem asserts that $\mathbb{P}_i\{\mathcal{H} < \infty\} = q^i$.

In this report, we observe asymptotic properties of trajectories of noncritical, $m \neq 1$, GWB system on the event $\{n < \mathcal{H} < k\}$. We use the following conditioned probability measure:

$$\mathbb{P}_i^{\mathcal{H}(n,k)}\{\cdot\} := \mathbb{P}_i\left\{\cdot\,\middle|\,n < \mathcal{H} < k\right\} \quad \text{and} \quad \mathbb{P}_i^{\mathcal{H}(n)}\{\cdot\} := \mathbb{P}_i\left\{\cdot\,\middle|\,n < \mathcal{H} < \infty\right\}$$

for any $n, k \in \mathbb{N}$. The first function denotes the probability distribution of the system awaiting its death in the last $k$ steps. The second one is the probability distribution of a system that will survive at the current moment but will eventually die.

## 2  Invariant Measure

In this step, we are interested in invariant properties of the state space $\mathcal{S}$ of GWB system. Recall that the set of nonnegative numbers $\{\mu_k, k \in \mathbb{N}\}$ is called an invariant measure with respect to transition probabilities $P_{ij}(\cdot)$, if it satisfied to the functional equation $\mu_j = \sum_i \mu_i P_{ij}(1)$. If $\sum_j \mu_j < \infty$, then it is called an invariant distribution. The monotone ratio limit property, proved in [2, Ch.I, §7], asserts that

$$\frac{\mathsf{p}_j(n)}{\mathsf{p}_1(n)} \uparrow \pi_j < \infty \qquad \text{as} \quad n \to \infty, \qquad (2.1)$$

and an appropriate GF $\mathcal{P}(s) := \sum_{j \in \mathcal{S}} \pi_j s^j$ converges for $s \in [0,1)$ and it satisfies to the following Abel functional equation

$$\mathcal{P}(f(n;s)) = \beta^n \mathcal{P}(s) + \frac{1-\beta^n}{1-\beta}\mathcal{P}(p_0) \qquad (2.2)$$

for all $n \in \mathbb{N}$. This equation expresses the invariance property of the positive numbers $\{\pi_j, j \in \mathcal{S}\}$ defined by monotone ratio limit property (2.1) with respect to the transition probabilities $P_{ij}(t)$.

In what follows we consider the noncritical case, i.e. $m \neq 1$.

*Remark 1.* It was stated in the monograph [2, Ch.I, §7, Th 2] that the series $\sum_{j\in S} \pi_j$ converges if $m < 1$ and diverges if $m > 1$. However, the statement about the divergence of this series in the case of $m > 1$ is incorrect. Faulty thinking led to this incorrect conclusion during the proof. Below we will prove that without any assumption $\sum_{j\in S} \pi_j$ converges in the case of $m > 1$, and under extra conditions, we find an upper bound depending on the extended Kolmogorov constant for this series.

**Theorem 1.** *If the system is supercritical, $m > 1$ then the series $\sum_{j\in S} \pi_j$ converges. Moreover, if $f''(1-) < \infty$ for $m < 1$ then*

$$\sum_{j\in S} \pi_j \leq \frac{q^3}{\mathcal{K}_q^2} < \infty.$$

*Proof.* Let us recall the discussions of the proof in [2, Ch.I, §7, Th 2] in which erroneous conclusions led to incorrect statements. First, putting $s = 1$ in (2.2), we obtain that

$$\sum_{j\in S} \pi_j = \frac{1}{1-\beta} \mathcal{P}(p_0).$$

Secondly, since $f(n;s) \to q$ as $n \to \infty$ uniformly in $s \in [0,r]$, the same formula entails the following relation:

$$\sum_{j\in S} \pi_j q^j = \frac{1}{1-\beta} \mathcal{P}(p_0).$$

Then

$$\sum_{j\in S} \pi_j = \mathcal{P}(q). \qquad (2.3)$$

Since $\sum_{j\in S} \pi_j s^j$ converges for $s \in [0,1)$, it is undoubtedly $\mathcal{P}(q) < \infty$.

Now, let $f''(1-) < \infty$ for $m < 1$. The Eq. (2.2) entails

$$\beta^n \pi_j = \sum_{k\in S} \pi_k P_{kj}(n)$$

for all $n \in \mathbb{N}$ and $j \in S$. Since $\pi_1 = 1$, then

$$\beta^n = \beta^n \pi_1 = \sum_{k\in S} \pi_k P_{k1}(n).$$

For $s = 0$, the formula (1.2) immediately implies that

$$P_{k1}(n) = k \mathsf{p}_0^{k-1}(n) \mathsf{p}_1(n).$$

Then, it follows that

$$\beta^n = \sum_{k\in S} \pi_k k \mathsf{p}_0^{k-1}(n) \mathsf{p}_1(n) = \mathsf{p}_1(n) \mathcal{P}'(\mathsf{p}_0(n))$$

$$> \frac{\mathsf{p}_1(n)}{\mathsf{p}_0(n)} \sum_{k\in S} \pi_k \mathsf{p}_0^k(n) = \frac{\mathsf{p}_1(n)}{\mathsf{p}_0(n)} \mathcal{P}(\mathsf{p}_0(n)).$$

Therefore
$$\frac{1}{\mathsf{p}_0(n)}\mathcal{P}(\mathsf{p}_0(n)) < \mathcal{P}'(\mathsf{p}_0(n)),$$
where
$$\mathcal{P}'(\mathsf{p}_0(n)) = \frac{\beta^n}{\mathsf{p}_1(n)}$$
for all $n \in \mathbb{N}$. Since $\lim_{n\to\infty}\mathsf{p}_0(n) = q$, taking the limit at $n \to \infty$ in both of the last two relations yields the following ones, respectively:
$$\mathcal{P}(q) \le q\mathcal{P}'(q), \tag{2.4}$$
and
$$\mathcal{P}'(q) = \frac{q^2}{\mathcal{K}_q^2}. \tag{2.5}$$
In obtaining the last statement we used the relation (1.5). Combining now relations (2.3)–(2.5) we obtain that
$$\sum_{j \in \mathcal{S}} \pi_j \le \frac{q^3}{\mathcal{K}_q^2} < \infty.$$

The theorem is proved.

Another invariant distribution can be constructed as follows. If $p_1 > 0$ and $q > 0$, then for all $j \in \mathcal{S}$ and $k \in \mathbb{N}$
$$\nu_j(k) := \lim_{n \to \infty} \mathbb{P}^{\mathcal{H}(n+k)}\{Z(n) = j\} = \frac{q^j - f^j(k;0)}{\beta^k \sum_{k \in \mathcal{S}} \pi_k q^k} \pi_j \tag{2.6}$$
is a probability function. For $\nu_j := \nu_j(0)$, the GF
$$\mathcal{V}(s) := \sum_{j \in \mathcal{S}} \nu_j s^j$$
has the form of $\mathcal{V}(s) = \mathcal{P}(qs)/\mathcal{P}(q)$; see [2, Ch.I, §8,§14].

**Theorem 2.** *The set of numbers $\{\nu_j(k), j \in \mathcal{S}\}$ forms an invariant distribution with respect to the probabilities $P_{ij}(\cdot)$, and for an appropriate GF $\mathcal{V}_k(s) := \sum_{j \in \mathcal{S}} \nu_j(k) s^j$ the following relation holds:*
$$\beta^k \mathcal{V}_k(s) = \mathcal{V}(s) - \mathcal{V}\left(\frac{\mathsf{p}_0(k)}{q}s\right), \tag{2.7}$$
*where $\mathcal{V}(s) = \sum_{j \in \mathcal{S}} \nu_j s^j$. If $f''(1-) < \infty$ for $m < 1$, then*
$$\mathcal{V}(s) = \frac{s}{1 + (1-s)q\gamma_q} \tag{2.8}$$
*and $\gamma_q = f''(q)/2\beta(1-\beta)$.*

*Proof.* Consider
$$\mathcal{V}_k(n;s) := \sum_{j \in \mathcal{S}} \mathbb{P}^{\mathcal{H}(n+k)}\{Z(n)=j\}s^j$$
and denoting $f_q(s) := f(qs)/q$, we rewrite
$$\mathcal{V}_k(n;s) = \frac{f(n+k,qs) - f(n+k,0)}{q - f(n+k,0)}$$
$$= 1 - \mathcal{G}(n; f_q(k,s)) \frac{1 - f_q(n,0)}{1 - f_q(n+k,0)},$$
where
$$\mathcal{G}(n;x) = \frac{1 - f_q(n,x)}{1 - f_q(n,0)}$$
and $f_q(n,s)$ is the $n$-fold iteration of $f_q(s)$. We note that $f'_q(1) = \beta < 1$, i.e. $f_q(s)$ is the GF of subcritical GW system. From the arguments in [2, Ch.I, §8, Th 1] it follows that there exists a limit function
$$\mathcal{G}(x) := \lim_{n \to \infty} \mathcal{G}(n;x)$$
such that $\mathcal{G}(f(x)) = \beta \mathcal{G}(x)$, iteration of this produces
$$\mathcal{G}(f_q(k,s)) = \beta^k \mathcal{G}(x). \tag{2.9}$$

On the other hand, it is easy to check that
$$\frac{1 - f_q(n,0)}{1 - f_q(n+k,0)} \longrightarrow \frac{1}{\beta^k} \quad \text{as} \quad n \to \infty.$$

Then
$$\mathcal{V}_k(s) = \lim_{n \to \infty} \mathcal{V}_k(n;s) = 1 - \mathcal{G}(f_q(k,s)) \frac{1}{\beta^k}. \tag{2.10}$$

A combination of (2.9) and (2.10) implies $\mathcal{V}_k(s) = 1 - \mathcal{G}(s)$ and hence
$$\mathcal{V}_k(f_q(s)) = 1 - \beta \mathcal{G}(s).$$
This leads to the Abel functional equation
$$1 - \mathcal{V}_k(f_q(s)) = \beta \cdot (1 - \mathcal{V}_k(s)), \tag{2.11}$$
which expresses the invariance property of the limiting-measure $\{\nu_j(k), j \in \mathcal{S}\}$ with respect to the probabilities $P_{ij}(\cdot)$. Setting now $s = 1$ in (2.11), we immediately obtain that
$$\sum_{j \in \mathcal{S}} \nu_j(k) = 1.$$

To prove (2.7), we rewrite (2.6) as follows:

$$\nu_j(k) = \frac{\nu_j}{\beta^k}\left[1 - \frac{p_0^j(k)}{q^j}\right]. \qquad (2.12)$$

where

$$\nu_j = \frac{\pi_j q^j}{\sum_{k \in \mathcal{S}} \pi_k q^k}.$$

Multiplying both sides of equality (2.12) by $s^j$ and summing over $j \in \mathcal{S}$, we obtain Eq. (2.7), where $\mathcal{V}(s) = \sum_{j \in \mathcal{S}} \nu_j s^j$. The form of (2.8) is in [8].

The theorem is proved.

Since the event $\{n < \mathcal{H} < n+k\}$ is complementary to $\{n+k < \mathcal{H} < \infty\}$, the following result is a logical continuation of the construction of the invariant distribution $\{\nu_j(k), j \in \mathcal{S}\}$ mentioned above.

**Theorem 3.** *There exists* $\lim_{n \to \infty} \mathbb{P}^{\mathcal{H}(n,n+k)}\{Z_n = j\} =: \sigma_j(k)$ *and*

$$\sigma_j(k) = \frac{\pi_j p_0^j(k)}{\sum_{i \in \mathcal{S}} \pi_i p_0^i(k)}.$$

*An appropriate GF* $\mathcal{C}_k(s) := \sum_{j \in \mathcal{S}} \sigma_j(k) s^j$ *has the form of*

$$\mathcal{C}_k(s) = \frac{\mathcal{P}(p_0(k)s)}{\mathcal{P}(p_0(k))}$$

*for any* $k \in \mathbb{N}$. *Moreover*, $\mathcal{C}_k(s) \to \mathcal{V}(s)$ *as* $k \to \infty$.

*Proof.* First, the conditional probability $\mathbb{P}^{\mathcal{H}(n,n+k)}\{Z_n = j\}$ we represent in the following form:

$$\mathbb{P}^{\mathcal{H}(n,n+k)}\{Z_n = j\} = \frac{\mathbb{P}\{Z_n = j, Z_n > 0, Z_{n+k} = 0\}}{\mathbb{P}\{Z_n > 0, Z_{n+k} = 0\}}$$

$$= \frac{\mathsf{P}_j(n) P_{j0}(k)}{\sum_{i \in \mathcal{S}} \mathsf{P}_i(n) P_{i0}(k)} = \frac{\mathsf{P}_j(n) \mathsf{p}_0^j(k)}{\sum_{i \in \mathcal{S}} \mathsf{P}_i(n) \mathsf{p}_0^i(k)}.$$

We divide the denominator and numerator of the last expression by the function $\mathsf{p}_1(n)$. Then letting $n \to \infty$, by the monotone ratio limit property (2.1), it follows that

$$\mathbb{P}^{\mathcal{H}(n,n+k)}\{Z_n = j\} \longrightarrow \sigma_j(k) \quad \text{as} \quad n \to \infty.$$

Now we will write out the appropriate GF:

$$\mathcal{C}_k(s) = \sum_{j \in \mathcal{S}} \sigma_j(k) s^j$$

$$= \sum_{j \in \mathcal{S}} \frac{\pi_j \mathsf{p}_0^j(k)}{\sum_{i \in \mathcal{S}} \pi_i \mathsf{p}_0^i(k)} s^j = \frac{\sum_{j \in \mathcal{S}} \pi_j \mathsf{p}_0^j(k) s^j}{\sum_{i \in \mathcal{S}} \pi_i \mathsf{p}_0^i(k)} = \frac{\mathcal{P}(\mathsf{p}_0(k)s)}{\mathcal{P}(\mathsf{p}_0(k))}.$$

The theorem is proved.

## 3 Conclusion

The paper considers discrete-time Galton-Watson branching systems. We study the asymptotic behavior of the transition probabilities of these systems. We observe the properties of the population size distribution in the system on positive trajectories. We present a detailed analysis of discrete-time stochastic branching systems with eventual extinction. One of the interesting results of this paper is Theorem 1, which corrects an error missed in the classical monograph by K.Athreya and P.Ney [2, Ch.I, §7, Th 2]. Recall that in this monograph the statement of the above-mentioned theorem was confirmed incorrectly.

By focusing on their limiting structures and pathways to extinction, we provide a framework for understanding the dynamics of collapsing populations. Future research could extend this work by:

1. Incorporating spatial dynamics to study extinction in distributed systems.
2. Exploring stochastic control mechanisms to delay extinction.
3. Applying the findings to more complex branching systems, such as age-dependent or continuous-time processes.

This framework offers valuable insights into the dynamics of systems constrained by eventual collapse, contributing to both theoretical and applied fields.

## A  Appendix (Some Delving into the Theory)

In the classical branching systems, individuals reproduce independently, and their offspring form the next generation. However, real-world populations often experience immigration, where new individuals are added to the population at each generation. So we can delve into the limiting behavior of a more complex stochastic process, called a discrete-time branching system allowing immigration. Based on the need to solve real problems, it often becomes necessary to investigate the conditions under which such systems converge to a stationary distribution and to analyze the properties of this limiting distribution. Papers [10,15,22,26,27] provide important results on the structural and limiting properties of branching systems allowing immigration in both discrete and continuous-time cases.

The discrete-time case model description is as follows. Let $X(n)$ denote the population size at generation $n$. The evolution of the system is governed by the following recurrent equation:

$$X(n+1) = \sum_{i=1}^{X(n)} \zeta(n,i) + I(n),$$

where

- $\zeta(n,i)$ represents the number of offspring of the $i$-th individual in the $n$-th generation.

- $I(n)$ denotes the number of immigrants arriving in the $n$-th generation.

It is assumed that

- Independence: The offspring numbers $\zeta(n,i)$ and the number of immigrants $I(n)$ are mutually independent.
- Stationarity: The offspring distribution and the immigration distribution remain constant over time.

The long-term behavior of the system $X(n)$ depends on the reproductive rate and the immigration rate.

- **Subcritical Case**: If the expected number of offspring per individual is less than one, the population will eventually die out, regardless of the immigration rate.
- **Critical Case**: If the expected number of offspring per individual is equal to one, the long-term behavior depends on the immigration rate. If the immigration rate is positive, the population will eventually stabilize to a non-degenerate stationary distribution.
- **Supercritical Case**: If the expected number of offspring per individual is greater than one, the population will grow exponentially, and the immigration process will have a negligible effect on long-term behavior.

In the critical case with positive immigration, the system converges to an invariant distribution. This distribution can be characterized by its probability generating function (PGF), denoted by $G(s)$. The PGF satisfies the Schröder-type functional equation. The invariant distribution can be obtained by solving this functional equation. Analytical solutions are often challenging, but numerical methods can be used to approximate the distribution.

Branching systems allowing immigration have numerous applications in various fields, including:

- **Population Biology**: Modeling the dynamics of populations with immigration, such as fish populations in a lake with regular stocking.
- **Epidemiology**: Studying the spread of infectious diseases with immigration of infected individuals.
- **Queueing Theory**: Analyzing queueing systems with customer arrivals and departures.
- **Finance**: Modeling the evolution of stock prices with random shocks.
- **Insurance**: Pricing insurance products and managing risk portfolios.

Thus, discrete-time stochastic branching systems allowing immigration provide a powerful framework for modeling the dynamics of populations with both internal growth and external input. The limiting behavior of these processes is influenced by the reproductive rate and the immigration rate. In the critical case with positive immigration, the process converges to a stationary distribution, which can be characterized by its probability generating function. Understanding the properties of this limiting distribution is crucial for various applications in diverse fields.

Further research directions are

- **Non-stationary Immigration**: Investigate the case where the immigration rate varies over time.
- **Dependent Offspring**: Relax the assumption of independent offspring and consider models with correlated offspring distributions.
- **Continuous-Time Models**: Extend the analysis to continuous-time branching processes with immigration.
- **Applications in Specific Domains**: Explore specific applications, such as epidemiology, ecology, and finance, and develop tailored models.

By delving deeper into these areas, researchers can gain valuable insights into the complex dynamics of real-world populations.

Beyond the limiting distribution, understanding the convergence rate of this distribution is crucial. Various techniques, such as coupling methods and martingale theory, can be employed to derive explicit bounds on the rate of convergence. These bounds provide quantitative information about how quickly the process approaches its stationary state. Moreover, the long-time behavior of the process can be characterized by various metrics, such as the expected population size, variance, and higher-order moments. Asymptotic expansions for these quantities can be derived using techniques like generating function analysis and Tauberian theorems.

In certain scenarios, the system may exhibit phase transitions as parameters, such as the reproduction rate or the immigration rate, are varied. These phase transitions can lead to dramatic changes in long-term behavior, including shifts from extinction to survival or from stable to oscillatory dynamics.

# References

1. Asmussen, S., Hering, H.: Branching Processes. Birkhäuser, Boston (1983)
2. Athreya, K.B., Ney, P.E.: Branching Processes. Springer, New York (1972)
3. Bingham, N., Goldie, C., Teugels, J.: Regular Variation. Cambridge (1987)
4. Bellman, R., Harris, T.E.: On the theory of age-dependent stochastic branching processes. Proc. Natl. Acad. Sci. USA **34**, 601–604 (1948)
5. Bellman, R., Harris, T.E.: On age-dependent binary branching processes. Ann. Math. **55**, 280–295 (1952)
6. Goldstein, M.I.: Critical age-dependent branching processes: single and multitype. Z. Wahrscheinlichkeitstheorie verw. Geb. **17**, 74–88 (1971)
7. Harris, T.E.: The Theory of Branching Processes. Springer, Berlin (1963)
8. Imomov, A.A., Murtazaev, M.: Renewed limit theorems for noncritical Galton-Watson branching systems. J. Theor. Probab. **37**(3), 2843–2858 (2024)
9. Imomov, A.A., Murtazaev, M.: On the Kolmogorov constant explicit form in the theory of Discrete-time Stochastic Branching Systems. J Appl. Prob. **61**(3), 927–941 (2024)
10. Imomov, A.A., Tukhtaev, E.E.: On asymptotic structure of critical Galton-Watson branching processes allowing immigration with infinite variance. Stoch. Model. **39**(1), 118–140 (2023)
11. Imomov, A.A., Meyliev, A.K.: On the asymptotic structure of non-critical Markov stochastic branching processes with continuous time. Vestn. Tomsk. Gos. Univ. Mat. Mekh. **69**, 22–36 (2021)

12. Imomov, A.A.: On a limit structure of the Galton-Watson branching processes with regularly varying generating functions. Prob. and Math. Stat. **39**(1), 61–73 (2019)
13. Imomov, A.A., Tukhtaev, E.E.: On application of slowly varying functions with remainder in the theory of Galton-Watson branching process. J. Sib. Fed. Univ. Math. Phys. **12**(1), 51–57 (2019)
14. Imomov, A.A.: On conditioned limit structure of the Markov branching process without finite second moment. Malays. J. Math. Sci. **11**(3), 393–422 (2017)
15. Imomov, A.A.: On long-term behavior of continuous-time Markov Branching Processes allowing immigration. J. Sib. Fed. Univ. Math. Phys. **7**(4), 443–454 (2014)
16. Jagers, P.: Branching Processes with Biological Applications. JW & Sons, Pitman Press (1975)
17. Jagers, P.: Age-dependent branching processes allowing immigration. Theory Probab. Appl. **13**, 225–236 (1968)
18. Kaplan, N., Pakes, A.G.: Supercritical age-dependent branching processes with immigration. Stoch. Proc. and their Appl. **2**, 371–389 (1974)
19. Karpenko, A.V., Nagaev, S.V.: Limit theorems for the total number of descendants for the Galton-Watson branching process. Theory Probab. Appl. **38**, 433–455 (1994)
20. Kennedy, D.P.: The Galton-Watson process conditioned on the total progeny. Jour. Appl. Prob. **12**, 800–806 (1975)
21. Haccou, P., Jagers, P., Vatutin, V.: Branching Processes: Variation, Growth, and Extinction of Populations. Cambridge University Press, Cambridge (2007)
22. Li, J., Chen, A., Pakes, A.G.: Asymptotic properties of the Markov branching process with immigration. J. Theor. Probab. **25**, 122–143 (2012)
23. Nagaev, S.V., Wachtel, V.: The critical Galton-Watson process without further power moments. J Appl. Prob. **44**(3), 753–769 (2007)
24. Pakes, A.G.: Critical Markov branching process limit theorems allowing infinite variance. Adv. Appl. Prob. **42**, 460–488 (2010)
25. Pakes, A.G.: Revisiting conditional limit theorems for the mortal simple branching process. Bernoulli **5**(6), 969–998 (1999)
26. Pakes, A.G.: Limit theorems for the simple branching process allowing immigration, I. The case of finite offspring mean. Adv. Appl. Prob. **11**, 31–62 (1979)
27. Pakes, A.G.: Some new limit theorems for critical branching processes allowing immigration. Stoch. Proc. and Appl. **3**, 175–185 (1975)
28. Seneta, E.: Regularly varying functions in the theory of simple branching processes. Adv. Appl. Prob. **6**(3), 408–420 (1974)
29. Slack, R.S.: A branching process with mean one and possible infinite variance. Wahrscheinlichkeitstheor. und Verv. Geb. **9**, 139–145 (1968)
30. Sevastyanov, B.A.: Branching Process. Nauka, Moscow (1971). (Russian)
31. Weiner, H.: Asymptotic probabilities in a critical age-dependent branching process. J Appl. Prob. **9**(4), 476–485 (1966)
32. Weiner, H.: On age-dependent branching processes. J Appl. Prob. **9**(4), 383–402 (1966)
33. Zolotarev, V.M.: More exact statements of several theorems in the theory of branching processes. Theory Probab. Appl. **2**(2), 245–253 (1957)

# A Power-Law Adjustable Coefficient Dynamic Pricing Model Considering Shortages and Leftovers

Anna Kitaeva[✉] and Yu Cao

National Research Tomsk State University, Tomsk, Russia
kit1157@yandex.ru

**Abstract.** We consider a single product during a sales period with a fixed duration and price sensitive intensity of a compound Poisson customer's flow. A model with a power adjustable coefficient of dynamic retail price control through the intensity of the demand is considered accommodating scenarios of shortages and leftovers. We consider the diffusion approximation of the stock level process to find the probabilistic characteristics of the selling process, and a linear approximation of the intensity-of-price dependence is used to find the expected revenue. Numerical examples illustrate the theoretical results.

**Keywords:** Dynamic Pricing · Price Sensitive Demand · Compound Poisson Demand · Diffusion Approximation · Shortages · Leftovers

## 1 Introduction and Problem Statement

The advent and continuous advancement of Information Technologies (IT) has significantly contributed to various fields, leading to transformative changes across industries, such as healthcare, finance, and manufacturing, providing tools for real-time data processing, predictive analytics, and automated decision-making processes, see, e.g., [1]. In finance, algorithmic trading powered by IT has enhanced market efficiency and trading speed [2].

In the realm of Supply Chain Management (SCM), IT has played a pivotal role in addressing market instabilities, optimizing resource allocation, and achieving dynamic economic strategies such as dynamic pricing. The utilization of IT in SCM enables firms to adapt rapidly to changing market conditions, thereby maintaining competitive advantage through responsive pricing mechanisms as discussed in [3].

The rapid growth in demand, coupled with accelerated product life cycles and frequent product updates, has exacerbated market uncertainties [4]. Such uncertainties pose significant risks for supply chain retailers, often resulting in scenarios where products sold out in advance, causing shortages, or conversely, where products remain unsold, leading to excessive inventory and resource wastage.

These challenges are further compounded by the environmental impact associated with resource-intensive production and disposal of surplus inventory.

Dynamic pricing, empowered by IT, allows firms to dynamically adjust prices based on current supply and demand conditions, thus playing a critical role in managing shortages and leftovers more effectively [5]. The use of advanced analytics can also allow retailers to optimize stock levels and set prices dynamically to respond to market uncertainties, providing a significant increase in retail revenues; see, e.g., key paper by Gallego and van Ryzin [6], and [7,8].

Shortages are a widely recognized concern in dynamic pricing research. Generally, it is assumed that unsatisfied demand is completely backlogged or completely lost. Backlog systems have received more attention in literature than lost sales inventory systems, probably due to the complexity of the lost sales systems' analysis, see, [9]. It seems reasonable to argue that settings with lost sales are more sensitive to stockouts than settings with backlogs, the lost sales assumption better models demand elasticity and opportunity costs associated with stockouts, making it a more realistic representation of consumer behavior. Here, we assume that the occurring shortages are lost sales.

Leftovers are also a topic that has received substantial coverage in the field of dynamic pricing. Perishable products deteriorate over time, becoming partially or wholly unsuitable for consumption at the end of their shelf-life; see, for example, [10,11]. At the end of the period, unsold units can be fully recycled, [12]; carried over to the next period for sale at a discounted price, [13]; or eliminated if they are entirely unsuitable for consumption, [14]. Here, we assume that any unsold units will be carried out to secondary market to sell or be disposed.

In [15], a zero-ending inventory price control model with a weight function depending on a power-law coefficient was studied without taking into account the possibility of shortages on practice for large values of the adjustable coefficient. In this paper, we build on previous research by also looking at cases of stockouts and leftovers, that is, we consider scenarios where the product is unavailable when the fixed-duration's sales session is nearing end or is not sold out completely during that session, resulting in stock shortages or excess inventory respectively.

Let us introduce the model's assumptions and notations. We consider a supply chain consisting of a vendor and customers. The vendor is a monopolist and seeks to maximize the revenue, ordering fixed lot $Q_0$ per unit price $d$ and selling it within a fixed period $T$. Replenishment during the period is impossible.

If shortages happen, unmet demand at the end of the period is lost. Leftovers are sold on the secondary market at salvage value per unit $s$, $s < d$ or they are disposed, in this case $s$ takes a negative value.

The demand is modeled as a Poisson process with intensity $\lambda(c)$, where $c = c(t)$ is a dynamic retail price per unit, the orders are independent identically distributed continuous random variables with the first and second moments $a_1$ and $a_2$ respectively.

We are going to consider the following model of expiration date-based price control

$$a_1 \lambda(c(t)) = \frac{Q(t)}{T(1 - t/T)^\gamma}, \qquad (1)$$

where coefficient $\gamma \neq 1$. The idea is the same as in [16], that is, we require that the instantaneous rate of a product's sale at time $t$ (left hand side of the equation) and the average rate at $[0, T-t]$ adjusted by coefficient $\gamma$ be equal to each other.

The probabilistic characteristics of the selling process and the expected revenues are obtained in the framework of a diffusion approximation of the stock level process. By employing the approximation, we are able to derive analytically tractable expressions that allow us to obtain solutions to the problem in a suitable form. Specifically, we assume that the stock level process $Q(\cdot)$ can be approximately described by the following stochastic differential equation:

$$dQ(t) = -a_1 \lambda(c(t))dt + \sqrt{a_2 \lambda(c(t))}dw(t),$$

where $w(\cdot)$ is the Wiener process.

It follows that the stock level process satisfies the following equation:

$$dQ(t) = -\frac{Q(t)}{T(1-t/T)^\gamma}dt + \sqrt{\frac{a_2}{a_1}\frac{Q(t)}{T(1-t/T)^\gamma}}dw(t). \quad (2)$$

Consider the influence of coefficient $\gamma$ on probabilistic characteristics of process $Q(\cdot)$.

## 2 Probabilistic Characters of the Stock Level Process

### 2.1 Expectation and Variance of Process $Q(\cdot)$

Let us denote expectation $E\{Q(t)\} = \overline{Q}(t) = \overline{Q}$. From (2) we have

$$d\overline{Q}(t) = -\frac{\overline{Q}(t)}{T(1-t/T)^\gamma}dt \quad (3)$$

with the initial condition $\overline{Q}(0) = Q_0$.

It follows that $\overline{Q}(t) = Q_0 \exp\left\{\frac{(1-t/T)^{1-\gamma} - 1}{1-\gamma}\right\}$.

Note that $\overline{Q}(T) = 0$ for $\gamma > 1$ and $\overline{Q}(T) = Q_0 \exp\left(-\frac{1}{1-\gamma}\right)$ for $\gamma < 1$. Thus, the average inventory level takes zero value before the end of the period for $\gamma > 1$, and there are leftovers for $\gamma < 1$. Let us denote $\alpha(x) = \exp\left\{\frac{(1-x)^{1-\gamma} - 1}{1-\gamma}\right\}$, $x \in [0, 1)$. Figure 1 shows the plots of function $\alpha(x)$ for $\gamma = 0.3, 0.5, 1, 1.5, 2$.

Denote $E\{Q^2(t)\} = \overline{Q^2}(t) = \overline{Q^2}$. Applying Ito's formula and averaging, we get

$$d(Q^2) = \left(-\frac{2Q^2(t)}{T(1-t/T)^\gamma} + \frac{a_2 Q(t)}{a_1 T(1-t/T)^\gamma}\right)dt + 2Q(t)\sqrt{\frac{a_2 Q(t)}{a_1 T(1-t/T)^\gamma}}dw(t)$$

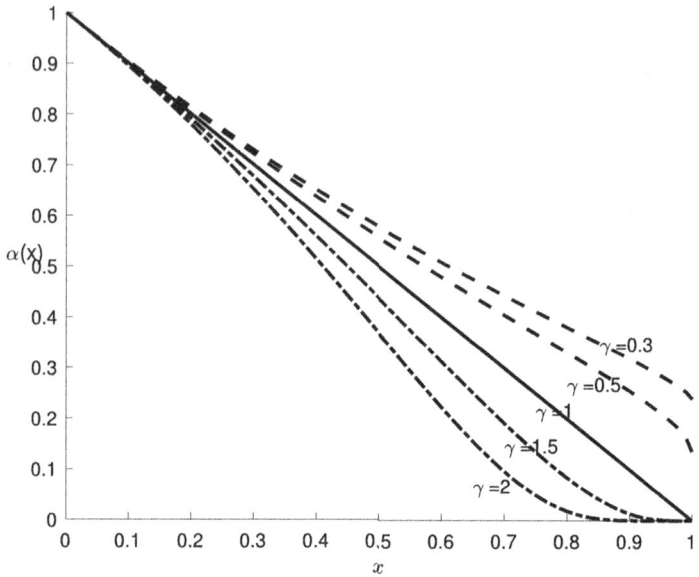

**Fig. 1.** $\overline{Q}(\cdot)/Q_0 = \alpha(\cdot)$ dependence of $t/T$ for $\gamma = 0.3, 0.5, 1, 1.5, 2$.

and

$$\frac{d\overline{Q^2}}{dt} = -2\frac{\overline{Q^2}}{T(1-t/T)^\gamma} + \frac{a_2 Q_0 \alpha(t/T)}{a_1 T(1-t/T)^\gamma} \quad (4)$$

subject to $\overline{Q^2}(0) = Q_0^2$.

Solution of (4) $\overline{Q^2} = Q_0^2 \alpha^2(t/T) + \frac{a_2 Q_0}{a_1}\alpha(t/T)(1 - \alpha(t/T))$. It follows that $Var\{Q(t)\} = V_Q(t) = V_Q = \frac{a_2 Q_0}{a_1}\alpha(t/T)(1 - \alpha(t/T))$. Note that $V_Q(T) = 0$ for $\gamma > 1$ and $V_Q(T) = \frac{a_2 Q_0}{a_1}\exp\left(-\frac{1}{1-\gamma}\right)\left(1 - \exp\left(-\frac{1}{1-\gamma}\right)\right)$ for $\gamma < 1$.

Thus, the lot is completely sold during the period almost surely for $\gamma > 1$, and for $\gamma < 1$, leftovers are obtained; $\overline{Q}(T) = Q_0 \exp\left(-\frac{1}{1-\gamma}\right)$.

## 2.2 Probability Density Function of Process $Q(\cdot)$

Consider the Laplace transform of the probability density function (PDF) of $Q(\cdot)$, $\Phi(p,t) = E\{\exp(-pQ(t))\}$. By applying Ito's formula, we get from (1)

$$d(e^{-pQ}) = \left(\frac{Qpe^{-pQ}}{T(1-t/T)^\gamma} + \frac{a_2 Q p^2 e^{-pQ}}{2a_1 T(1-t/T)^\gamma}\right)dt \\ - pe^{-pQ}\sqrt{\frac{a_2 Q}{a_1 T(1-t/T)^\gamma}}dw(t). \quad (5)$$

After averaging (5) with respect to $Q = Q(\cdot)$, it follows

$$T(1-t/T)^\gamma \frac{\partial \Phi(p,t)}{\partial t} + p\left(1 + \frac{a_2}{2a_1}p\right)\frac{\partial \Phi(p,t)}{\partial p} = 0. \tag{6}$$

Solving (6) by the method of characteristic, we get $\Phi(p,t) = \varphi\left(\frac{p}{p+\beta}\alpha(t/T)\right)$, where $\varphi(\cdot)$ is an unknown function and parameter $\beta = 2a_1/a_2$. From the initial condition it follows that $\Phi(p,0) = \exp(-pQ_0)$, so $\varphi(z) = \exp\left(-\frac{Q_0\beta z}{1-z}\right)$, where $z = p/(p+\beta)$. Finally, we get

$$\Phi(p,t) = \exp\left(-\frac{Q_0\beta p\alpha(t/T)}{p+\beta - p\alpha(t/T)}\right). \tag{7}$$

To test the formula let us find

$$\lim_{\gamma \to 1} \exp\left\{-\frac{Q_0\beta p\alpha(t/T)}{p+\beta - p\alpha(t/T)}\right\} = \exp\left\{-\frac{\beta p(T-t)Q_0}{pt+\beta T}\right\}.$$

The result is the same as in [16].

Using inverse Laplace transform, we obtain PDF of $Q(\cdot)$

$$f(x,t) = \exp\left(-\frac{\beta\alpha Q_0}{1-\alpha}\right)\left\{\delta(x) + \exp\left(-\frac{\beta Q_0}{1-\alpha}\right) \right. \\ \left. \times \sqrt{\frac{\beta^2\alpha Q_0}{(1-\alpha)^2 x}} I_1\left(2\sqrt{\frac{\beta^2\alpha Q_0 x}{(1-\alpha)^2}}\right)\right\}, \tag{8}$$

where $\alpha = \alpha(t/T)$, $I_1(\cdot)$ is the modified Bessel function of the first kind and first order, $\delta(\cdot)$ is the Dirac delta function. The term containing the delta function in (8) arises because at some point of time a customer can buy all inventories on hand.

## 2.3 Selling Duration

From (8) it follows that the cumulative distribution function (CDF) of the length of time $\tau$ it takes to sell lot $Q_0$

$$F_\tau(t) = \exp\left(-\frac{\beta Q_0\alpha(t/T)}{1-\alpha(t/T)}\right). \tag{9}$$

Figure 2 shows the plot of $F_\tau(t)$ for $Q_0/a_1 = 300, a_2/a_1^2 = 4/3, \gamma = 0.8, 1.5, 2, 3$.

From (9) it follows that $F_\tau(T) = 1$ for $\gamma > 1$, and the shortages can occur in practice for large values of $\gamma$.

Expectation of the sales duration

$$E\{\tau\} = \int_0^T (1-F_\tau(t))\,dt$$
$$= T\left(1 - e^{\beta Q_0}\int_0^{1-\alpha(1)} \frac{\exp(-\beta Q_0/x)\left((1-\gamma)\ln(1-x)+1\right)^{\frac{\gamma}{1-\gamma}}}{1-x}dx\right).$$

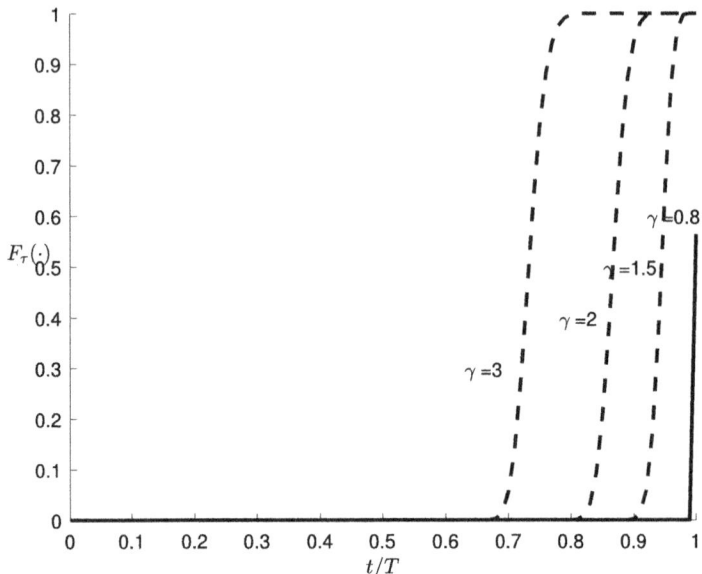

**Fig. 2.** $F_\tau(\cdot)$ dependence on $t/T = x$ for $\gamma = 0.8, 1.5, 2, 3$.

Let us consider the time point $T_1$, $T_1 = Tx_0$, where $x_0$ is the the solution of equation $E\{Q(T_1)\} = Q_0\alpha(x_0) = a_1$, that is, $x_0 = x_0(\gamma, a_1/Q_0) = 1 - [(1-\gamma)\ln(a_1/Q_0) + 1]^{\frac{1}{1-\gamma}}$, as the moment of shortages occurrence. In simulation, we take the smallest value that satisfies $Q(T_1) < a_1$ as $T_1$.

In Table 1 values of $x_0$ for $\gamma = 1.05, 1.2, 1.5, 2$ and $a_1/Q_0 = 0.005, 1/300$, $0.0001, a_1/Q_0 \approx 0$ in MATLAB package are presented. In Fig. 3 plots of $x_0$ dependence on $1 < \gamma \le 2$ for the same values of $a_1/Q_0$ are given; function $x_0(\gamma, a_1/Q_0)$ is monotonically decreasing, and $\lim_{\gamma \to 1+} = 1 - a_1/Q_0$.

**Table 1.** Values of $x_0$ for $\gamma = 1.05, 1.2, 1.5, 2$.

| $a_1/Q_0$ | $\gamma = 1.05$ | $\gamma = 1.2$ | $\gamma = 1.5$ | $\gamma = 2$ |
|---|---|---|---|---|
| 0.005 | 0.9909 | 0.9730 | 0.9249 | 0.8412 |
| 1/300 | 0.9934 | 0.9778 | 0.9326 | 0.8508 |
| 0.0001 | 0.9995 | 0.9944 | 0.9667 | 0.8999 |
| $\approx 0$ (MATLAB) | 0.9999 | 0.9963 | 0.9711 | 0.9078 |

According to the numerical results, for $a_1/Q_0 < 1/300$ and $\gamma = 2$ about 15% of the cycle period can be taken up by the shortage intervals.

In Fig. 4 the results of illustrative simulation of $T_1$ are presented, where $x_0$ satisfies $\alpha(x_0) = a_1/Q_0 = 1/300$. For generating a non-homogeneous Poisson

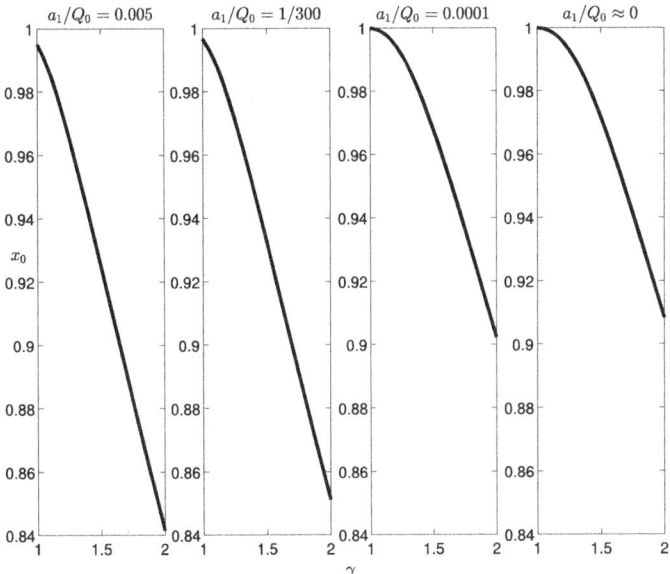

**Fig. 3.** $x_0$ dependence on $1 < \gamma \le 2$ for $a_1/Q_0 = 0.005, 1/300, 0.0001, a_1/Q_0 \approx 0$ in MATLAB.

process, the thinning algorithm is applied. This method is used throughout the rest of the paper as well as the number of simulation iterations being 1000, and we take average values for each variable value. Here $1 < \gamma \le 2$, $T = 10$, $Q_0 = 1500$ and $a_1 = 5$, purchases are uniformly distributed over $(0, 10)$ or exponentially distributed. The black curve is the theoretical result, while the solid red curve represents results of simulation in case of uniformly distributed purchases and the dotted blue curve represents the results in case of exponentially distributed purchases.

## 3 The Expected Revenue and Its Optimization

Let us consider linear approximation of the intensity-of-price dependence

$$\lambda(c) = \lambda_0 - \lambda_1 \frac{c(t) - c_0}{c_0}, \qquad (10)$$

where $c_0$ is a stationary price corresponding stationary intensity $\lambda_0$ and parameter $\lambda_1 > 0$ characterizes the sensitivity of $\lambda(\cdot)$ to the relative price's deviations from the stationary price. Linear dependence of the customers' flow intensity on the price is common in literature; see, for example, [17].

From (1) and (10) we get

$$c(t) = c_0 \left( 1 + \frac{\lambda_0}{\lambda_1} - \frac{Q(t)}{a_1 \lambda_1 T (1 - t/T)^\gamma} \right).$$

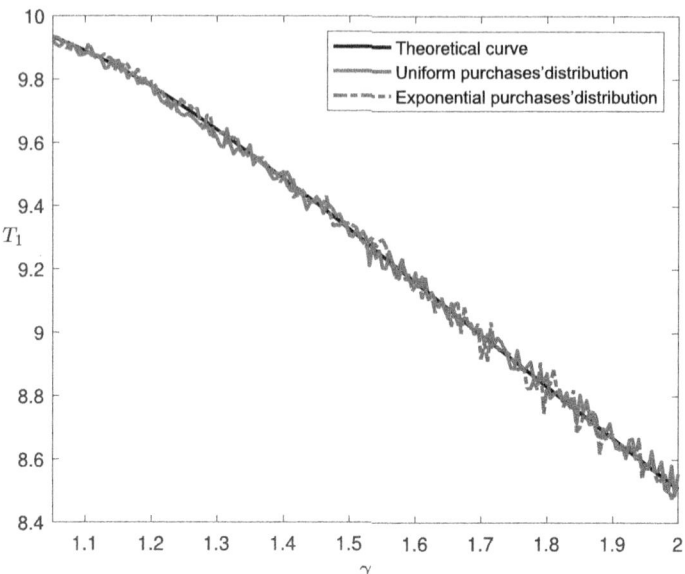

**Fig. 4.** $T_1$ Simulation, $T = 10, Q_0 = 1500$, purchases are Uniform$(0, 10)$ or Exp$(5)$.

The average revenue at time unit

$$E\left\{c(t)a_1\lambda(c)\right\} = c_0 E\left\{\left(1 + \frac{\lambda_0}{\lambda_1} - \frac{1}{a_1\lambda_1}\frac{Q(t)}{T(1-t/T)^\gamma}\right)\frac{Q(t)}{T(1-t/T)^\gamma}\right\}$$

$$= c_0\left(1 + \frac{\lambda_0}{\lambda_1}\right)\frac{\overline{Q}(t)}{T(1-t/T)^\gamma} - \frac{c_0}{a_1\lambda_1}\frac{\overline{Q^2}(t)}{T^2(1-t/T)^{2\gamma}}.$$

Next, we will derive the cycle revenues for shortages and leftovers incorporating corresponding penalties.

### 3.1 The Expected Revenue Considering Shortages

We consider the setting where unsatisfied demand is lost and assume the demand intensity during shortages' period $T - T_1$ is static, and according to $T_1$ definition $\lambda(c(T_1)) \approx \dfrac{1}{T(1-T_1/T)^\gamma}$ and $c(T_1) \approx c_0\left(1 + \dfrac{\lambda_0}{\lambda_1} - \dfrac{1}{\lambda_1 T(1-T_1/T)^\gamma}\right)$.

Let us consider the average shortage penalty as follows

$$\overline{S}_s = -a_1 c(T_1)\lambda(c(T_1))(T - T_1)$$

$$\approx -a_1 c_0\left(1 + \frac{\lambda_0}{\lambda_1} - \frac{1}{\lambda_1 T(1-T_1/T)^\gamma}\right)(1 - T_1/T)^{1-\gamma}.$$

Thus, the expected revenue considering shortages

$$\overline{S}_{\gamma>1} = \int_0^{T_1} E\{c(t)\lambda(t)\}\,\mathrm{d}t + \overline{S}_s = c_0 Q_0 \left\{ 1 + \frac{\lambda_0}{\lambda_1} - \frac{1}{\lambda_1 T} \right.$$
$$\left. \times \int_{1-T_1/T}^{1} \left( \frac{Q_0 a_1 - a_2}{a_1^2} \alpha^2 (1-z) + \frac{a_2}{a_1^2} \alpha(1-z) \right) \frac{\mathrm{d}z}{z^{2\gamma}} \right\} \qquad (11)$$
$$- a_1 c_0 \left( 1 + \frac{\lambda_0}{\lambda_1} - \frac{1}{\lambda_1 T (1-T_1/T)^{\gamma}} \right) (1 - T_1/T)^{1-\gamma}.$$

Weighted expected revenue $\dfrac{\overline{S}_{\gamma>1}}{a_1 c_0}$ depends on four dimensionless system's parameters $Q_0/a_1, \lambda_0/\lambda_1, \lambda_1 T, a_2/a_1^2$, except $\gamma$. In Fig. 5 the results of illustrative simulation of weighted revenue's dependence on $\gamma$ are presented for different sets of the system's parameters. The black curves represent the theoretical results, and the red curves are the simulation results.

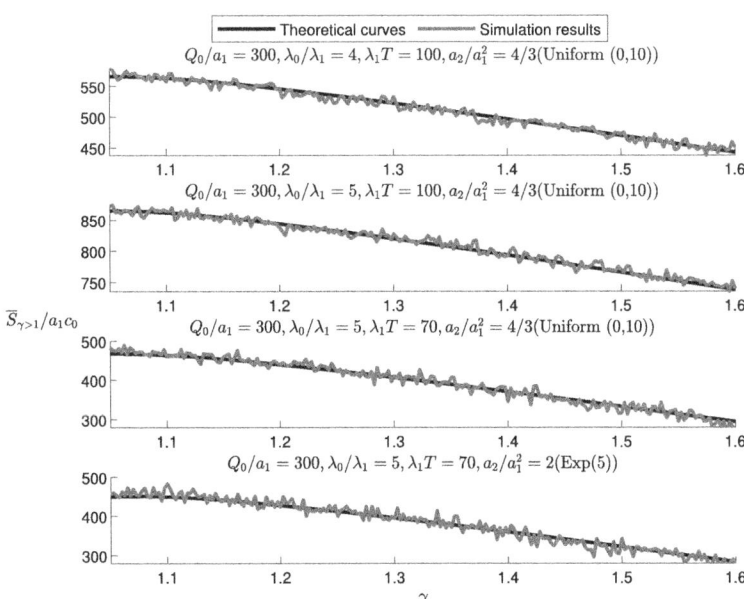

**Fig. 5.** $\dfrac{\overline{S}_{\gamma>1}}{a_1 c_0}$ dependence on $\gamma$ for different sets of the system's parameters.

An increase in ratio $\lambda_0/\lambda_1$, similar to an increase in the coefficient $\lambda_1 T$, leads to an increase in price within the sales period, which leads to a significant increase in the revenue; compare the first three subplots in Fig. 5. Ratio $a_2/a_1^2$ determines the coefficient of variation of purchases. An increase in ratio $a_2/a_1^2$ does not decline the revenue significantly, a higher $a_2/a_1^2$ induce larger dispersion and increase the optimal coefficient $\gamma$ required to achieve maximum revenue; compare the last two subplots in Fig. 5.

The revenue is a concave function with respect to a lot size. Figure 6 depicts weighted revenue $\dfrac{\overline{S}_{\gamma>1}}{a_1 c_0}$ dependence on $Q_0/a_1$ for $\gamma = 1.2, 1.4, 1.6$; $\lambda_0/\lambda_1 = 4$, $\lambda_1 T = 100, a_2/a_1^2 = 4/3$.

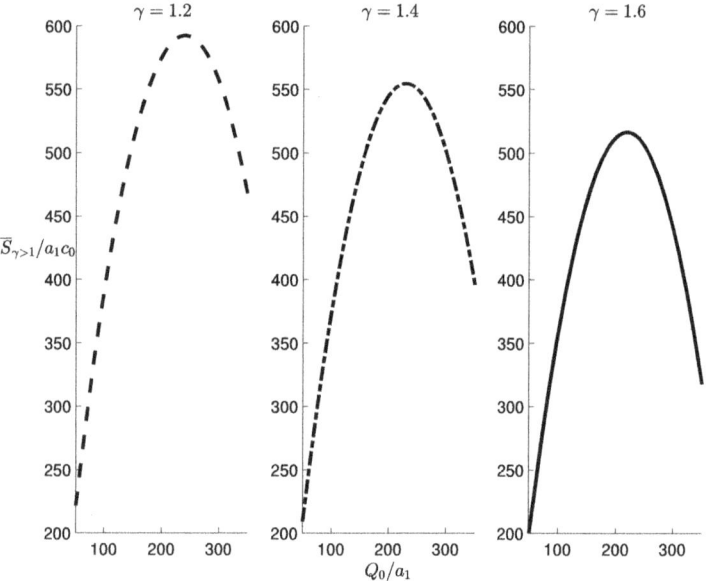

**Fig. 6.** $\dfrac{\overline{S}_{\gamma>1}}{a_1 c_0}$ dependence on $Q_0/a_1$ for $\gamma = 1.2, 1.4, 1.6$.

The large $\gamma$, the less optimal $Q_0/a_1$ and the corresponding maximal revenue due to increasing value of penalties for stockout in the revenue.

For $\gamma$ close to 1 from above, we can find the ratio $Q_0/a_1 = R$ giving maximum of the weighted revenue considering $T_1/T \approx 1$ and neglecting the last term in (11). Let us denote $\dfrac{\overline{S}_{\gamma\to 1+}}{a_1 c_0} = \overline{SW}_{\gamma\to 1+}$, then $\dfrac{\partial \overline{SW}_{\gamma\to 1+}}{\partial R} = 1 + \dfrac{\lambda_0}{\lambda_1} - \dfrac{p_1(2R - a_2/a_1^2) + p_2 a_2/a_1^2}{\lambda_1 T}$, where $p_1 = \displaystyle\int_0^1 \alpha^2(1-z)\dfrac{\mathrm{d}z}{z^{2\gamma}}$, $p_2 = \displaystyle\int_0^1 \alpha(1-z)\dfrac{\mathrm{d}z}{z^{2\gamma}}$.

It follows that $R_{\gamma\to 1+opt} = \dfrac{(1+\lambda_0/\lambda_1)\lambda_1 T + (p_1 - p_2)a_2/a_1^2}{2p_1}$.

$R_{\gamma\to 1+opt}$ for different sets of system's parameters for $\gamma = 1 + 1 \times 10^{-10}$ are shown in Table 2.

In Fig. 7 the dependence of optimal $\gamma$ on $R$ is presented; $\gamma_{opt}$ maximizes $\overline{SW}_{\gamma>1}$ for $\lambda_0/\lambda_1 = 4, \lambda_1 T = 100, a_2/a_1^2 = 4/3$. For $R \gg 1$, $\gamma_{opt}$ is close to 1. In

**Table 2.** $R_{\gamma \to 1+opt}$ for different sets of system's parameters, $\gamma = 1 + 1 \times 10^{-10}$.

| $\lambda_0/\lambda_1$ | 4 | 5 | 5 | 5 |
|---|---|---|---|---|
| $\lambda_1 T$ | 100 | 100 | 70 | 70 |
| $a_2/a_1^2$ | 4/3 | 4/3 | 4/3 | 2 |
| $R_{\gamma \to 1+opt}$ | 225.1 | 275.1 | 185.1 | 172.6 |

case of multiplicative adjustable coefficient, we obtain the same result for large lots, see [18].

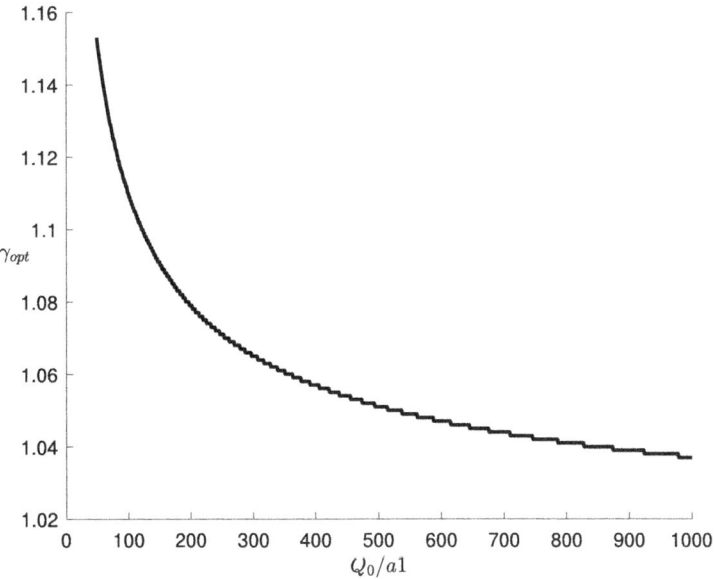

**Fig. 7.** $\gamma_{opt} = \arg\max_{\gamma} \overline{S}_{\gamma>1}$ dependence on $R = Q_0/a_1$; $\lambda_0/\lambda_1 = 4$, $\lambda_1 T = 100$, $a_2/a_1^2 = 4/3$.

### 3.2 The Expected Revenue Considering Leftovers

If $\gamma < 1$ then from (9) it follows $F_\tau(T) = \exp\left(-\dfrac{\beta Q_0 \exp(-1/(1-\gamma))}{1 - \exp(-1/(1-\gamma))}\right) < 1$, leftovers are possible. Expectation of the leftovers $\overline{Q}(T) = Q_0 \exp\left(-\dfrac{1}{1-\gamma}\right)$.

We evaluate the value of unsold products at the end of the sales cycle as follows:
$$\overline{S}_l = \eta d Q_0 \exp\left(-\dfrac{1}{1-\gamma}\right),$$

where coefficient $\eta$ is related to the possible spoilage of the product during the sales period, $\eta \leq 1$; see [19].

Then, the expected revenue considering leftovers

$$\overline{S}_{\gamma<1} = \int_0^T E\left\{c(t)\lambda(t)\right\} dt + \widetilde{S}_l = c_0 Q_0 \left\{1 + \frac{\lambda_0}{\lambda_1} - \frac{1}{\lambda_1 T}\right.$$
$$\left. \times \int_0^1 \left(\frac{Q_0 a_1 - a_2}{a_1^2}\alpha^2(1-z) + \frac{a_2}{a_1^2}\alpha(1-z)\right)\frac{dz}{z^{2\gamma}}\right\} + \eta d Q_0 \exp\left(-\frac{1}{1-\gamma}\right).$$

Weighted revenue $\dfrac{\overline{S}_{\gamma<1}}{a_1 c_0}$ increases monotonically with respect to $\gamma$. The results of the weighted revenues simulation for different sets of the system's parameters are presented in Fig. 8. The black curves represent the theoretical results, and the red curves are the simulation results.

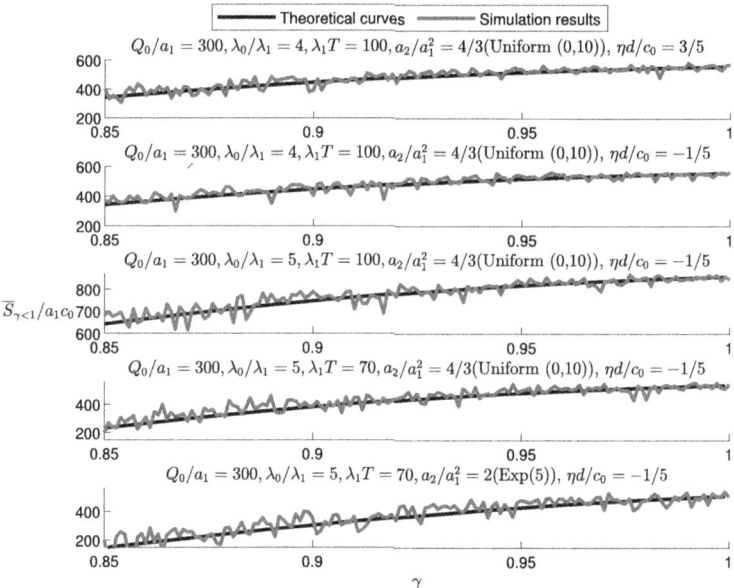

**Fig. 8.** $\dfrac{\overline{S}_{\gamma<1}}{a_1 c_0}$ dependence on $\gamma$ for different sets of the system's parameters.

Similar to the shortages scenario, increasing coefficients $\lambda_0/\lambda_1$ or $\lambda_1 T$ yields a higher demand rate and the corresponding price, thereby enhances revenue; an increasing coefficient $a_2/a_1^2$ leads to the revenue's decreasing and also increases dispersion. Decreasing coefficient $\eta d/c_0$ corresponds to a decrease in the value of leftovers or an increase in handling costs. Both factors contribute to a decrease in revenue, but the decrease is insignificant due to the small quantity of leftovers.

The revenue is concave with respect to the stock level as in case of shortages. Figure 9 depicts weighted revenue $\overline{SW}_{\gamma<1} = \dfrac{\overline{S}_{\gamma<1}}{a_1 c_0}$ dependence on $R = Q_0/a_1$ for $\gamma = 0.85, 0.9, 0.95$; $\lambda_0/\lambda_1 = 4, \lambda_1 T = 100, a_2/a_1^2 = 4/3, \eta d/c_0 = -1/5$.

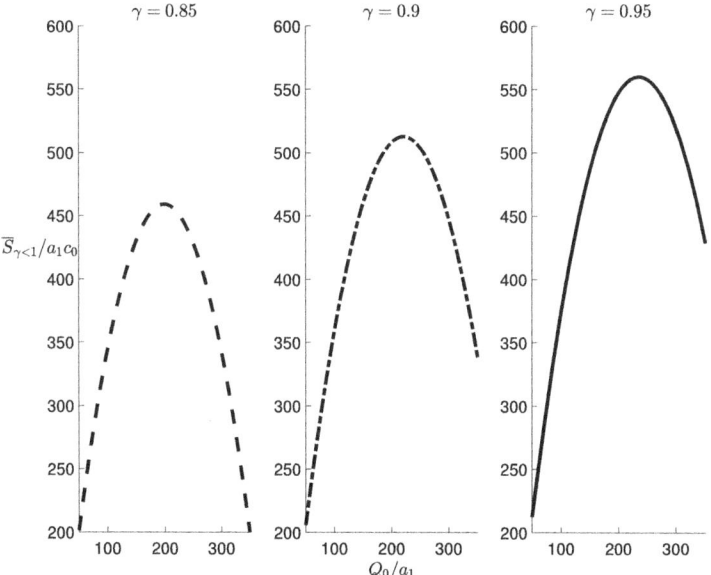

**Fig. 9.** $\dfrac{\overline{S}_{\gamma<1}}{a_1 c_0}$ dependence on $Q_0/a_1$ for $\gamma = 0.85, 0.9, 0.95$.

Here we obtain the same picture as previously, that is, tending $\gamma$ to one we increase the optimal lot size and the corresponding revenue.

We get the following equation for optimal $R = Q_0/a_1$:

$$\dfrac{\partial \overline{SW}_{\gamma<1}}{\partial R} = 1 + \dfrac{\lambda_0}{\lambda_1} - \dfrac{p_1(2R - a_2/a_1^2) + p_2 a_2/a_1^2}{\lambda_1 T} + p_3 = 0,$$

where $p_3 = \dfrac{\eta d}{c_0} \exp\left(-\dfrac{1}{1-\gamma}\right)$. Thus, we get

$$R_{\gamma<1 opt} = \dfrac{(1 + p_3 + \lambda_0/\lambda_1)\lambda_1 T + (p_1 - p_2)a_2/a_1^2}{2p_1}.$$

Note that $R_{\gamma\to 1-opt} = R_{\gamma\to 1+opt}$. In Table 3 $R_{\gamma\to 1-opt}$ for different sets of system's parameters and $\gamma = 1 - 1 \times 10^{-10}$ are given.

**Table 3.** $R_{\gamma\to 1-opt}$ for different sets of system's parameters, $\gamma = 1 - 1 \times 10^{-10}$.

| $\lambda_0/\lambda_1$ | 4 | 5 | 5 | 5 | 5 |
|---|---|---|---|---|---|
| $\lambda_1 T$ | 100 | 100 | 70 | 70 | 70 |
| $a_2/a_1^2$ | 4/3 | 4/3 | 4/3 | 2 | 2 |
| $\eta d/c_0$ | -1/5 | -1/5 | -1/5 | -1/5 | 3/5 |
| $R_{\gamma\to 1-opt}$ | 226.1 | 276.1 | 186.1 | 174.2 | 174.2 |

## 4 Conclusion

In a stable environment, that is, $Q_0/a_1 \approx \lambda_0 T$ the basic model of the intensity of price dependence works well. The dynamic pricing strategy proposed in [16] almost surely compensates for random demand fluctuations described by diffusion process. The models with the linear and power coefficients proposed in [15, 18] can adjust the intensity-of-price dependence to the real demand characteristics.

The basic model and the model with the linear adjustable coefficient guarantee almost surely zero-ending inventory at the end of sales cycle, just like the power-law model with $\gamma > 1$. The last model assumes strong intensity-of-price dependence, that leads to possibility of stockouts for large values of the adjustable coefficient, this needs to be taken into account in practice. Here we propose a heuristic method of stockouts' consideration based on the time when the expected stock level falls below the average purchases' quantity. The results of the sales process simulation show good correspondence between the theoretical and simulated stockouts' intervals, as well as between the corresponding theoretical and simulated revenues. Note that the formula for $R_{\gamma\to 1+opt}$ can be used for the optimal lot size calculation for any $\gamma > 1$, if we do not consider the penalties for stockouts, that is, in the framework of [15].

Large values of the adjustable coefficient can lead to a significant reduction in revenue. In practice, according to numerical results, models with $\gamma > 1.2$ should be used with caution. On the other hand, using the optimal $\gamma > 1$ allows us to increase the revenue compared to the basic model with $\gamma = 1$.

The possibility of leftovers in case of $\gamma < 1$ is also considered in the framework of the linear intensity-of-price dependence approximation; see (10). Based on the results achieved within this approximation for $0.85 < \gamma < 1$, we do not recommend using the power model for $\gamma < 0.95$. It seems that the linear approximation does not work well for small $\gamma$, so we plan to study the case of leftovers in detail without considering this approximation.

## References

1. Singh, P., Khoshaim, L., Nuwisser, B., et al.: How Information Technology (IT) is shaping consumer behavior in the digital age: a systematic review and future research directions. Sustainability **16**(4), 1556 (2024)

2. Oyeniyi, L.-D., Ugochukwu, C.-E., Mhlongo, N.-Z.: Analyzing the impact of algorithmic trading on stock market behavior: a comprehensive review. World J. Ada. Eng. Technol. Sci. **11**(2), 437–453 (2024)
3. Sundram, V.-P.-K., Chhetri, P., Bahrin, A.-S.: The consequences of information technology, information sharing and supply chain integration, towards supply chain performance and firm performance. J. Int. Logist. Trade **18**(1), 15–31 (2020)
4. Hofstra, N., Spiliotopoulou, E., de Leeuw, S.: Ordering decisions under supply uncertainty and inventory record inaccuracy: an experimental investigation. Decis. Sci. **55**(3), 303–318 (2024)
5. Chen, X., Simchi-Levi, D.: Coordinating inventory control and pricing strategies with random demand and fixed ordering cost. Manuf. Serv. Oper. Manag. **5**(1), 59–62 (2003)
6. Gallego, G., van Ryzin, G.: Optimal dynamic pricing of inventories with stochastic demand over finite horizons. Manag. Sci. **40**(8), 999–1020 (1994)
7. Zhao, W., Zheng, Y.-S.: Optimal dynamic pricing for perishable assets with non-homogeneous demand. Manag. Sci. **46**, 375–388 (2000)
8. Sahay, A.: How to reap higher profits with dynamic pricing. MIT Sloan Manag. Rev. **48**(4), 53–60 (2007)
9. Wang, X., Disney, S.-M., Wang, J.: Exploring the oscillatory dynamics of a forbidden returns inventory system. Int. J. Prod. Econ. **147**, 3–12 (2014)
10. Nahmias, S.: Perishable inventory theory: a review. Oper. Res. **30**(4), 680–708 (1982)
11. Wang, C., Chen, X.: Joint order and pricing decisions for fresh produce with put option contracts. J. Oper. Res. Soc. **69**(3), 474–484 (2018)
12. Fu, K., Gong, X., Liang, G.: Managing perishable inventory systems with product returns and remanufacturing. Prod. Oper. Manag. **28**(6), 1366–1386 (2019)
13. Chintapalli, P.: Simultaneous pricing and inventory management of deteriorating perishable products. Ann. Oper. Res. **229**, 287–301 (2015)
14. Santos, M.-C., Agra, A., Poss, M.: Robust inventory theory with perishable products. Ann. Oper. Res. **289**(2), 473–494 (2020)
15. Kitaeva, A.-V., Stepanova, N.-V., Zhukovskiy, O.-I.: Profit optimization with a power-law adjustable coefficient for zero ending inventories dynamic pricing model, stochastic demand, and fixed lifetime product. IFAC-PapersOnLine **55**(10), 1793–1797 (2022)
16. Kitaeva, A.-V., Stepanova, N.-V., Zhukovskaya, A.-O.: Zero ending inventory dynamic pricing model under stochastic demand, fixed lifetime product, and fixed order quantity. IFAC-PapersOnLine **52**(13), 2482–2487 (2019)
17. Li, S., Zhang, J., Tang, W.: Joint dynamic pricing and inventory control policy for a stochastic inventory system with perishable products. Int. J. Prod. Res. **53**(10), 2937–2950 (2015)
18. Kitaeva, A.-V., Stepanova, N.-V., Zhukovskaya, A.-O.: Profit optimization for zero ending inventories dynamic pricing model under stochastic demand and fixed lifetime product. IFAC-PapersOnLine **53**(2), 10505–10510 (2020)
19. Agi, M.-A., Soni, H.-N.: Joint pricing and inventory decisions for perishable products with age-, stock-, and price-dependent demand rate. J. Oper. Res. Soc. **71**(1), 85–99 (2020)

# Inferences of Modularity for Graphs Evolved by the Clustering Attachment Model

Natalia Markovich[(✉)] and Maksim Ryzhov

V.A. Trapeznikov Institute of Control Sciences, Russian Academy of Sciences, Profsoyuznaya Str. 65, 117997 Moscow, Russia
nat.markovich@gmail.com

**Abstract.** The clustering attachment (CA) is an evolution model of random graphs. In contrast to preferential attachment, the CA leads to a light-tailed node degree distribution and clusters of exceedances of the modularity over a sufficiently high threshold. The modularity shows the connectivity of nodes and serves to divide graphs into communities. An extremal index approximates the mean cluster size of exceedances over a high threshold. Considering the change of the modularity at each evolution step, the extremal index of the modularity random sequence indicates the consecutive large connectivity of nodes. It reflects the community appearance during the evolution. Our simulation shows how parameters of the CA model impact on the bursts of the modularity. The comparison of the CA evolution without and with node and edge deletion is provided.

**Keywords:** Evolution · Clustering attachment · Modularity · Clustering coefficient

## 1 Introduction

The study of random networks and graphs and their evolution attracts interest of many researchers [1–3]. The temporal burstiness of (human) communities is one of the problem under the investigation [4]. Random graphs are used as models to approximate the random networks.

Let $G_t = (V_t, E_t)$ be an undirected graph at evolution step $t \geq 0$, where $V_t$ and $E_t$ are sets of vertices and edges, respectively. The evolution of the graph is represented as a sequence of graphs $G_0, G_1, ...$, where $G_0$ is the initial graph from which the evolution begins. One can interpret $t$ as discrete time moments.

We focus at a clustering attachment (CA) as the evolution model. The CA was proposed in [4] and developed in [5,6] for undirected graphs. By [4], the

---

The authors were supported by the Russian Science Foundation RSF, project number 24-21-00183.

attachment of a new node (vertex) to one of existing nodes by a new edge may happen at time $t$ with probability[1]

$$P_{CA}(i,t) \propto c_{i,t}^\alpha + \epsilon, \qquad \alpha, \epsilon \geq 0 \qquad (1)$$

proportional to the clustering coefficient of node $i$

$$c_{i,t} = \begin{cases} 0, & k_{i,t} = 0 \text{ or } k_{i,t} = 1, \\ 2\Delta_{i,t}/(k_{i,t}(k_{i,t}-1)), & k_{i,t} \geq 2, \end{cases} \qquad (2)$$

$c_{i,t} \in [0,1]$. The latter measures the tendency of the node $i$ to form triangles of the nearest nodes in its neighborhood. $\Delta_{i,t}$ is the number of triangles of connected nodes involving node $i$ and $k_{i,t}$ is the degree of node $i$ both at time $t$. $\alpha$ and $\epsilon$ are parameters of the CA. By the CA a new node may only be attached to an existing node $i$ involved in some number of triangles $\Delta_{i,t}$ at time $t$ for $\epsilon = 0$. If $\epsilon > 0$ dominates the term $c_{i,t}^\alpha$, then $c_{i,t}$ does not impact much in the attachment probability $P_{CA}(i,t)$. The CA leads to specific phenomena such as light-tailed distributed node degrees and bursts of the modularity sequence that is built by the graph evolution. The first phenomenon is observed in [4,6] by simulation study.

The community detection constitutes an important part of the analysis of random graphs. One of the popular methods of the community detection is to find a partition of the vertex set that maximizes the modularity. The modularity $Q(t)$ of an undirected graph shows how many edges exist within communities and between them:

$$Q(t) = \frac{1}{2\|E_t\|} \sum_{ij} \left[ A_{ij} - \frac{k_{i,t} k_{j,t}}{2\|E_t\|} \right] \mathbf{1}(i,j), \qquad (3)$$

where $A_{ij}$ denote elements of the adjacency matrix $A$ of the graph $G_t$, $\|E_t\| = \frac{1}{2}\sum_{ij} A_{ij}$ is the number of edges in $G_t$, $k_{i,t}$ is the degree of vertex $i$ (i.e. the number of neighbors of $i$) at evolution step $t$, $\mathbf{1}(i,j)$ is equal to 1 when nodes $i$ and $j$ in $G_t$ belong to the same community and zero otherwise, [7]. By [8,9] (3) can be reduced to

$$Q(t) = \sum_{i=1}^{J(t)} \left[ \frac{L_{C_i^t,t}}{\|E_t\|} - \left( \frac{k_{C_i^t,t}}{2\|E_t\|} \right)^2 \right], \qquad (4)$$

where the sum iterates over all communities $\{C_i^t\}$, $J(t)$ is the total number of communities in the graph $G_t$, $L_{C_i^t,t}$ is the number of intra-community links for community $C_i^t$ and $k_{C_i^t,t}$ is the sum of degrees of the nodes in the community $C_i^t$. The modularity reaches a minimum value when it is possible to bipartite a graph [8]. This fact is used in developing community identification or the graph cut algorithm.

We aim to study clusters of exceedances of the modularity sequence arisen during the CA evolution. To this end, we estimate the extremal index (EI) that

---
[1] $x \propto y$ means that there is a non-zero constant $C$ such that $x = Cy$.

approximates the mean cluster size, i.e. the number of exceedances over a sufficiently high threshold per cluster [16]. Considering the change of the modularity at each evolution step, the EI of the modularity random sequence indicates the consecutive large connectivity of nodes. It reflects the community appearance during the evolution. The CA models with and without node and edge deletion are studied.

In our simulation, we use the Greedy Modularity Maximization Algorithm (GMMA) to partition a graph into communities [8]. The GMMA is a hierarchical agglomeration algorithm for detecting community structure which is faster than many competing algorithms. Its running time on a network with $n$ vertices and $m$ edges is $O(md \log(n))$, where $d$ is a depth of the "dendrogram" describing a hierarchical decomposition of the network into communities. Typically, $m \sim n$, $d \sim \log n$ hold for real-world networks.

The paper is organized as follows. Related work is given in Sect. 2. In Sect. 3 the modularity sequences generated by the CA without and with node and edge deletion are investigated. In Sect. 4 the estimation of the EI of the modularity is considered. The exposition is finalized with conclusions.

## 2 Related Work

A stationary sequence $\{Y_n\}_{n\geq 1}$ of random variables (r.v.s) with cumulative distribution function $F(x)$ and $M_n = \max_{1 \leq j \leq n} Y_j$ is said to have the EI $\theta \in [0, 1]$ if for each $0 < \tau < \infty$ there is a sequence of real numbers $u_n = u_n(\tau)$ such that

$$\lim_{n\to\infty} n(1 - F(u_n)) = \tau \quad \text{and} \quad \lim_{n\to\infty} P\{M_n \leq u_n\} = e^{-\tau\theta} \quad (5)$$

hold [10]. The EI plays a key role in the extreme value analysis since it allows us to get a limit distribution of a sample maximum of r.v.s when the latter are dependent, i.e. $P\{M_n \leq u_n\} = F^{n\theta}\{u_n\} + o(1)$ as $n \to \infty$. The EI is equal to one for independent identically distributed r.v.s. The converse is incorrect. Moreover, the EI measures the local clustering tendency of high threshold exceedances of a underlying random sequence. As closer to zero $\theta$, as stronger is the local dependence. The reciprocal value $1/\theta$ approximates the mean number of exceedances over a sufficiently high threshold per cluster, i.e. the mean cluster size. The cluster can be determined as a block of data with at least one exceedance over $u$.

To estimate $\theta$ we use the intervals estimator [11] that is one of the most accurate and simple estimators. Taking the exceedance times $1 \leq S_1 < ... < S_{N_u} \leq n$ of the underlying sequence $\{Y_n\}_{n\geq 1}$, the observed interexceedance times are $T_i = S_{i+1} - S_i$ for $i \in \{1, ..., N_u - 1\}$. $N_u = \sum_{i=1}^{n} \mathbf{1}\{Y_i > u\}$ is the number of observations exceeding a predetermined high threshold $u$, $L \equiv L(u) = N_u - 1$. The intervals estimator is defined as

$$\hat{\theta}_n(u) = \begin{cases} \min(1, \hat{\theta}_n^1(u)), & \text{if } \max\{T_i : 1 \leq i \leq L\} \leq 2, \\ \min(1, \hat{\theta}_n^2(u)), & \text{if } \max\{T_i : 1 \leq i \leq L\} > 2, \end{cases} \quad (6)$$

where

$$\hat{\theta}_n^1(u) = \frac{2(\sum_{i=1}^{L} T_i)^2}{L \sum_{i=1}^{L} T_i^2}, \quad \hat{\theta}_n^2(u) = \frac{2(\sum_{i=1}^{L} (T_i - 1))^2}{L \sum_{i=1}^{L} (T_i - 1)(T_i - 2)}. \tag{7}$$

The intervals estimator requires a choice of $u$ as a single parameter. Usually, a high quantile of $\{Y_n\}$ is selected as $u$. We find $u$ in Sect. 4 by discrepancy method proposed in [12] that is a data-driven method.

## 3 Modularity Evolution by Clustering Attachment with and without Node and Edge Deletion

Let us consider the CA evolution when a newly appended node attaches to a fixed number $m_0 \geq 2$ of existing nodes by $m_0$ new edges at each evolution step. Following [6], we use further the normalized probabilities to append a new edge between an existing node $i$ and a new node $v \notin V_{t-1}$ at evolution step $t$

$$P_{CA}(i,t) = \frac{c_{i,t}^\alpha + \epsilon}{\sum_{j \in V_t} c_{j,t}^\alpha + \|V_t\|\epsilon}, \quad i \in V_t, \ \alpha > 0 \tag{8}$$

and

$$P_{CA}(i,t) = \frac{\mathbf{1}\{c_{i,t} > 0\} + \epsilon}{\sum_{j \in V_t} \mathbf{1}\{c_{j,t} > 0\} + \|V_t\|\epsilon}, \quad i \in V_t, \ \alpha = 0. \tag{9}$$

It is assumed in (8) and (9) that not all $c_{j,t}$, $j \in V_t$ and $\epsilon$ are equal to zero at the same time [6]. Note that (8) is the same as (1) with $C = \left(\sum_{j \in V(t)} c_{j,t}^\alpha + \|V(t)\|\epsilon\right)^{-1}$.

We analyse bursts of the modularity $Q(t)$ at time $t$ during the CA evolution. The latter bursts may happen in clusters of consecutive exceedances of the normalized modularity

$$\psi(t) = Q(t)/\langle Q \rangle - 1 \tag{10}$$

over a high threshold $u$. $\langle Q \rangle$ denotes the average of the modularity values over some evolution period. The clustering of $\psi(t)$ is enhanced when $\alpha$ in (1) grows [4]. A cluster structure of $\psi(t)$ and the average clustering coefficient

$$\overline{C}_t = \frac{1}{\|V_t\|} \sum_{i \in V_t} c_{i,t} \tag{11}$$

against evolution steps $t$ for two pairs of the CA parameters $(\alpha, \epsilon) \in \{(1,0),(1,1)\}$ and three strategies of node and edge deletion are observed in Fig. 1, where $m_0$ is taken equal to $\{2,3\}$. The pairs $(\alpha, \epsilon) \in \{(0,0),(0,1)\}$ and the corresponding modularity series having a visual clustering structure were shown in Fig. 2, [13].

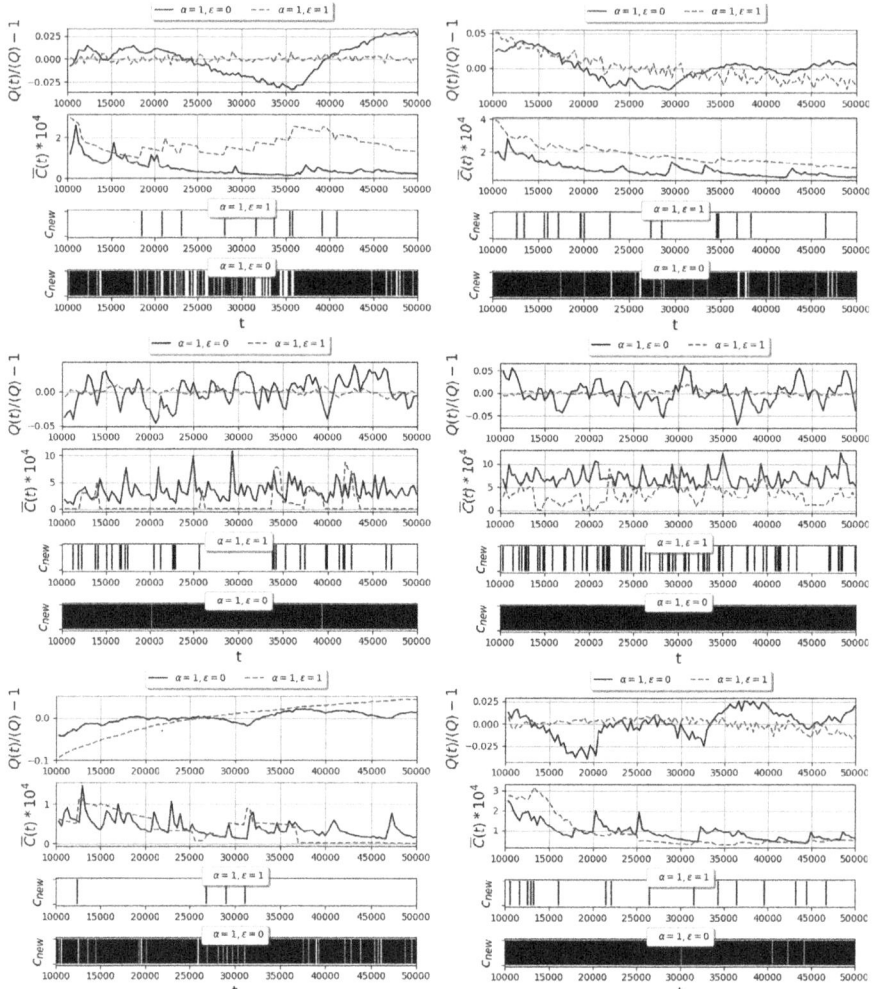

**Fig. 1.** The evolution of the normalized graph modularity (10) and the average clustering coefficient (11) against the CA evolution steps, and spike trains denoting a point process of the evolution steps when newly appended nodes create a triangle with existing nodes: without node and edge deletion (top), with uniform node deletion (middle) and with uniform edge deletion (bottom), where the left column relates to $m_0 = 2$ and the right one to $m_0 = 3$.

The case $\epsilon = 0$ is specific. If, in addition, node $i$ does not belong to any triangle of nodes and $c_i = 0$ holds, then $P_{CA}(i,t) = 0$ follows by (8) or (9) assuming not all $c_j = 0, j \in V_t$. The latter implies that new nodes cannot be attached to node $i$ that reflects on the modularity by (3). If nodes are isolated, then the attachment to them is unlike since their triangle counts are zero-valued.

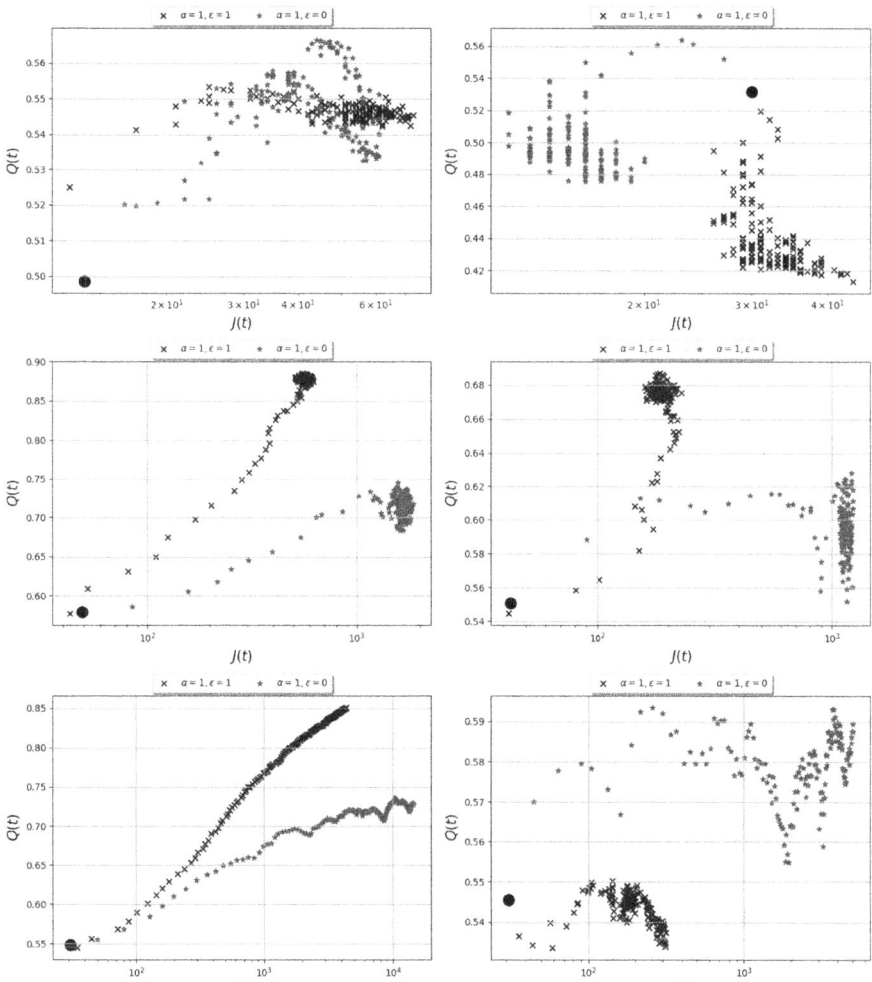

**Fig. 2.** The graph modularity $Q(t)$ (3) against the number of communities $J(t)$ found by the GMMA algorithm for the CA evolution with $m_0 = 2$ (left column) and $m_0 = 3$ (right column): without node and edge deletion (top line), with uniform node deletion (middle line) and with uniform edge deletion (bottom line) and for $(\alpha, \epsilon) \in \{(1,1), (1,0)\}$. Black circles represent the modularity of the initial graph.

The CA evolution without node and edge deletion, with the uniform node and edge deletion are compared in Fig. 1. $\langle Q \rangle$ denotes the modularity average over evolution steps in the interval $t \in [10^4, 5 \cdot 10^4]$.

The average clustering coefficient $\overline{C}_t$ tends to decay as $t$ grows for evolution without node and edge deletion and with edge deletion, see Fig. 1 (top, bottom). Since the node deletion leads to the graph with a fixed number of nodes, $\overline{C}_t$ becomes larger and due to an increasing of the number of isolated nodes it tends to a stable and relatively large value in comparison with the other two strategies, Fig. 1 (middle). A stable value of $\overline{C}$ in the case of node deletion indicates on the geometry in the graph by [14]. The geometry means here the communication of nodes with their neighbor nodes, e.g., in a close geolocation. Spike trains in Fig. 1 indicate the evolution steps with the creation of a new triangle. Figure 1 shows that the case $(\alpha, \epsilon) = (1, 1)$ leads to significantly smaller triangle counts than the case $(\alpha, \epsilon) = (1, 0)$ for any removal strategy.

Figure 2 helps us to understand why the modularity $Q(t)$ tends to decrease or increase in Fig. 1 for sufficiently large steps of the evolution. The most specific case is provided by the increasing modularity in Fig. 1 (bottom, left). At the beginning of the evolution, the number of communities tends to increase due to a large amount of small communities with a few nodes. Later, the small communities obtain more links to large communities and became finally parts of the latter ones. This leads to a stabilization of the number of communities and the modularity.

Figure 2 shows the number of communities $J(t)$ corresponding to the modularity $Q(t)$ represented in Fig. 1. The graph obtained by the CA with parameters $m_0 = 2$, $(\alpha, \epsilon) = (1, 0)$ starting from a triangle of connected nodes and containing $5 \cdot 10^3$ nodes is used as an initial graph from which the evolution begins. Apart of the CA with $m_0 = 2$ and the uniform edge deletion in Fig. 2 (bottom, left), the modularity and the number of communities are stabilized at certain values. In the latter figure, the modularity grows as the numbers of communities increases due to the appearance of isolated nodes and nodes with small degrees considered as communities. For the case of an exception, the evolution steps are not enough to reach a stability due to the enlargement of communities.

In contrast to $m_0 = 2$, the CA with $m_0 = 3$ in Fig. 2 (bottom, right) does not lead to a large number of communities. This is because the appearance of isolated nodes and nodes with small degrees is unlikely since three edges are appended and one edge is deleted at each step. Then the modularity in Fig. 2 (bottom, right) is smaller and with a smaller number of communities than one in Fig. 2 (bottom, left). The modularity is smaller in Fig. 2 (middle, right) than in Fig. 2 (middle, left) with about the same number of communities since the

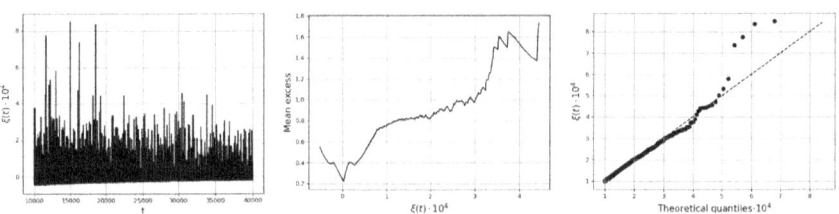

**Fig. 3.** Residuals $\{\xi_t \cdot 10^4\}$, $\xi_t = \psi(t) - \psi(t-1)$ for the CA evolution with $(\alpha, \epsilon) = (1, 1)$ and the edge deletion against $t$ (left), their mean excess function (middle) and their QQ-plot against Pareto type IV quantiles (see (12)) calculated by exceedances $\xi_t > 10^{-4}$ (right).

amount of edges with $m_0 = 3$ is larger. The normalized modularity $\{\psi(t)\}$ in (10) may be close to zero (i.e. $Q$ is close to $\langle Q \rangle$) or it may increase without visual clusters as in Fig. 1 (bottom, left) for the case $(\alpha, \epsilon) = (1, 1)$. In the latter case, new nodes are attached with a nearly uniform probability to existing nodes since $\epsilon$ in (8) dominates $(c_{i,t})^\alpha$ and new triangles appear rarely.

Let us focus on the increasing normalized modularity $\psi(t)$ for $(\alpha, \epsilon) = (1, 1)$ in Fig. 1 (left column, bottom). Figure 3 (left) shows deviations $\xi_t = \psi(t) - \psi(t-1)$ at time interval $t \in \Omega = \{10^4, 4 \cdot 10^4\}$ that contain several bursts. The sample mean excess function

$$e_n(u) = \sum_{i=1}^{n} (\xi_i - u) \mathbf{1}\{\xi_i > u\} / \sum_{i=1}^{n} \mathbf{1}\{\xi_i > u\}$$

in Fig. 3 (middle), where $\mathbf{1}\{\cdot\}$ denotes an event indicator function, has a linear behaviour that indicates a Pareto distribution of $\xi_t$, $t \in \Omega$. The QQ-plot in Fig. 3 (right) represents the sample quantiles of $\xi_t$ against Pareto type IV quantiles. There are outliers that indicate a heavier distribution tail than the examined shifted Pareto type IV [15]

$$P\{X > x\} = \begin{cases} \left[1 + \left(\frac{x-\mu}{\sigma}\right)^\gamma\right]^{-\alpha}, & \text{if } x > \mu, \\ 1, & \text{if } x \leq \mu, \end{cases} \quad (12)$$

with parameters $(\sigma, \alpha, \gamma, \mu) = (7, 29785.7, 1.116, 1.0003 \cdot 10^{-4})$ calculated by the maximum likelihood method. The program of this method has been taken in Mathematica software. The large value of $\alpha$ indicates that $\xi_t$ is distributed with a rather light heavy tail.

## 4 Evolution of the Extremal Index of the Modularity

The EI estimates of the modularity against values of the parameter $\alpha$ in (8) without node and edge deletion and with uniform node or edge deletion are shown in Fig. 4. For $\epsilon = 0$ the EI's decay as $\alpha$ increases is represented. This decay is as slower as larger is $m_0$. The EI's tend to be stable and close to 1 for $\epsilon = 1$. The case $\hat{\theta} \approx 1$ may imply the lack of clusters of modularity exceedances and that the exceedances over a sufficiently high threshold may occur singly. For $\alpha \in [0, 1]$ and $\epsilon = 0$ the EI's are also stable and close to 1, see Fig. 4, since $c_{i,t}^\alpha$ and thus the attachment probability $P_{CA}(i,t)$ are larger than that for $\alpha > 1$. If a node or an edge is deleted at each step of the evolution, then the EI estimates coincide for different $m_0$ and their values depend only on $\epsilon$, see Fig. 4 (top right, bottom). The smaller EI's, the stronger is the clustering (or a local dependence) of modularity sequences.

For $\epsilon = 0$ the EI's are closer to zero when a node or an edge is deleted than the case without any deletion. Then the clustering of the modularity is stronger. The local dependence of the modularity is weaker (and the mean cluster size is smaller) when the CA is provided without the node and edge deletion. The latter is in agreement with the clustering of the modularity in Fig. 1. Let us recall that the maximum of the underlying random sequence does not exceed a sufficiently high threshold almost surely, if $\theta = 0$ by the definition of the EI (5).

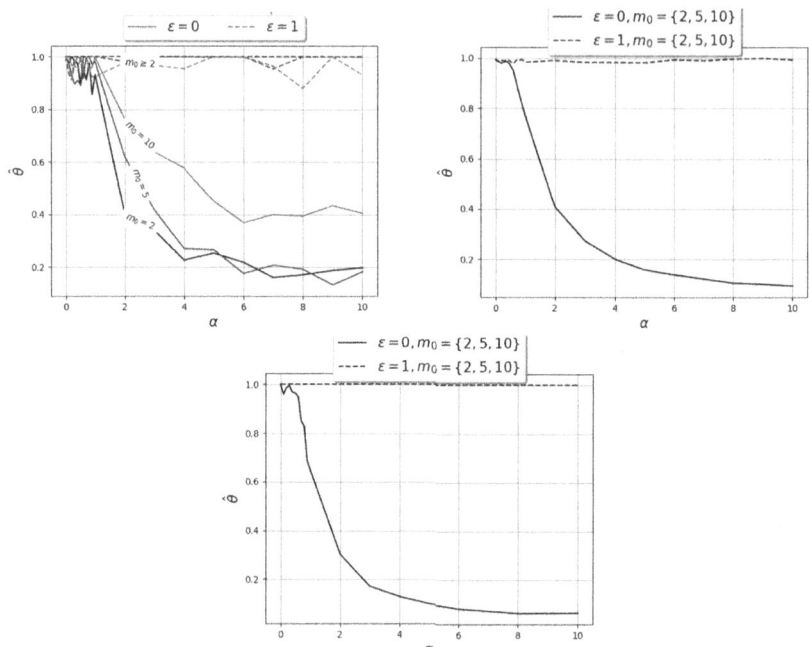

**Fig. 4.** Intervals estimates (6), (7) of the EI's $\hat{\theta}$ of the normalized modularity $\psi(t) = Q(t)/\langle Q \rangle - 1$ averaged over 100 simulated graphs evolved by the CA without the node and edge deletion (top left), with the uniform node deletion (top right) and with the uniform edge deletion (bottom) against the parameter $\alpha$ in (8).

## 5 Conclusion

The CA evolution model with possible node and edge deletion is studied by simulation. The study is focused at bursts of the modularity sequence that can be characterized by the extremal index. The extremal index of the modularity is estimated assuming the latter random sequence is stationary distributed. The cases $\epsilon = 0$ and $\epsilon = 1$, where $\epsilon$ is the CA parameter, differ significantly. For $\epsilon = 0$ the extremal index decays to some stable value and thus, the burstiness of the modularity (its local dependence) grows as the CA parameter $\alpha$ increases. The burstiness is stronger when the node or edge deletion takes place. For $\epsilon = 1$ the extremal index tends to one. It may imply single independent bursts of the modularity.

Regarding the relation of the modularity and the number of communities it is shown that both tend to stability with the evolution. The reason is that during the evolution communities are enlarged and hence, their number decreases. The tendency and stability values are different for $\epsilon = 0$ and $\epsilon = 1$ and also strongly depend on $m_0$ and the strategy to delete or not delete nodes and edges.

# References

1. Bollobás, B.: Random Graphs, 2nd edn. Cambridge University Press, Cambridge (2001)
2. van der Hofstad, R.: Random Graphs and Complex Networks, vol. 1. Cambridge University Press, Cambridge (2017)
3. Newman, M.: Networks. Oxford University Press, Oxford, New York (2018)
4. Bagrow, J., Brockmann, D.: Natural emergence of clusters and bursts in network evolution. Phys. Rev. X **3**(2), 021016 (2012)
5. Markovich, N.M., Vaičiulis, M.: Investigation of triangle counts in graphs evolving by clustering attachment. Autom. Remote Control **85**(11), 1095–1107 (2024)
6. Markovich, N.M., Ryzhov, M.S., Vaičiulis, M.: Inferences for random graphs evolved by clustering attachment. arXiv: 2403.00551v1, pp. 1–25 (2024, submitted)
7. Dugué, N., Perez, A.: Directed Louvain: maximizing modularity in directed networks. In: Research Report Université d'Orléans. hal-01231784, pp. 1–15 (2016). http://dx.doi.org/10.13140/RG.2.1.4497.0328CrossRef
8. Clauset, A., Newman, M.E., Moore, C.: Finding community structure in very large networks. Phys. Rev. E **70**(6), 066111 (2004). https://doi.org/10.1103/PhysRevE.70.066111CrossRef
9. Hagberg, A.A., Schult, D.A., Swart, P.J.: Exploring network structure, dynamics, and function using NetworkX. In: Varoquaux, G., Vaught, T., Millman, J. (eds.) Proceedings of the 7th Python in Science Conference (SciPy2008), (Pasadena, CA USA), pp. 11–15. Scientific Research An Academic Publisher (2008)
10. Leadbetter, M.R., Lingren, G., Rootzén, H.: Extremes and Related Properties of Random Sequence and Processes. Springer, New York (1983)
11. Ferro, C., Segers, J.: Inference for clusters of extreme values. J. R. Statist. Soc. B. **65**, 545–556 (2003)
12. Markovich, N.M., Rodionov, I.V.: Threshold selection for extremal index estimation. J. Nonparametric Stat. **36**(3), 527–546 (2023). https://doi.org/10.1080/10485252.2023.2266050CrossRef
13. Markovich, N.M., Ryzhov, M.S.: Clusters of exceedances for evolving random graphs. In: Vishnevskiy, V., Samouylov, K., Kozyrev, D. (eds.) DCCN 2022. LNCS, vol. 13766, pp. 67–74. Cham, Springer (2022)
14. Michielan, R., Litvak, N., Stegehuis, C.: Detecting hyperbolic geometry in networks: why triangles are not enough. Phys. Rev. E **106**(5), 054303 (2022)
15. Dutang, C., Goulet, V., Langevin, N.: Feller-pareto and related distributions: numerical implementation and actuarial applications. J. Stat. Softw. **103**(6), 1–22 (2022)
16. Beirlant, J., Goegebeur, Y., Teugels, J., Segers, J.: Statistics of Extremes: Theory and Applications, p. 498. Wiley, Chichester (2004)

# Queueing Model with Correlated Arrival Process, Simultaneous Service of a Finite Number of Customers, and Pre-processing, Post-processing and Co-processing of Customers

Alexander Dudin[(✉)] and Olga Dudina

Department of Applied Mathematics and Computer Science, Belarusian State University, 220030 Minsk, Belarus
{dudin,dudina}@bsu.by

**Abstract.** In this paper, we analyse a queueing model with a Markov arrival process and the possibility of simultaneous service of a finite number of customers defined by the multidimensional Markov chain. Service in this system is implemented in cooperation with another, infinite-server, queueing system. The variants when service in the infinite-server system is implemented before, after, and in parallel with service in the main system are considered. An explicit form of the generator of the multidimensional Markov chains describing the analysed variants of the queueing system is obtained. It is shown that in all these variants, the system is stable. The way for solving the problem of the stationary distribution computation in these variants is outlined as well. The results can be used for optimization of the choice of the system parameters.

**Keywords:** simultaneous service · tandem queue · $MAP$ · stationary distribution · ergodicity

## 1 Introduction

In this paper, we consider three variants of symbiosis of two queueing systems. One system, which we call system 1, is the main system of the symbiosis and provides essential service. The system can process simultaneously a finite number of customers. Their service process is described by a multidimensional Markov chain ($MC$). This chain is defined by: (i) the transition intensities at the moments when service of some customer is finished or transition occurs without any service completion, and (ii) the transition probabilities of the chain at the moments of a new service beginning. A more detailed description of the notion of the simultaneous service process, which includes the variants of service by several independent servers, limited processor sharing, etc., is given in the description of the model in the next section. Note that group departure of customers after service at the main system is not assumed in this model.

The second system, which we call system 2, is the auxiliary system. This system has an infinite number of servers. The service time distribution is assumed to be exponential. These servers prepare customers for service in the main system, provide some additional service after the finish of the essential service, or provide service instead of the servers of the main system when it is temporarily overloaded, e.g., provide an opportunity for self-service of customers when all the servers of the main system are busy.

We suppose that the input flow of customers to this symbiosis of two systems is defined by the Markov arrival process ($MAP$). The $MAP$ is essentially a more general process than the very popular in the literature stationary Poisson process, renewal process with $PH$ distribution (for definition, see, e.g., [1,2]) of inter-arrival times. It is known that the $MAP$ is suitable for description of the arrival processes with correlated inter-arrival times and times having large variance, which is typical, e.g., to certain information flows in telecommunications and contact centers. For more information about the $MAP$ see, e.g., [3–9].

In a unified way, we sequentially consider three variants of the structure of the symbiosis.

In the first variant, system 2 implements a pre-processing of a customer. This means that, before entering system 1, a customer is processed by system 2 and then transits to system 1, if the capacity of this system is not exhausted. In the contrary case, the customer makes a random choice between the options to be lost (leave the system without service in system 1) or return for service in system 2. In other words, in the first variant, the considered symbiosis is the dual tandem of queueing systems 2 and 1. A similar system with pre-processing in an infinite-server system was considered in [10]. There, it was assumed that the heterogeneous marked $MAP$ flow of customers enters the tandem consisting of the infinite-server system and the system with a finite number of identical servers. One type of customers has a low priority. Customers of this type are lost if all servers at the second stage of the tandem are busy at the moment of service completion of the low-priority customer at the first stage. To reduce the share of the lost customers after service at the first stage, low-priority customers are not admitted for service at the first phase when the number of busy servers at the second phase exceeds some fixed threshold value. In the model considered in this paper, we assume only one type of customers, the possibility of a customer repeated service at stage 1 when the main system is not available. The service process assumed in [10] is an important particular case of the service process assumed in this paper. The stationary behavior of the model is analysed in [10] where customer loss at both stages is possible. The problem of the optimal choice of the threshold is numerically solved.

In the second variant, system 2 implements post-processing of a customer. This means that, after service in system 1, the customer proceeds for service in system 2. In other words, in the second variant, the considered symbiosis is the dual tandem of queueing systems 1 and 2. Customer loss can occur at the first stage when the maximally allowed number of customers receive service at a customer arrival moment.

In the third variant, the main service is provided by system 1. Servers of system 2 do not start processing arriving customers if there are idle servers in system 1. Only the customers arriving when the capacity of the system is exhausted 1 start service in system 2. In other words, in the third variant, the considered symbiosis is the set of the parallel systems in which system 2 provides service only to the overflow of customers arriving at system 1. Interruption of service in an infinite-server system in case service by the main system becomes available is not permitted.

All these three variants can be considered as the simple examples of queueing networks with two multi-server nodes. To the best of our knowledge, these queueing networks were not considered in the existing literature for the case of the general Markov description of simultaneous service of a finite number of servers.

The outline of the results presentation is as follows. In Sect. 2, the description of a simultaneous service process and of the arrival process is done. In Sects. 3–5, the first, second, and third variants of the symbiosis structure are analysed, respectively. For all variants, the explicit block matrix form of the generator of the multidimensional Markov chain describing the behavior of the system is presented. Formulas for computation of some performance measures of the system are derived. Section 6 contains some concluding remarks.

## 2 Description of a Simultaneous Service Process and Arrival Process

We suppose that the main system can provide service simultaneously to up to $N$ customers. This includes, as particular cases, many possible scenarios of service.

The most well-known and well-investigated scenario is that the service is provided by $N$ independent, identical servers. The service time of a customer by a server has an exponential distribution or more general phase-type ($PH$) distribution, see, e.g., [1,2]. The assumption that the service time has an exponential distribution drastically simplifies the analysis of the corresponding queueing system because there is no need to account for the residual service times in all busy servers. The assumption that the service time has the phase-type distribution essentially complicates the analysis but allows to account for not only the mean service time but also its variance and higher moments.

The evident disadvantage of the system resource sharing via the organization of independent servers is the underutilization of the resource. The part of the resource allotted to the currently idle servers is not used. This disadvantage disappears if all the resources are permanently shared by all customers presenting in the system. In the simplest settings, this means that the total possible service rate (the capacity of bandwidth) in the system, say, $\mu$, is uniformly shared by all customers presenting in the system. If $i$ customers reside in the system, then each of them receives service at rate $\frac{\mu}{i}$, $i > 0$. Such a discipline is called processor sharing ($PS$), see, e.g., [11,12]. The mechanism of $PS$ is not applicable in many real-world systems, e.g., telecommunication systems, in which service at a too low rate may not be acceptable for customers. Therefore, the discipline of limited processor sharing ($LPS$) appeared, see, e.g., [13–15]. This discipline assumes that not more than $N$ customers can obtain service at the same time, with the rate inversely proportional to the number of concurring customers. The number $N$ is sometimes called the multi-programming, concurrency level, or multiplicity. Besides the undesirability of service at a too low rate, sometimes the service at a too high rate is also undesirable. This led to the consideration of the so-called mixed service discipline, for references, see, e.g., [16]. It assumes that a limited number of customers (corresponding to so-called non-elastic traffic) receives service at a constant rate, as in the multi-server system, while another finite group of customers is processed with the use of the $LPS$ discipline.

The range of existing and possible queueing models with different mechanisms of simultaneous service of customers is quite wide. The consideration of simultaneous service mechanism in this paper is to present a general framework for uniform consideration of various queueing systems with such existing and future service disciplines.

We consider a queueing system with the possibility to provide service simultaneously to up to $N$ customers. The description of the service process is as follows. Let the number of customers in service at the moment $t$ be $n_t = \overline{0, N}$. This notation means that $n_t$ admits values in the set $\{0, 1, \ldots, N\}$. Let $\eta_t$ be the auxiliary finite-state process such that the two-dimensional process $\{n_t, \eta_t\}$ behaves between the successive admissions of customers for service to the system as a Quasi-Death-Process that is the partial case of the Quasi-Birth-and-Death ($QBD$) Process; for definition and more details see [1].

Under the fixed state $n$ of the process $n_t$, $n = \overline{0, N}$, the process $\eta_t$ admits values from the set of cardinality $s_n$. The Quasi-Death-Process $\{n_t, \eta_t\}$ under the fixed value $n$ of the process $n_t$ is completely defined by two matrices, $\mathbf{S}_n^0$ and $\mathbf{S}_n^-$. The non-diagonal entries of the square matrix $\mathbf{S}_n^0$, $n = \overline{0, N}$, of size $s_n$ describe the transition rates of the process $\eta_t$ within its state space that do not lead to the end of a customer service. The diagonal entries of the matrix $\mathbf{S}_n^0$ are negative. The module of such an entry defines the rate of the process $\eta_t$ exit from the corresponding state. The matrix $\mathbf{S}_n^-$, $n = \overline{1, N}$, of size $s_n \times s_{n-1}$ describes transition rates of the process $\eta_t$ when one of $n$ customers processed in the system departs.

Note that the matrix $\mathbf{S}_n^0$ is the sub-generator and the matrix $\mathbf{S}_n^0 + \mathbf{S}_n^-$ is the generator. This implies, in particular, that $\mathbf{S}_n^0 \mathbf{e} + \mathbf{S}_n^- \mathbf{e} = \mathbf{0}^T$, $n = \overline{1, N}$, where $\mathbf{e}$ and $\mathbf{0}$ are column and row vectors of 1's and 0's of a suitable size, respectively, and $^T$ denotes transposition of the vector.

Let the number of customers receiving service in the system be equal to $n$, $n = \overline{0, N-1}$, and a new customer is admitted to the system for service. The number of customers in service increases to $n+1$, and the process $\eta_t$ makes the transitions with the probabilities that constitute the stochastic matrix $\mathbf{S}_n^+$, $n = \overline{0, N-1}$, of size $s_n \times s_{n+1}$.

In the simplest case, when service is provided by $N$ independent servers, the service time in which has the exponential distribution with the rate, say, $\varepsilon$, the process $\eta_t$ has only one state, and the matrices $\mathbf{S}_n^0$, $\mathbf{S}_n^-$, and $\mathbf{S}_n^+$ turn to the scalars equal to $-\varepsilon n, \varepsilon n$, and 1, correspondingly.

Here we suggest that the matrices $\mathbf{S}_n^0$, $\mathbf{S}_n^-$, and $\mathbf{S}_n^+$ are known and implement analysis of the tandems with the service process in the main system defined by these matrices.

As it was mentioned above, the arrival process of customers to the considered kinds of symbiosis of two queueing systems is the $MAP$, see [3–9]. Arrivals can occur only at the transition moments of the so-called underlying $MC$. This process $\nu_t$, $t \geq 0$, has the state space $\{1, 2, \ldots, W\}$ and the generator $D$. This generator is split into two matrices $D_0$ and $D_1$ as $D = D_0 + D_1$ where the entries of the matrix $D_1$ define the rates of transitions of the process $\nu_t$ inside the set $\{1, 2, \ldots, W\}$ which are accompanied by the customers arrival. The matrix $D_0$ is the subgenerator, and the generator $D$ is assumed to be irreducible.

The invariant probability vector $\boldsymbol{\theta}$ of the $MC$ $\nu_t$ is calculated as the unique solution to the system $\boldsymbol{\theta} D = \mathbf{0}$, $\boldsymbol{\theta}\mathbf{e} = 1$. The mean arrival rate $\lambda$ of the $MAP$ is calculated as $\lambda = \boldsymbol{\theta} D_1 \mathbf{e}$. Formulas for calculation of the higher moments of inter-arrival times and the coefficient of their correlation can be found, e.g., in [4,5,8].

Below we present the analysis of performance of the described above variants of symbiosis of systems 1 and 2.

## 3 Scenario with Customers Pre-processing

Let us assume that the queueing system with simultaneous service of customers describes the operation of the second stage of the considered tandem queue. The first stage of this tandem contains an unlimited number of independent, identical servers. All arriving in the $MAP$ customers are admitted to this infinite-server system (system 2) and immediately begin service. The service time of an arbitrary customer at an arbitrary server of this system has the exponential distribution with the rate $\mu$. Service in the infinite-server system can be interpreted as some kind of customer's preparation for the service in the main system or reservation of certain resources mandatory for further use in the main service.

After the end of service at the first stage, with the probability $p$, $0 < p \leq 1$, the customer transits for service at the second stage (main server) or, with the complementary probability, it is discovered that the main service is not required. The customer finishes service and departs from the tandem system. For example, during the search of information about some entities containing in the relational database via some available index, the user may find the required information and will not need to be processed using the tables of the database. Or, trying to connect to an agent of some information system, the user receives complete service from the IVR (Interactive Voice Response) machine and skips talking to the agent. Or during the preliminary inspection of a car it is found that the car does not need attention of a highly qualified repairman. In [10], the infinite-server system corresponds to several seconds of free connection in a cellular mobile network, after which some users drop a connection to avoid payment at the main system.

According to the given above description of the operation of the second stage of a tandem, maximum $N$ customers can be processed at this stage. A customer that finishes the pre-processing (service at the first stage) and decides to transit to the main service when the number of customers at the second stage is less than $N$ transits to the second stage and starts service. If this number is equal to $N$, the customer cannot enter the second stage. With the probability $q$, $q > 0$, this customer is lost. With the complementary probability, the customer returns for the repeated service at the first stage. Duration of the repeated services also has the exponential distribution with the rate $\mu$.

The structure of the described tandem with customers preprocessing is illustrated by Fig. 1.

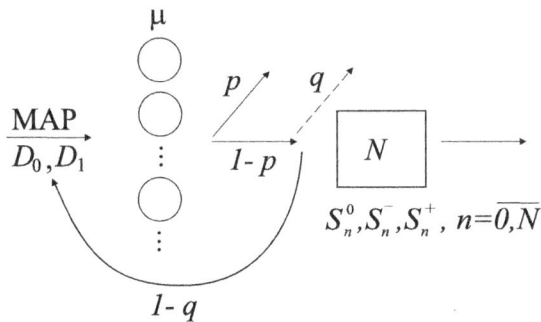

**Fig. 1.** Structure of the tandem with customers preprocessing.

Let us analyse the described tandem queueing system. Its behavior is described by the four-dimensional process

$$\zeta_t = \{i_t, n_t, \nu_t, \eta_t\},\ i_t \geq 0,\ n_t = \overline{0, N},\ \nu_t = \overline{1, W},\ \eta_t = \overline{1, s_{n_t}},\ t \geq 0,$$

where, at the moment $t$, $i_t$ is the number of customers at the first stage, $n_t$ is the number of customers at the second stage, $\nu_t$ is the state of the underlying process of arrivals, and $\eta_t$ is the state of the underlying process of customers processing at the second stage. Due to the Markov description of all four components of the process $\zeta_t$, it is the $MC$.

To simplify analysis of four-dimensional $MC$ $\zeta_t$, let us enumerate its components in the lexicographic order and call as the sub-level $(i, n)$ the set of the states of the $MC$ such as the values of the first two components, $(i_t, n_t)$ are $(i, n)$. Level $i$ is the set $((i, 0), (i, 1), \ldots, (i, N))$ of sub-levels $(i, n)$, $n = \overline{0, N}$.

Let $\mathcal{Q}_{i,j}^{(n,n')}$ be the matrix consisting of the transition rates from the states that belong to the sub-level $(i, n)$ to the states that belong to the sub-level $(j, n')$, $i, j \geq 0$, $|i - j| \leq 1$, $n, n' \in \{0, 1, \ldots, N\}, |n - n'| \leq 1$. By default, the diagonal entries of the matrix $\mathcal{Q}_{i,i}^{(n,n)}$ are negative, and their moduli define the rates of the exit from the corresponding state of the $MC$ $\zeta_t$. The matrix $\mathcal{Q}_{i,j}$ consisting of the blocks $\mathcal{Q}_{i,j}^{(n,n')}$, $n, n' \in \{0, 1, \ldots, N\}, |n - n'| \leq 1$, defines transition rates between the levels $i$ and $j$. The infinite-size matrix $\mathcal{Q}$ consisting of the blocks $\mathcal{Q}_{i,j}$ is the generator of the $MC$ $\zeta_t$.

**Lemma 1.** The generator $\mathcal{Q}$ of the $MC$ $\zeta_t$ is the block tri-diagonal matrix with the blocks $\mathcal{Q}_{i,j}$ that have the following form:

- The matrix $\mathcal{Q}_{i,i}$ has the diagonal blocks of the form

$$\mathcal{Q}_{i,i}^{(n,n)} = -i\mu I_{Ws_n} + D_0 \oplus \mathbf{S}_n^0, \; n = \overline{0, N-1},$$

$$\mathcal{Q}_{i,i}^{(N,N)} = -i\mu(1 - p(1-q))I_{Ws_N} + D_0 \oplus \mathbf{S}_N^0,$$

and the subdiagonal blocks of the form

$$\mathcal{Q}_{i,i}^{(n,n-1)} = I_W \otimes \mathbf{S}_n^-, \; n = \overline{1, N}.$$

- The matrix $\mathcal{Q}_{i,i+1}$ is the diagonal matrix with the blocks of the form

$$\mathcal{Q}_{i,i+1}^{(n,n)} = D_1 \otimes I_{s_n}, \; n = \overline{0, N}.$$

- The matrix $\mathcal{Q}_{i,i-1}$ has the diagonal blocks of the form

$$\mathcal{Q}_{i,i-1}^{(n,n)} = i\mu(1-p)I_{Ws_n}, \; n = \overline{0, N-1},$$

$$\mathcal{Q}_{i,i-1}^{(N,N)} = i\mu(pq + 1 - p)I_{Ws_N},$$

and the updiagonal blocks of the form

$$\mathcal{Q}_{i,i-1}^{(n,n+1)} = i\mu p I_W \otimes \mathbf{S}_n^+, \; n = \overline{0, N-1},$$

where $\otimes$ and $\oplus$ are symbols of Kronecker product and sum of matrices, see, e.g., [17].

It can be verified that the $MC$ $\zeta_t$ belongs to the class of asymptotically quasi-Toeplitz Markov chains ($AQTMC$), see [8,18]. Because, with the positive probability $q$ a customer departs from tandem when the second system is full ($N$ customers receive a service there), using the results from [18], by analogy with [24] the following statement can be proven.

**Theorem 1.** The $MC$ $\zeta_t$ is ergodic for any set of system parameters.

This implies that the so-called stationary probabilities of the system states exist:
$$\pi(i, n, \nu, \eta) = \lim_{t \to \infty} P\{i_t = i, n_t = n, \nu_t = \nu, \eta_t = \eta\},$$

$$i \geq 0, n = \overline{0, N}, \; \nu = \overline{1, W}, \; \eta = \overline{1, s_n}.$$

Let us combine the probabilities of the states that belong to the sublevel $(i, n)$ into the row vectors $\boldsymbol{\pi}(i, n)$. Denote also

$$\boldsymbol{\pi}_i = (\boldsymbol{\pi}(i, 0), \boldsymbol{\pi}(i, 1), \ldots, \boldsymbol{\pi}(i, N)), \; i \geq 0.$$

It is well-known that the row vectors $\pi_i$, $i \geq 0$, satisfy the following system of equilibrium equations:

$$(\pi_0, \pi_1, \ldots, \pi_i, \ldots)\mathcal{Q} = \mathbf{0}, \quad (\pi_0, \pi_1, \ldots, \pi_i, \ldots)\mathbf{e} = 1.$$

Since it is evident that the generator $\mathcal{Q}$ does not possess the quasi-Toeplitz property, i.e., the form of the block $\mathcal{Q}_{i,j}$ depends not only on the difference $j - i$ but also on $i$, the problem of solving the infinite system of equilibrium equations is very difficult. To solve this system, we use the numerically stable algorithms for solving such equations presented in [18–21]. These algorithms suggest the construction and solution of another infinite system equations based on the use of the notion of the so-called censored $MCs$ (see, e.g., [22,23]) with different censoring levels.

More details about concrete the realization of these algorithms, including recommendations about the proper choice of truncation levels in recursions, which are used, and computation of infinite sums can be found in [18–21].

Having computed the stationary probabilities of the system states, it is possible to compute a variety of performance measures of the system.

The mean number $L^{(\infty)}$ of busy servers at an arbitrary moment in the infinite-server system is computed as

$$L^{(\infty)} = \sum_{i=1}^{\infty} i\pi \mathbf{e}.$$

The mean customers departure rate $R^{(\infty)}$ from the infinite-server system is computed as

$$R^{(\infty)} = \mu L^{(\infty)}.$$

The mean number $L^{(main)}$ of busy servers in the main system is computed as

$$L^{(main)} = \sum_{i=0}^{\infty} \sum_{n=1}^{N} n\pi_i \begin{pmatrix} \mathbf{0}_{Ws_0}^T \\ \mathbf{0}_{Ws_1}^T \\ \vdots \\ \mathbf{0}_{Ws_{n-1}}^T \\ \mathbf{e}_{Ws_n} \\ \mathbf{0}_{Ws_{n+1}}^T \\ \vdots \\ \mathbf{0}_{Ws_N}^T \end{pmatrix}.$$

The mean total number $L^{(total)}$ of customers processed at an arbitrary moment in the tandem is computed as

$$L^{(total)} = \sum_{i=0}^{\infty} \sum_{n=1}^{N} (i+n)\boldsymbol{\pi}_i \begin{pmatrix} \mathbf{0}_{Ws_0}^T \\ \mathbf{0}_{Ws_1}^T \\ \vdots \\ \mathbf{0}_{Ws_{n-1}}^T \\ \mathbf{e}_{Ws_n} \\ \mathbf{0}_{Ws_{n+1}}^T \\ \vdots \\ \mathbf{0}_{Ws_N}^T \end{pmatrix}.$$

The mean customer departure rate $R^{(main)}$ from the main system is computed as

$$R^{(main)} = \sum_{i=0}^{\infty} \boldsymbol{\pi}_i \begin{pmatrix} \mathbf{0}_{Ws_0}^T \\ (I_W \otimes \mathbf{S}_1^-)\mathbf{e} \\ (I_W \otimes \mathbf{S}_2^-)\mathbf{e} \\ \vdots \\ (I_W \otimes \mathbf{S}_N^-)\mathbf{e} \end{pmatrix}.$$

The loss probability of an arbitrary customer

$$P_{loss} = 1 - \frac{R^{(main)}}{pR^{(\infty)}}.$$

## 4 System with Customers Post-processing

Now consider the scenario when at first a customer receives service in the main system and then it is processed by the infinite server-system. E.g., the customer received service in the main system has to be packed, cooled down, or to reduce radiation level, etc.

As above, we assume that the customers arrive at the system in the $MAP$ given by the matrices $D_0$ and $D_1$. If the arriving customer meets less than $N$ customers in the main system, it starts service immediately. Otherwise, the customer is lost. As above, the simultaneous service process in the main system is defined by the set of the matrices $\mathbf{S}_n^0$, $\mathbf{S}_n^-$, and $\mathbf{S}_n^+$, $n = \overline{0, N}$. After service completion in the main system, with the probability $\delta$, $0 \leq \delta < 1$, the customer departs from the considered tandem without service in the infinite-server system. With the complementary probability, the customer transits for service in the infinite-server system. The service time in this system has the exponential distribution with the rate $\mu$.

The structure of the described tandem with customers post-processing is illustrated by Fig. 2.

Operation of the considered tandem system is also described by the $MC$ of the form

$$\zeta_t = \{i_t, n_t, \nu_t, \eta_t\}, \ i_t \geq 0, n_t = \overline{0, N}, \ \nu_t = \overline{1, W}, \ \eta_t = \overline{1, s_{n_t}}, \ t \geq 0,$$

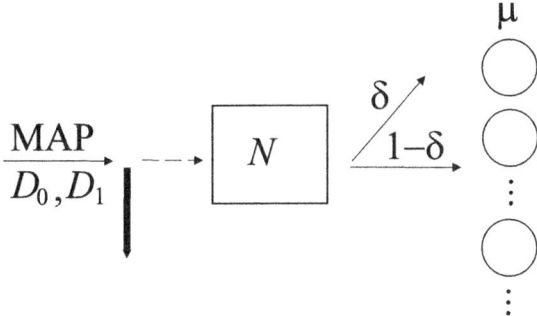

**Fig. 2.** Structure of the tandem with post-processing.

where, at the moment $t$, $i_t$ is the number of customers at the infinite-server system (at the second stage of the tandem), $n_t$ is the number of customers at the first stage of the tandem (system with simultaneous service up to $N$ customers), $\nu_t$ is the state of the underlying process of arrivals, and $\eta_t$ is the state of the underlying process of customers processing at the first stage.

**Lemma 2.** The generator $\mathcal{Q}$ of the MC $\zeta_t$ is the block tri-diagonal matrix with the blocks $\mathcal{Q}_{i,j}$ that have the following form:

- The block-tridiagonal matrix $\mathcal{Q}_{i,i}$ has the non-zero blocks of the following form:
  the diagonal blocks are defined as

  $$\mathcal{Q}_{i,i}^{(n,n)} = -i\mu I_{Ws_n} + D_0 \oplus \mathbf{S}_n^0, \; n = \overline{0, N-1},$$

  $$\mathcal{Q}_{i,i}^{(N,N)} = -i\mu I_{Ws_N} + (D_0 + D_1) \oplus \mathbf{S}_N^0,$$

  the subdiagonal blocks are defined as

  $$\mathcal{Q}_{i,i}^{(n,n-1)} = \delta I_W \otimes \mathbf{S}_n^-, \; n = \overline{1, N}.$$

  the updiagonal blocks are defined as

  $$\mathcal{Q}_{i,i}^{(n,n+1)} = D_1 \otimes \mathbf{S}_n^+, \; n = \overline{0, N-1}.$$

- The matrix $\mathcal{Q}_{i,i+1}$ has the non-zero blocks only below the diagonal that have the form

  $$\mathcal{Q}_{i,i+1}^{(n,n-1)} = (1-\delta)I_W \otimes \mathbf{S}_n^-, \; n = \overline{1, N}.$$

- The matrix $\mathcal{Q}_{i,i-1}$ has the diagonal blocks of the form

  $$\mathcal{Q}_{i,i-1}^{(n,n)} = i\mu I_{Ws_n}, \; n = \overline{0, N}.$$

It can be shown that the $MC$ $\zeta_t$ is always stable. The vectors $\boldsymbol{\pi}_i$, $i \geq 0$, of the stationary probabilities can be computed with the use of the numerically stable algorithms recommended in the previous section.

Formulas for computation of the mean number $L^{(\infty)}$ of busy servers at the infinite-server system, the mean number $L^{(main)}$ of busy servers in the main system, and the mean total number of customers in the tandem $L^{(total)}$, as well as mean departure rates $R^{(main)}$ and $R^{(\infty)}$ are the same as in the previous section.

The loss probability $P_{loss}^{(main)}$ of the customers in the main system is calculated by the formulas

$$P_{loss}^{(main)} = \frac{1}{\lambda} \sum_{i=0}^{\infty} \boldsymbol{\pi}_i \begin{pmatrix} \mathbf{0}_{Ws_0}^T \\ \mathbf{0}_{Ws_1}^T \\ \vdots \\ \mathbf{0}_{Ws_{N-1}}^T \\ (D_1 \otimes I_{s_N})\mathbf{e} \end{pmatrix}$$

or

$$P_{loss}^{(main)} = 1 - \frac{R^{(main)}}{\lambda}.$$

The presence of two different formulas for $P_{loss}^{(main)}$ is helpful for control of the accuracy of formula derivation and computer implementation.

## 5   System with Customers Co-processing

Now consider the system where all customers are serviced by the main system if it is not overloaded at their arrival moment. While the customers arriving when the number of customers in the main system is equal to $N$ are lost with the probability $(1-p)$ or receive service in the infinite-server system with the complementary probability. The description of arrival and service processes at both systems is the same as in the previous sections.

The structure of the described tandem with customers co-processing is illustrated by Fig. 3.

Operation of the considered tandem system is also described by the $MC$ of form

$$\zeta_t = \{i_t, n_t, \nu_t, \eta_t\}, \; i_t \geq 0, n_t = \overline{0, N}, \; \nu_t = \overline{1, W}, \; \eta_t = \overline{1, s_{n_t}}, \; t \geq 0,$$

where, at the moment $t$, $i_t$ is the number of customers at the infinite-server system (overhead system), $n_t$ is the number of customers at the main system, $\nu_t$ is the state of the underlying process of arrivals, and $\eta_t$ is the state of the underlying process of customers processing at the main system.

**Lemma 3.** *The generator $\mathcal{Q}$ of the $MC$ $\zeta_t$ is the block tri-diagonal matrix with the blocks $\mathcal{Q}_{i,j}$ that have the following form:*

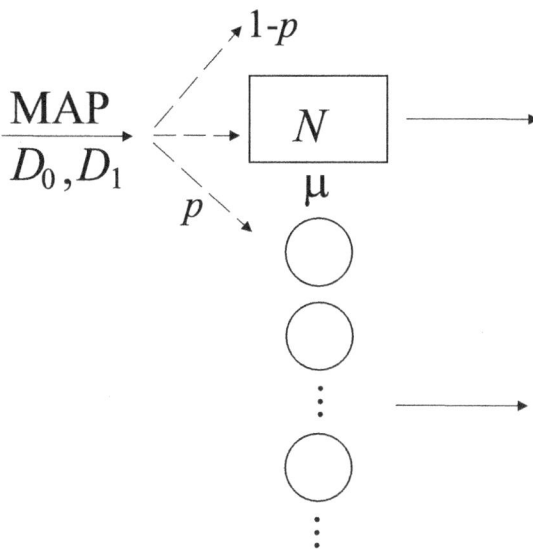

**Fig. 3.** Structure of the tandem with co-processing.

- The block-tridiagonal matrix $\mathcal{Q}_{i,i}$ has the non-zero blocks of the following form:
  the diagonal blocks are defined as
  $$\mathcal{Q}_{i,i}^{(n,n)} = -i\mu I_{WI_{s_n}} + D_0 \oplus \mathbf{S}_n^0, \ n = \overline{0, N-1},$$
  $$\mathcal{Q}_{i,i}^{(N,N)} = -i\mu I_{Ws_N} + D_0 \oplus \mathbf{S}_N^0 + (1-p)D_1 \otimes I_{s_N},$$
  the subdiagonal blocks are defined as
  $$\mathcal{Q}_{i,i}^{(n,n-1)} = \delta I_W \otimes \mathbf{S}_n^-, \ n = \overline{1, N}.$$
  the updiagonal blocks are defined as
  $$\mathcal{Q}_{i,i}^{(n,n+1)} = D_1 \otimes \mathbf{S}_n^+, \ n = \overline{0, N-1}.$$

- The matrix $\mathcal{Q}_{i,i+1}$ has only one non-zero block
  $$\mathcal{Q}_{i,i+1}^{(N,N)} = pD_1 \otimes I_{s_N}.$$

- The matrix $\mathcal{Q}_{i,i-1}$ has the diagonal blocks of the form
  $$\mathcal{Q}_{i,i-1}^{(n,n)} = i\mu I_{Ws_n}, \ n = \overline{0, N}.$$

It can be shown that the $MC$ $\zeta_t$ is always stable. The vectors $\boldsymbol{\pi}_i$, $i \geq 0$, of the stationary probabilities can be computed with the use of the numerically stable algorithms recommended in Sect. 3.

Formulas for computation of the mean number $L^{(\infty)}$ of busy servers at the infinite-server system and the mean number $L^{(main)}$ of busy servers in the main system, and the mean total number of customers in the tandem $L^{(total)}$, as well as the mean departure rates $R^{(main)}$ and $R^{(\infty)}$ are the same as in the previous sections.

The probability $P^{(overhead)}$ of the event that an arbitrary arriving customer cannot enter the main system (and will receive service at the infinite-server system) is calculated as

$$P^{(overhead)} = \frac{1}{\lambda} \sum_{i=0}^{\infty} \pi_i \begin{pmatrix} \mathbf{0}_{Ws_0}^T \\ \mathbf{0}_{Ws_1}^T \\ \vdots \\ \mathbf{0}_{Ws_{N-1}}^T \\ (D_1 \otimes I_{s_N})\mathbf{e} \end{pmatrix}$$

or

$$P^{(overhead)} = 1 - \frac{R^{(main)}}{\lambda}.$$

The share of customers that receive service in the infinite-server system is calculated as

$$P^{(\infty)} = (1-p)P^{(overhead)}$$

or

$$P^{(\infty)} = \frac{\mu L^{(\infty)}}{\lambda}.$$

The existence of two different formulas for computation of the share $P^{(\infty)}$ is useful for control of the correctness of formulas for computation and their computer implementation.

The probability that an arbitrary customer will not receive service in the considered symbiosis of systems is calculated as

$$P_{loss} = pP^{(overhead)}.$$

The probability $P^{(emp-nonemp)}$ that at an arbitrary moment the main system is empty while the infinite-server system is not empty is calculated as

$$P^{(emp-nonemp)} = \sum_{i=1}^{\infty} \pi_i \begin{pmatrix} \mathbf{e}_{Ws_0} \\ \mathbf{0}^T_{Ws_1} \\ \vdots \\ \mathbf{0}^T_{Ws_{N-1}} \\ \mathbf{0}^T_{Ws_N} \end{pmatrix}.$$

It is worth noting that, due to the existence of the alternative options of receiving service at the main system or the infinite-server system and the possibility of payment of a high price for the use of each server during a unit of time or very slow service in the infinite-server system, various optimization problems can be formulated. E.g., the optimization problem can be of the form: to find a minimal capacity $N$ of the main system such that the value $P^{(\infty)}$ will not exceed the fixed in advance values, say 0.05. Such and similar optimization problems can be numerically solved based on the algorithmic and analytical results presented in this section.

## 6 Conclusion

In this paper, we analysed three variants of cooperation of a service system with a general description of simultaneous service process of a finite number of customers (main system) and an infinite-server system. The arrival flow of customers is described by the $MAP$. The service time in the infinite-server system has the exponential distribution. Analysed variants are: (i) service in the infinite-server system precedes service in the main system; (ii) service in the infinite-server system follows service in the main system; (iii) service in the infinite-server system is provided only to the customers who arrived when the main system is overloaded. In all variants, the generator of the multidimensional $MC$ that describes the behavior of the system is derived, the existence of a stationary distribution is stated, the algorithm for computation of this distribution is outlined, and formulas for computation of the basic performance measures are presented.

## References

1. Neuts, M.F.: Matrix-Geometric Solutions in Stochastic Models: An Algorithmic Approach. Courier Corporation, North Chelmsford (1994)
2. O'Cinneide, C.A.: Phase-type distributions: open problems and a few properties. Stoch. Model. **15**, 731–757 (1999)
3. Gonzalez, M., Lillo, R.E., Ramirez Cobo, J.: Call center data modeling: a queueing science approach based on Markovian arrival processes. Qual. Technol. Quant. Manag. **2024** (2024). https://doi.org/10.1080/16843703.2024.2371715
4. Chakravarthy, S.R.: The batch Markovian arrival process: a review and future work. Adv. Prob. Theory Stochast. Process. **1**(1), 21–49 (2001)

5. Chakravarthy, S.R.: Introduction to Matrix-Analytic Methods in Queues 1: Analytical and Simulation Approach - Basics. ISTE Ltd, London and John Wiley and Sons, New York (2022)
6. Chakravarthy, S.R.: Introduction to Matrix-Analytic Methods in Queues 2: Analytical and Simulation Approach - Queues and Simulation. ISTE Ltd, London and John Wiley and Sons, New York (2022)
7. Lucantoni, D.M.: New results on the single server queue with a batch Markovian arrival process. Commun. Stat. Stochast. Models **7**(1), 1–46 (1991)
8. Dudin, A.N., Klimenok, V.I., Vishnevsky, V.M.: The Theory of Queuing Systems with Correlated Flows. Springer, Cham (2020)
9. Vishnevskii, V.M., Dudin, A.N.: Queueing systems with correlated arrival flows and their applications to modeling telecommunication networks. Autom. Remote. Control. **78**(8), 1361–1403 (2017). https://doi.org/10.1134/S000511791708001X
10. Kim, C., Dudin, S.: Priority tandem queueing model with admission control. Comput. Ind. Eng. **61**(1), 131–140 (2011)
11. Yashkov, S.F.: Processor-sharing queues: some progress in analysis. Queueing Syst. **2**, 1–17 (1987)
12. Yashkov, S.F., Yashkova, A.S.: Processor sharing: a survey of the mathematical theory. Autom. Remote Control **68**, 1662–1731 (2007)
13. Yamazaki, G., Sakasegawa, H.: An optimal design problem for limited processor sharing systems. Manag. Sci. **33**(8), 1010–1019 (1987)
14. Dudin, A.N., Dudin, S.A., Dudina, O.S., Samouylov, K.E.: Analysis of queueing model with processor sharing discipline and customers impatience. Oper. Res. Perspect. **5**, 245–255 (2018)
15. Telek, M., Van Houdt, B.: Response time distribution of a class of limited processor sharing queues. ACM SIGMETRICS Perform. Eval. Rev. **45**(3), 143–155 (2018)
16. Dudin, A., Dudin, S., Dudina, O.: Analysis of a queueing system with mixed service discipline. Methodol. Comput. Appl. Probab. **25**(2), 57 (2023)
17. Graham, A.: Kronecker Products and Matrix Calculus with Applications. Courier Dover Publications, Mineola (2018)
18. Klimenok, V.I., Dudin, A.N.: Multi-dimensional asymptotically quasi-Toeplitz Markov chains and their application in queueing theory. Queueing Syst. **54**, 245–259 (2006)
19. Dudin, S., Dudin, A., Kostyukova, O., Dudina, O.: Effective algorithm for computation of the stationary distribution of multi-dimensional level-dependent Markov chain with upper block-Hessenberg structure of the generator. J. Comput. Appl. Math. **366**, 112425 (2020)
20. Dudina, O., Kim, C., Dudin, S.: Retrial queuing system with Markovian arrival flow and phase-type service time distribution. Comput. Ind. Eng. **66**(2), 360–373 (2013)
21. Dudin, S., Dudina, O.: Retrial multi-server queuing system with PHF service time distribution as a model of a channel with unreliable transmission of information. Appl. Math. Model. **65**, 676–695 (2019)

22. Kemeny, J.G., Snell, J.L., Knapp, A.W.: Denumerable Markov Chains: With a Chapter of Markov Random Fields by David Griffeath, vol. 40. Springer, Cham (2012)
23. Bini, D.A., Latouche, G., Meini, B.: Numerical Methods for Structured Markov Chains. OUP Oxford (2005)
24. Dudin, A., Dudin, S., Klimenok, V., Dudina, O.: Stability of queueing systems with impatience, balking and non-persistence of customers. Mathematics **12**(14) (2024)

# Reliability Analysis of a k-out-of-n System with External Service and Non-preemptive Priority Under N-Policy with Multiple Server Vacations

Binumon Joseph[1] and K. P. Jose[2](✉)

[1] Government Engineering College Idukki, Painavu 685603, Kerala, India
[2] PG & Research Department of Mathematics, St. Peter's College, Kolenchery 682311, Kerala, India
kpjspc@gmail.com

**Abstract.** This paper examines the reliability of a k-out-of-n system, where a single server takes several vacations. In addition to servicing defective system components, the server offers its services to external customers. Poisson distribution is used to describe the arrival of externally failing customers and the failure rate of internal components. The service time of internal components and external customers follows an exponential distribution. N policy governs the non-preemptive external services. When the number of defective system components exceeds N, the server's vacation will end and internal service will resume. The Matrix Geometric method is used to analyse the model. A suitable cost function and several system performance measures are established.

**Keywords:** Reliability · k-out-of-n system · Multiple vacation · N-Policy · Matrix-Analytic Method

## 1 Introduction

There is always some risk of component failure in any system. If the system can continue to function without any problems, even if certain components fail, the system will be more dependable. A k-out-of-n system has at least k functional components out of a total of n components, to ensure functioning efficiently. Recently, k-out-of-n systems have been extensively studied in the literature, in the context of system optimisation and reliability computation. A k-out-of-n system with an unreliable server that takes multiple vacations under (N,T)policy has been studied by Chakravarthy et al. [3]. There is a chance of component failure, and the failure time has an exponential distribution. Phase-type distribution governs both the component's service time and the failing server's repair

---

The authors acknowledge the financial support provided by FIST Program, Dept. of Science and Technology, Govt. of India, to the PG & Research Dept. of Mathematics through SR/FST/College-2018- XA 276(C).

time. Chakravarthy et al. [2] examined a system with an unreliable server that serves a damaged machine from a pool of N machines, where a server's failure time is exponentially distributed. There is a phase-type distribution to both the repair time of the server and the service time of a broken machine.

When a server is idle, it might be utilised to serve customers from outside to increase revenue from commercial operations. The server's experience will also improve with such outside customer service. Krishnamoorthy et al. [8] conducted an analysis of a k-out-of-n system that provides MAP arrival service to external clients. Phase-type distributions are used to serve defective system components and for external customers. Dudin et al. [4] conducted an analysis of a k-out-of-n system that utilized idle time to serve external customers. If the server is busy, external clients with BMAP arrival are sent to an orbit. Krishnamoorthy et al. [9] analysed the reliability of a k-out-of-n system serving external clients and derived several performance measures using the Matrix geometric technique. The switching of servers between internal and external clients is controlled by N-policy. Under N policy, Joseph and Jose [7] examined a k-out-of-n system that had server vacation and offered service to outside clients. The service time of internal components and external customers follows an exponential distribution. Thresiamma and Jose [12] minimise the long-term cost in a production inventory system with positive service time by applying the N policy. Wu et al. [14] investigated a single-vacation, k-out-of-n: G repairable system whose vacation and repair times are distributed according to general distributions. Several reliability measures including availability, failure rate, and mean time to first failure of the system are developed in a steady-state by employing the supplementary variable technique.

Jain and Jain [6] examined a problem involving machine repair, that involved several servers and an asynchronous server vacation. In the study of a standby system with a single repairman, Yang et al. [15] developed a working vacation, in which the repairman performs service at a reduced rate while on vacation. Wang et al. [13] examined a repairable system with non-identical components under phase-type distributed multiple vacations of a single server. Liu et al. [10] presented a mixed redundancy technique in the reliability study of a multistate system under phase-type distributed multiple vacations of a single server. The number of failed components in the system at any given time was calculated in Eryilmaz [5] by looking at a k-out-of-n system with different component types and non-identical failure distributions.

This paper examines a k-out-of-n system with a single server that serves both external clients and failed internal components. In order to balance the primary and external services and preserve system reliability, a non-preemptive N-policy is implemented. If the number of failed internal components reaches N, during the service of an external failed component, then after completing the ongoing service of the external failed component, the server changes its service to internal service. If the system is free from external failed components and the number of internal failed components is less than N, the server takes multiple vacations.

The rest of the paper is structured as follows: In Sect. 2, the mathematical model is defined and examined. The steady-state probability vector and stability requirements of the system are determined in Sect. 3. System performance measures are introduced in Sect. 4. The numerical analysis of the model is illustrated in Sect. 5, to show how N-policy and outside client service affect system reliability. Section 6 gives conclusion.

## 2 Mathematical Modelling and Analysis of the Problem

A k-out-of-n system starts functioning with all its components working well. During the operation of the system, the system components are subject to failure. The failure time of a system component follows an exponential distribution with parameter $\lambda_s/i$, where $i$ is the number of working components in the system. Then the average failure rate of system components per unit time is $\lambda_s$. A single server, service the failed system components, and the service time of failed components follows an exponential distribution with parameter $\mu_s$. The idle time of the server is effectively utilised by offering service to the external customers, who are from outside the system. The arrival of external customers to the system follows an exponential distribution with parameter $\lambda_e$. The service time of external customers is exponentially distributed with parameter $\mu_e$. As a result of the service, the server gains more experience from varied external service scenarios, its performance will improve. Additionally, this will raise revenue without sacrificing the system's reliability. The service of failed internal system components and external failed customers are controlled by N-policy. The server starts the service of internal failed system components when the number of internal failed components reaches the level N. If the server is busy with the service of external customers when the number of internal failed system components reaches the level N, the server changes its service from external service to internal service after completing the ongoing external service.

The server goes on vacation when the system is free from external customers and the number of internal failed components is less than N. The vacation time is exponentially distributed with parameter $\nu$. The server takes a vacation after the service of internal failed components, even when external customers are waiting for service. After completing one server vacation, if the number of failed internal components is less than N and there are no external customers waiting for service, then the server goes for another vacation. Additionally, if the server is on vacation, the vacation is interrupted when the failure count of internal components reaches N, and the server promptly repairs each of the N internal failures one by one. External customers do not join the system for service when the server is occupied with internal components. Also, when the server is busy with external customers and the number of internal failed system components is N or more, a newly arrived external customer does not join the system. If not, the external customers are added to a queue of infinite length.

Let $E(t)$ represent the number of external components in the system, $I(t)$ be the number of internal failed components and $S(t)$ the status of the server at time $t$.

$$S(t) = \begin{cases} 0, & \text{if the server is in vacation,} \\ 1, & \text{if the server services the internal failed components,} \\ 2, & \text{if the server services the external customers.} \end{cases}$$

Then $\{X(t), t \geq 0\}$ where $X(t) = (E(t), S(t), I(t))$ is a continuous time Markov chain with the state space $\Omega = \{(i_1, 0, i_2)/i_1 \geq 0, 0 \leq i_2 \leq N-1\} \cup \{(i_1, 1, i_2)/i_1 \geq 0, 1 \leq i_2 \leq n-k+1\} \cup \{(i_1, 2, i_2)/i_1 \geq 1, 0 \leq i_2 \leq n-k+1\}$.
In sequel, we use the following notations:

1. $I_n$ - $n^{th}$ order identity matrix.
2. $E_k$ - $k^{th}$ order square matrix defined as

$$E_k(i, j) = \begin{cases} 1, & \text{if } j = i+1; 1 \leq i \leq k-1, \\ , & \text{if } j = i; 1 \leq i \leq k, \\ , & \text{otherwise.} \end{cases}$$

3. $E'_k$ - transpose of $E_k$.
4. $r_k(i)$ - 1×k order row matrix with $i^{th}$ element is 1 and all other elements are zeros.
5. $c_k(i)$ - transpose of $r_k(i)$.
6. **e** - a column matrix of appropriate order with all elements 1.
7. $\otimes$ - Kronecker product of matrices.

The block tridiagonal infinitesimal generator matrix of $\{X(t), t \geq 0\}$ is

$$Q = \begin{pmatrix} B_1 & B_0 & & \\ B_2 & A_1 & A_0 & \\ & A_2 & A_1 & A_0 \\ & & \ddots & \ddots & \ddots \end{pmatrix}, \text{ where } B_1 = \begin{pmatrix} B_{11} & B_{12} \\ B_{13} & B_{14} \end{pmatrix}, B_{11} = \lambda_s E_N - \lambda_e I_N,$$

$B_{12} = \lambda_s \Big( C_N(N) \otimes r_{n-k+1}(N) \Big), \quad B_{13} = \mu_s \Big( C_{n-k+1}(1) \otimes r_N(1) \Big),$

$B_{14} = \lambda_s E_{n-k+1} + \mu_s E'_{n-k+1} + \lambda_s \Big( C_{n-k+1}(n-k+1) \otimes r_{n-k+1}(n-k+1) \Big),$

$B_0 = \begin{pmatrix} \lambda_e I_N & 0 & 0 \\ 0 & 0 & 0 \end{pmatrix}, B_2 = \begin{pmatrix} 0 & 0 \\ 0 & 0 \\ B_{21} & B_{22} \end{pmatrix}, B_{21} = \begin{pmatrix} \mu_e I_N \\ 0 \end{pmatrix},$

$B_{22} = \begin{pmatrix} 0 & 0 \\ 0 & \mu_e I_{(n-k+2-N)} \end{pmatrix}.$

$A_1 = \begin{pmatrix} A_{11} & A_{12} & A_{13} \\ A_{14} & A_{15} & A_{16} \\ A_{17} & A_{18} & A_{19} \end{pmatrix}, \text{ where } A_{11} = \lambda_s E_N - (\lambda_e + \nu) I_N, A_{13} = \begin{pmatrix} \nu I_N & 0 \end{pmatrix},$

$A_{12} = \lambda_s \Big( C_N(N) \otimes r_{n-k+1}(N) \Big), \quad A_{14} = \mu_s \Big( C_{n-k+1}(1) \otimes r_N(1) \Big),$

$A_{15} = \lambda_s E_{n-k+1} + \mu_s E'_{n-k+1} + \lambda_s \Big( r_{n-k+1}(n-k+1) \otimes C_{n-k+1}(n-k+1) \Big),$

$A_{16} = 0_{(n-k+1)\times(n-k+2)}$, $A_{17} = 0_{(n-k+2)\times N}$, $A_{18} = 0_{(n-k+2)\times(n-k+1)}$,

$$A_{19} = \begin{pmatrix} \lambda_s E_N - (\lambda_e + \mu_e)I_N & \lambda_1\left(C_N(N) \otimes r_{(n-k+2-N)}\right) \\ 0_{(n-k+2-N)\times N} & A_{191} \end{pmatrix}, \text{ where}$$

$A_{191} = \lambda_s E_{n-k+2-N} + \mu_s E'_{n-k+2-N} + \lambda_s\left(r_{n-k+2-N}(n-k+2-N) \otimes C_{n-k+2-N}(n-k+2-N)\right)$.

$$A_0 = \begin{pmatrix} \lambda_e I_N & 0 & 0 \\ 0 & 0 & 0 \\ 0 & 0 & A_{01} \end{pmatrix}, A_{01} = \begin{pmatrix} \lambda_e I_N & 0 \\ 0 & 0 \end{pmatrix}, A_2 = \begin{pmatrix} 0 & 0 & 0 \\ 0 & 0 & 0 \\ 0 & A_{21} & A_{22} \end{pmatrix},$$

$$A_{21} = \begin{pmatrix} 0 & 0 \\ 0 & \mu_e I_{(n-k+2-N)} \end{pmatrix}, A_{22} = \begin{pmatrix} \mu_e I_N & 0 \\ 0 & 0 \end{pmatrix}.$$

## 3 Stability Condition

Let $\boldsymbol{\Pi} = (\Pi_1, \Pi_2, \Pi_3)$ be the steady state probablity vector of the generator matrix $A = A_2 + A_1 + A_0 = \begin{pmatrix} A_{11}^* & A_{12}^* & A_{13}^* \\ A_{21}^* & A_{22}^* & 0 \\ 0 & A_{32}^* & A_{33}^* \end{pmatrix}$,

where $A_{11}^* = \lambda_s E_N - \nu I_N, A_{12}^* = A_{12}, A_{13}^* = A_{13}, A_{21}^* = A_{14}, A_{22}^* = A_{15}, A_{32}^* = A_{21}$, and $A_{33}^* = \begin{pmatrix} \lambda_s E_N & \lambda_s\left(C_N(N) \otimes r_{(n-k+2-N)}\right) \\ 0_{(n-k+2-N)\times N} & A_{191} \end{pmatrix}$. Then $\boldsymbol{\Pi} A = 0$ and $\boldsymbol{\Pi} e = 1$.

Let $\Pi$ is subdivided as $\Pi_1 = ((\pi_{(1,1)}, \pi_{(1,2)}, \pi_{(1,3)} \ldots, \pi_{(1,N)}), \Pi_2 = (\pi_{(2,1)}, \pi_{(2,2)}, \pi_{(2,3)} \ldots, \pi_{(2,n-k+1)}), \Pi_3 = (\pi_{(3,1)}, \pi_{(3,2)}, \pi_{(3,3)} \ldots, \pi_{(3,n-k+2)}))$.

From $\boldsymbol{\Pi} A = 0$ we obtain,

$$\Pi_1 A_{11}^* + \Pi_2 A_{21}^* = 0 \quad (1)$$
$$\Pi_1 A_{12}^* + \Pi_2 A_{22}^* + \Pi_3 A_{32}^* = 0 \quad (2)$$
$$\Pi_1 A_{13}^* + \Pi_3 A_{33}^* = 0 \quad (3)$$

From Eq. (1),

$$\pi_{(1,j)} = \pi_{(2,1)} \frac{\mu_s \lambda_s^{j-1}}{(\lambda_s + \nu)^j}, 1 \leq j \leq N. \quad (4)$$

Equation (3) gives

$$\Pi_3 = -\Pi_1 A_{13}^* A_{33}^{*-1}. \quad (5)$$

The matrix $A_{33}^*$ can be partetioned as $A_{33}^* = \begin{pmatrix} A_{331}^* & A_{332}^* \\ 0 & A_{334}^* \end{pmatrix}$. Hence $A_{33}^{*-1} = \begin{pmatrix} A_{331}^{*-1} & -A_{331}^{*-1} A_{332}^* A_{334}^{*-1} \\ 0 & A_{334}^{*-1} \end{pmatrix}$ (Chakravarthy [1]). Then the matrix $A_{13}^* A_{33}^{*-1}$ becomes,

$$A_{13}^*{A_{33}^*}^{-1}_{(i,j)} = \begin{cases} \dfrac{-\nu}{\lambda_s}, & 1 \le i \le N, 1 \le j \le N, i \le j, \\[2mm] \dfrac{-\nu(\lambda_s)^{j-N-1}}{(\lambda_s+\mu_e)^{j-N}}, & 1 \le i \le N, N+1 \le j \le (n-k+1), \\[2mm] \dfrac{-\nu(\lambda_s)^{n-k+1-N}}{(\lambda_s+\mu_e)^{n-k+1-N}\mu_2}, & 1 \le i \le N, j = (n-k+2), \\[2mm] 0, & \text{otherwise.} \end{cases} \qquad (6)$$

Substituting Eqs. (4) and (6) in (5) we obtain

$$\pi_{(3,j)} = \begin{cases} \pi_{(2,1)} \dfrac{\mu_s}{\lambda_s} \left[1 - \left(\dfrac{\lambda_s}{\lambda_s+\nu}\right)^j\right], & 1 \le j \le N, \\[2mm] \pi_{(2,1)} \left[1 - \left(\dfrac{\lambda_s}{\lambda_s+\nu}\right)^N\right] \dfrac{\mu_1 \lambda_s^{j-N-1}}{(\lambda_s+\mu_e)^{j-N}}, & N+1 \le j \le n-k+1, \\[2mm] \pi_{(2,1)} \left[1 - \left(\dfrac{\lambda_s}{\lambda_s+\nu}\right)^N\right] \dfrac{\mu_1 \lambda_s^{j-N-1}}{\mu_e(\lambda_s+\mu_e)^{j-N-1}}, & j = n-k+2. \end{cases} \qquad (7)$$

The above Markov chain is stable if and only if $\Pi A_0 e < \Pi A_2 e$. We partetioned $\Pi_3$ as $\Pi_{31}$ and $\Pi_{32}$, where $\Pi_{31} = (\pi_{(3,1)}, \pi_{(3,2)}, \ldots, \pi_{(3,N)})$ and $\Pi_{32} = (\pi_{(3,N+1)}, \pi_{(3,N+2)}, \ldots, \pi_{(3,n-k+2)})$. Then

$$\Pi A_0 e = \lambda_e [\Pi_1 e + \Pi_{31} e]. \qquad (8)$$

From Eq. (4),

$$\Pi_1 e = \pi_{(2,1)} \frac{\mu_s}{\nu} \left[1 - \left(\frac{\lambda_s}{\lambda_s+\nu}\right)^N\right]. \qquad (9)$$

$\Pi_{31} e$ can be obtained from Eq. (7) as

$$\Pi_{31} e = \pi_{(2,1)} \left[\frac{N\mu_s}{\lambda_s} - \frac{\mu_s}{\nu}\left(1 - \left(\frac{\lambda_s}{\lambda_s+\nu}\right)^N\right)\right]. \qquad (10)$$

$$\Pi A_0 e = \pi_{2,1} \frac{N\lambda_e \mu_s}{\lambda_s}. \qquad (11)$$

$$\Pi A_2 e = \mu_e [\Pi_{31} e + \Pi_{32} e]. \qquad (12)$$

$$\Pi_{32} e = \pi_{(2,1)} \frac{\mu_s}{\mu_e} \left[1 - \left(\frac{\lambda_s}{\lambda_s+\nu}\right)^N\right]. \qquad (13)$$

From Eq. (12),

$$\Pi A_2 e = \pi_{(2,1)} \mu_e \left[ \frac{N\mu_s}{\lambda_s} + \left( \frac{\mu_s}{\mu_e} - \frac{\mu_s}{\nu} \right) \left( 1 - \left( \frac{\lambda_s}{\lambda_s + \nu} \right)^N \right) \right]. \quad (14)$$

**Theorem 1.** *The system is stable if and only if*

$$N \left( \frac{\lambda_e - \mu_e}{\lambda_s} \right) < \left( \frac{\nu - \mu_e}{\nu} \right) \left[ 1 - \left( \frac{\lambda_s}{\lambda_s + \nu} \right)^N \right].$$

*Proof.* The continuous time Markov chain $\{X(t), t \geq 0\}$ is stable if and only if $\Pi A_0 e < \Pi A_2 e$.

Using Eq. (11) and Eq. (14), the stability condition $\Pi A_0 e < \Pi A_2 e$ become

$$N \left( \frac{\lambda_e - \mu_e}{\lambda_s} \right) < \left( \frac{\nu - \mu_e}{\nu} \right) \left[ 1 - \left( \frac{\lambda_s}{\lambda_s + \nu} \right)^N \right]. \quad (15)$$

### 3.1 Steady State Probability Vector

The Markov process $\{X(t), t \geq 0\}$ is a level-independent quasi birth death (QBD) process. The stationary distribution when it exists, has a matrix geometric solution. Let $\mathbf{x} = (x_0, x_1, x_2, \ldots)$ be the probability steady state vector of $Q$, the generator matrix of the process. Then $\mathbf{x}$ satisfies the equation $\mathbf{x}Q = 0$ and the normalizing condition $\mathbf{x}\mathbf{e} = 1$. Here $\mathbf{e}$ represents the column matrix of 1's with infinite order. Then

$$x_{i+1} = x_i \, R \; \forall \, i \geq 1,$$

where R is the minimal nonnegative solution of the matrix equation $A_0 + RA_1 + R^2 A_2 = 0$ [11]. The boundary probability vectors $x_0$ and $x_1$ are obtained from the equations

$$x_0 B_0 + x_1 B_2 = 0,$$
$$x_0 B_1 + x_1 (RA_2 + A_1) = 0.$$

Using normalization, $x_0 \mathbf{e} + x_1 (I - R)^{-1} \mathbf{e} = 1$, one can solve the equations for $x_0$ and $x_1$.

## 4  System Performance Measures

The important system performance measures are given below.

1. Portion of time the system is down,
$$P_F = \sum_{i=0}^{\infty} x(i, 1, n-k+1) + \sum_{i=1}^{\infty} x(i, 2, n-k+1).$$
2. Reliability of the system, $P_R = 1 - P_F$.

3. The average number of external customers in the queue,
$$N_Q = \sum_{i=0}^{\infty} i \sum_{j=0}^{N-1} x(i,0,j) + \sum_{i=0}^{\infty} i \sum_{j=1}^{n-k+1} x(i,1,j) + \sum_{i=1}^{\infty} (i-1) \sum_{j=0}^{n-k+1} x(i,2,j).$$
4. The average number of failed system components,
$$N_{IF} = \sum_{j=0}^{N-1} j \sum_{i=0}^{\infty} x(i,0,j) + \sum_{j=1}^{n-k+1} j \sum_{i=0}^{\infty} x(i,1,j) + \sum_{j=0}^{n-k+1} j \sum_{i=1}^{\infty} x(i,2,j).$$
5. Fraction of time the server in an external service, $P_{EB} = \sum_{i=1}^{\infty} \sum_{j=0}^{n-k+1} x(i,2,j).$
6. Probability that the server was found on vacation, $P_v = \sum_{i=0}^{\infty} \sum_{j=0}^{N-1} x(i,0,j).$
7. Expected rate of external customer loss,
$$E_{EL} = \lambda_e \sum_{i=0}^{\infty} \sum_{j=1}^{n-k+1} x(i,1,j) + \lambda_e \sum_{i=1}^{\infty} \sum_{j=N}^{n-k+1} x(i,2,j).$$
8. Average number of external customers waiting while the server is on vacation,
$$EE_v = \sum_{i=0}^{\infty} i \sum_{j=0}^{N-1} x(i,0,j).$$
9. Average number of internal components waiting while the server is on vacation, $EI_v = \sum_{j=0}^{N-1} j \sum_{i=0}^{\infty} x(i,0,j).$

## 5 Numerical Analysis

This section describes the numerical experiments that were conducted to examine the effects of various parameters on the performance measurements.

### 5.1 Effect of N Policy on Performance Measures

In Table 1, various performance measures are listed for different values of the policy level N. The value of $P_F$ increases with an increase in N. For lower values of N, the server spends more time serving the main components, and hence $N_Q$, the number of outside customers waiting for service decreases with an increase of N. Also, $P_{EB}$ increases with an increase of N. As N increases, the server spends more time on external services, and hence $E_{EL}$ decreases and $N_{IF}$ increases with an increase in N. In this model, the idle time is used for vacation and external service. Then an increase in server's busy period with external customers results in a decrease in the vacation time. So $P_v$ and $EE_v$ decrease and $EI_v$ increases with an increase of N.

**Table 1.** The effect of N policy level on performance measures for $\lambda_s = 5, \lambda_e = 4, \mu_s = 7, \mu_e = 5, n = 50, k = 20, \nu = 4$.

| N  | $P_F$    | $N_Q$  | $E_{EL}$ | $N_{IF}$ | $P_v$  | $P_{EB}$ | $EE_v$ | $EI_v$ |
|----|----------|--------|----------|----------|--------|----------|--------|--------|
| 2  | 0.000014 | 7.6796 | 3.1225   | 3.5794   | 0.1102 | 0.1755   | 0.7814 | 0.0433 |
| 3  | 0.000017 | 6.6748 | 3.0760   | 4.0733   | 0.1009 | 0.1848   | 0.5929 | 0.0757 |
| 5  | 0.000026 | 5.6952 | 3.0135   | 5.0454   | 0.0884 | 0.1973   | 0.3988 | 0.1271 |
| 7  | 0.000041 | 5.2302 | 2.9758   | 6.0169   | 0.0809 | 0.2048   | 0.3021 | 0.1744 |
| 10 | 0.000085 | 4.8776 | 2.9430   | 7.4836   | 0.0744 | 0.2114   | 0.2261 | 0.2482 |
| 13 | 0.000185 | 4.6938 | 2.9238   | 8.9580   | 0.0706 | 0.2153   | 0.1854 | 0.3263 |
| 16 | 0.000419 | 4.5834 | 2.9110   | 10.4320  | 0.0682 | 0.2178   | 0.1604 | 0.4074 |
| 19 | 0.000978 | 4.5103 | 2.9010   | 11.8930  | 0.0666 | 0.2198   | 0.1437 | 0.4906 |
| 22 | 0.002329 | 4.4581 | 2.8911   | 13.3190  | 0.0656 | 0.2218   | 0.1320 | 0.5763 |
| 25 | 0.005632 | 4.4181 | 2.8772   | 14.6600  | 0.0652 | 0.2246   | 0.1239 | 0.6663 |

## 5.2 System's Reliability Corresponding to Variation of Different System Parameters

From Table 2 and Table 3, it is clear that as the N policy level increases the reliability of the system slowly decreases. Also, it is observed that, the increase in the failure rate of the system components results in a decrease in the reliability of the system. At the same time the increase in the service rate of internal components helps to improve the system reliability. Due to the N policy, the arrival rate of external customers to the system and the service rate of external customers do not much affect the reliability of the system. Since the service priority is nonpreemptive, the increase in the service rate of external customers slightly improve the system reliability. By the assumption of vacation interruption and N policy, the vacation parameter does not affect the reliability of the system. The reliability of the system directly depends on the number of working components of the system. As n, the total number of system components increases, the number of working components at a time increases. This results in obtaining a much reliable system (Figs. 1, 2, 3, 4, 5 and 6).

## 5.3 Cost Function

As the N policy level increases, the system's reliability gradually decreases. From Table 1, it is evident that, for higher values of N, the system gains more external customers and has a smaller loss of external customers. The income from the external service decreases, and the expected loss rate of external consumers increases for smaller values N. It is necessary to evaluate the optimal figure for N in order to minimise system costs without affecting reliability. This section formulates and analyses a suitable cost function with respect to the fluctuation of N and various parameters. The cost per unit of time incurred if the system fails is shown by $C_1$. The holding cost of each external customer within the

**Table 2.** System's Reliability corresponding to different values of $\lambda_s, \lambda_e$ and $\nu$

| N | $P_R$ | | | $P_R$ | | | $P_R$ | | |
|---|---|---|---|---|---|---|---|---|---|
| | $\lambda_s=4$ | $\lambda_s=5$ | $\lambda_s=6$ | $\lambda_e=3$ | $\lambda_e=3.5$ | $\lambda_e=4$ | $\nu=4$ | $\nu=4.5$ | $\nu=5$ |
| 2 | 1.00000 | 0.99999 | 0.99850 | 0.99999 | 0.99999 | 0.99999 | 0.99999 | 0.99999 | 0.99999 |
| 3 | 1.00000 | 0.99998 | 0.99836 | 0.99998 | 0.99998 | 0.99998 | 0.99998 | 0.99998 | 0.99998 |
| 5 | 1.00000 | 0.99997 | 0.99806 | 0.99998 | 0.99997 | 0.99997 | 0.99997 | 0.99997 | 0.99997 |
| 6 | 1.00000 | 0.99997 | 0.99788 | 0.99997 | 0.99997 | 0.99997 | 0.99997 | 0.99997 | 0.99997 |
| 8 | 1.00000 | 0.99995 | 0.99745 | 0.99995 | 0.99995 | 0.99995 | 0.99995 | 0.99995 | 0.99995 |
| 9 | 1.00000 | 0.99993 | 0.99719 | 0.99994 | 0.99994 | 0.99993 | 0.99993 | 0.99993 | 0.99993 |
| 11 | 1.00000 | 0.99989 | 0.99659 | 0.99990 | 0.99989 | 0.99989 | 0.99989 | 0.99989 | 0.99989 |
| 12 | 1.00000 | 0.99986 | 0.99622 | 0.99987 | 0.99986 | 0.99986 | 0.99986 | 0.99986 | 0.99986 |
| 14 | 0.99999 | 0.99976 | 0.99534 | 0.99978 | 0.99977 | 0.99976 | 0.99976 | 0.99976 | 0.99976 |
| 15 | 0.99999 | 0.99968 | 0.99482 | 0.99971 | 0.99969 | 0.99968 | 0.99968 | 0.99968 | 0.99968 |
| 17 | 0.99997 | 0.99945 | 0.99353 | 0.99949 | 0.99947 | 0.99945 | 0.99945 | 0.99945 | 0.99945 |
| 18 | 0.99995 | 0.99926 | 0.99276 | 0.99932 | 0.99929 | 0.99926 | 0.99926 | 0.99927 | 0.99927 |
| 20 | 0.99987 | 0.99870 | 0.99086 | 0.99880 | 0.99875 | 0.99870 | 0.99870 | 0.99870 | 0.99870 |
| 21 | 0.99979 | 0.99826 | 0.98969 | 0.99840 | 0.99833 | 0.99826 | 0.99826 | 0.99826 | 0.99826 |
| 23 | 0.99942 | 0.99688 | 0.98683 | 0.99713 | 0.99700 | 0.99688 | 0.99688 | 0.99688 | 0.99688 |
| 24 | 0.99904 | 0.99581 | 0.98506 | 0.99614 | 0.99598 | 0.99581 | 0.99581 | 0.99581 | 0.99582 |

**Table 3.** System's Reliability corresponding to different values of $\mu_s, \mu_e$ and $n$.

| N | $P_R$ | | | $P_R$ | | | $P_R$ | | |
|---|---|---|---|---|---|---|---|---|---|
| | $\mu_s=6$ | $\mu_s=7$ | $\mu_s=8$ | $\mu_e=5$ | $\mu_e=6$ | $\mu_e=7$ | $n=45$ | $n=50$ | $n=55$ |
| 2 | 0.99926 | 0.99999 | 1.00000 | 0.99999 | 0.99999 | 0.99999 | 0.99992 | 0.99999 | 1.00000 |
| 3 | 0.99918 | 0.99998 | 1.00000 | 0.99998 | 0.99998 | 0.99999 | 0.99991 | 0.99998 | 1.00000 |
| 5 | 0.99899 | 0.99997 | 1.00000 | 0.99997 | 0.99998 | 0.99998 | 0.99986 | 0.99997 | 1.00000 |
| 6 | 0.99888 | 0.99997 | 1.00000 | 0.99997 | 0.99997 | 0.99997 | 0.99982 | 0.99997 | 0.99999 |
| 8 | 0.99860 | 0.99995 | 1.00000 | 0.99995 | 0.99995 | 0.99996 | 0.99972 | 0.99995 | 0.99999 |
| 9 | 0.99843 | 0.99993 | 1.00000 | 0.99993 | 0.99994 | 0.99995 | 0.99964 | 0.99993 | 0.99999 |
| 11 | 0.99801 | 0.99989 | 0.99999 | 0.99989 | 0.99990 | 0.99991 | 0.99941 | 0.99989 | 0.99998 |
| 12 | 0.99775 | 0.99986 | 0.99999 | 0.99986 | 0.99987 | 0.99988 | 0.99923 | 0.99986 | 0.99997 |
| 14 | 0.99711 | 0.99976 | 0.99997 | 0.99976 | 0.99979 | 0.99980 | 0.99870 | 0.99976 | 0.99995 |
| 15 | 0.99670 | 0.99968 | 0.99995 | 0.99968 | 0.99972 | 0.99974 | 0.99829 | 0.99968 | 0.99994 |
| 17 | 0.99569 | 0.99945 | 0.99989 | 0.99945 | 0.99951 | 0.99955 | 0.99701 | 0.99945 | 0.99990 |
| 18 | 0.99506 | 0.99926 | 0.99984 | 0.99926 | 0.99935 | 0.99940 | 0.99604 | 0.99926 | 0.99986 |
| 20 | 0.99346 | 0.99870 | 0.99963 | 0.99870 | 0.99885 | 0.99894 | 0.99299 | 0.99870 | 0.99976 |
| 21 | 0.99245 | 0.99826 | 0.99944 | 0.99826 | 0.99847 | 0.99858 | 0.99065 | 0.99826 | 0.99968 |
| 23 | 0.98986 | 0.99688 | 0.99872 | 0.99688 | 0.99725 | 0.99744 | 0.98334 | 0.99688 | 0.99942 |
| 24 | 0.98821 | 0.99581 | 0.99806 | 0.99581 | 0.99630 | 0.99656 | 0.97781 | 0.99581 | 0.99922 |

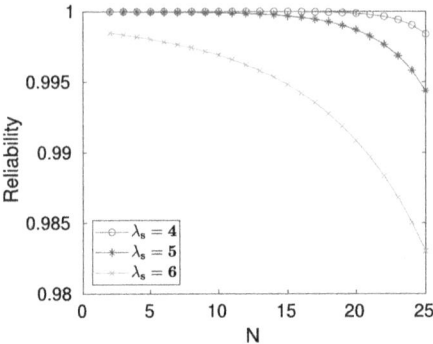

**Fig. 1.** Reliability and failure rate of internal components, $\lambda$.

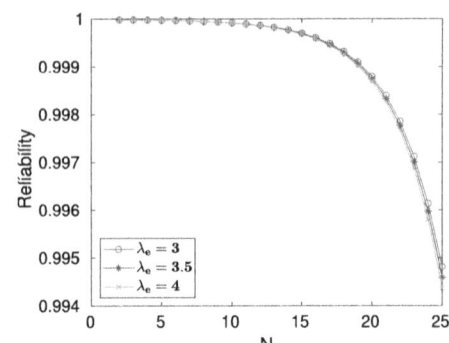

**Fig. 2.** Reliability and arrival rate of external components, $\lambda_e$.

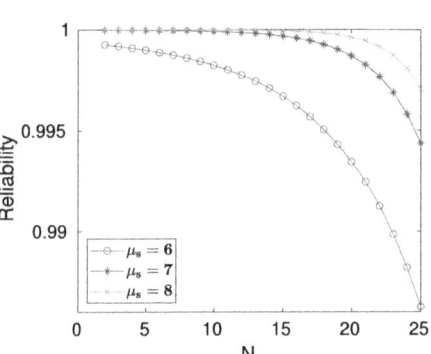

**Fig. 3.** Reliability and the service rate $\mu_s$ of internal components.

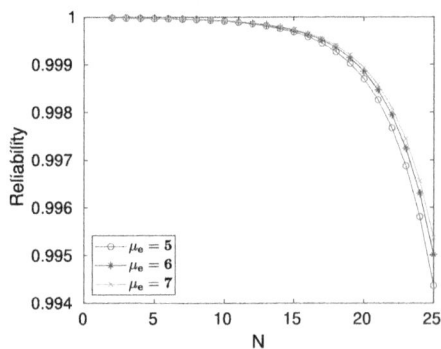

**Fig. 4.** Reliability and the service rate $\mu_e$, of external customers.

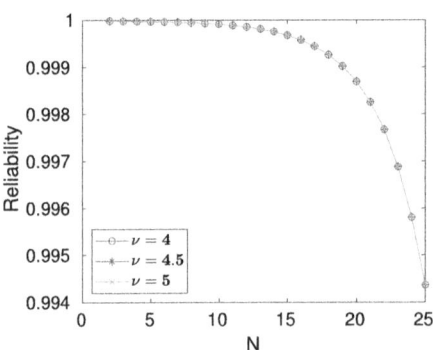

**Fig. 5.** Reliability and the vacation rate $\nu$.

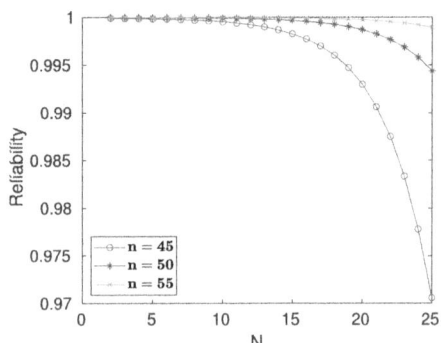

**Fig. 6.** Reliability and n, the total number of system components.

queue for one unit of time is denoted by $C_2$; the cost due to the loss of one external customer is represented by $C_3$; The holding cost of each failed system

**Table 4.** Variation of Cost corresponding to $\lambda_s, \lambda_e$ and $\nu$

| N | Cost | | | Cost | | | Cost | | |
|---|---|---|---|---|---|---|---|---|---|
| | $\lambda_s = 4$ | $\lambda_s = 5$ | $\lambda_s = 6$ | $\lambda_e = 3$ | $\lambda_e = 3.5$ | $\lambda_e = 4$ | $\nu = 4$ | $\nu = 4.5$ | $\nu = 5$ |
| 2 | 124950 | 143630 | 163670 | 78975 | 101000 | 143630 | 143630 | 124560 | 115130 |
| 3 | 117020 | 134880 | 154140 | 77864 | 98408 | 134880 | 134880 | 121690 | 114360 |
| 5 | 109360 | 126390 | 144980 | 76770 | 95641 | 126390 | 126390 | 118490 | 113530 |
| 7 | 106100 | 122670 | 140920 | 76539 | 94521 | 122670 | 122670 | 117110 | 113390 |
| 9 | 104670 | 120930 | 138970 | 76769 | 94219 | 120930 | 120930 | 116630 | 113650 |
| 11 | 104130 | 120180 | 138080 | 77262 | 94366 | 120180 | 120180 | 116660 | 114140 |
| 14 | 104160 | 120030 | 137770 | 78277 | 95048 | 120030 | 120030 | 117210 | 115140 |
| 17 | 104740 | 120500 | 138130 | 79487 | 96043 | 120500 | 120500 | 118110 | 116320 |
| 20 | 105610 | 121340 | 138870 | 80835 | 97239 | 121340 | 121340 | 119230 | 117640 |
| 23 | 106690 | 122460 | 139860 | 82340 | 98627 | 122460 | 122460 | 120550 | 119100 |
| 25 | 107560 | 123390 | 140630 | 83489 | 99705 | 123390 | 123390 | 121590 | 120210 |

**Table 5.** Variation of Cost corresponding to $\mu_s, \mu_e$ and $n$

| N | Cost | | | Cost | | | Cost | | |
|---|---|---|---|---|---|---|---|---|---|
| | $\mu_s = 6$ | $\mu_s = 7$ | $\mu_s = 8$ | $\mu_e = 5$ | $\mu_e = 6$ | $\mu_e = 7$ | $n = 45$ | $n = 50$ | $n = 55$ |
| 2 | 155020 | 143630 | 136120 | 143630 | 114040 | 103700 | 143640 | 143630 | 143630 |
| 3 | 146760 | 134880 | 127000 | 134880 | 109220 | 100300 | 134890 | 134880 | 134880 |
| 5 | 138930 | 126390 | 118010 | 126390 | 104610 | 97186 | 126400 | 126390 | 126390 |
| 7 | 135600 | 122670 | 113990 | 122670 | 102780 | 96137 | 122680 | 122670 | 122670 |
| 9 | 134110 | 120930 | 112060 | 120930 | 102140 | 95974 | 120950 | 120930 | 120930 |
| 11 | 133530 | 120180 | 111180 | 120180 | 102110 | 96257 | 120220 | 120180 | 120180 |
| 14 | 133550 | 120030 | 110900 | 120030 | 102650 | 97110 | 120120 | 120030 | 120020 |
| 17 | 134130 | 120500 | 111270 | 120500 | 103570 | 98237 | 120710 | 120500 | 120470 |
| 20 | 135040 | 121340 | 112010 | 121340 | 104720 | 99528 | 121840 | 121340 | 121260 |
| 23 | 136170 | 122460 | 113040 | 122460 | 106070 | 100970 | 123700 | 122460 | 122260 |
| 25 | 137040 | 123390 | 113910 | 123390 | 107100 | 102050 | 125620 | 123390 | 123020 |

component for one unit of time is represented by $C_4$; and the cost/unit of time if the server is on vacation is represented by $C_5$. Then the expected cost/unit time, $Cost = C_1 P_F + C_2 N_Q + C_3 E_{EL} + C_4 N_{IF} + C_5 P_v$.

The effect of N policy level on cost function is explained in Tables 4 and 5 and Figs. 7, 8, 9, 10, 11 and 12. It shows the existence of an optimum value for N. Up to a certain point, the cost decreases as N increases, but after a value of N, the cost increases as N increases. All these estimations are done by taking $C_1 = 200000, C_2 = 8000, C_3 = 25000, C_4 = 1000$ and $C_5 = 5000$.

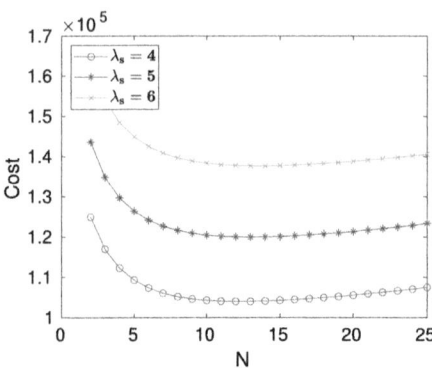

**Fig. 7.** Cost variation corresponding to failure rate of internal components, $\lambda_s$.

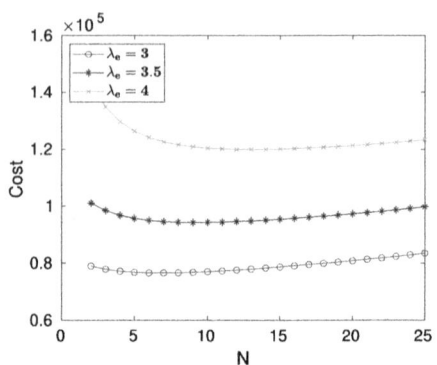

**Fig. 8.** Cost variation corresponding to arrival rate of external customers, $\lambda_e$.

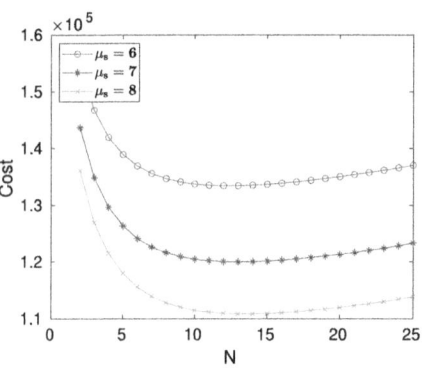

**Fig. 9.** Cost variation corresponding to service rate of internal components $\mu_s$.

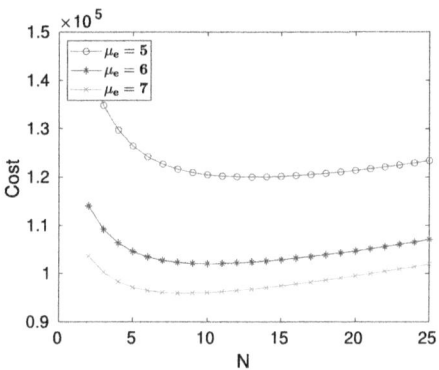

**Fig. 10.** Cost variation corresponding to service rate of external customers $\mu_e$.

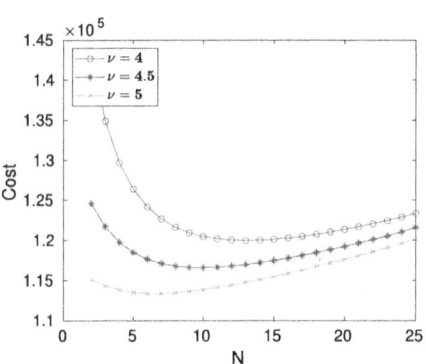

**Fig. 11.** Cost variation corresponding to the vacation parameter $\nu$.

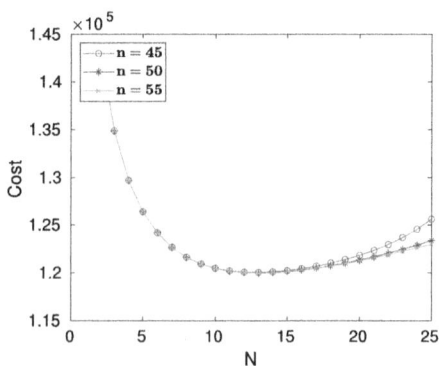

**Fig. 12.** Cost variation corresponding to n, the total number of system components.

## 6 Conclusion

This paper examined the reliability of a k-out-of-n system in which a single server took vacations. An N-policy was regulated to manage the external service. A numerical analysis was conducted to examine the impact of the N-policy on system reliability. The tabular and graphical representations indicated a clear trend: as N increased, system reliability decreased. For each value of N, variations in reliability in response to different system parameters were also analysed. The Matrix Geometric Method was applied to analyse the model, and a suitable cost function was derived, enabling the determination of an optimal N value that minimized total cost for a specified level of system reliability.

Providing services to external customers and taking a server vacation were effective ways to utilize server idle time and boost system income. However, the length of the vacation and external services needed careful management in systems that required a few components to operate to prevent adverse effects on system reliability. In the future, the impact of a working vacation on system reliability and the N-policy will be explored. Another possible extension of this study will involve the use of PH distributions for the service time.

## References

1. Chakravarthy, S.R.: Introduction to Matrix-Analytic Methods in Queues 1. Wiley, Hoboken (2022)
2. Chakravarthy, S.R., Agarwal, A.: Analysis of a machine repair problem with an unreliable server and phase type repairs and services. Naval Res. Logistics (NRL) **50**(5), 462–80 (2003)
3. Chakravarthy, S.R., Krishnamoorthy, A., Ushakumari, P.V.: A k-out-of-n reliability system with an unreliable server and phase type repairs and services: the (N, T) policy. J. Appl. Math. Stoch. Anal. **14**(4), 361–80 (2001)
4. Dudin, A.N., Krishnamoorthy, A., Narayanan, V.C.: Idle time utilization through service to customers in a retrial queue maintaining high system reliability. J. Math. Sci. **191**, 506–17 (2013)
5. Eryilmaz, S.: The number of failed components in a k-out-of-n system consisting of multiple types of components. Reliab. Eng. Syst. Saf. **175**, 246–250 (2018)
6. Jain, A., Jain, M.: Multi server machine repair problem with unreliable server and two types of spares under asynchronous vacation policy. Int. J. Math. Oper. Res. **10**(3), 286–315 (2017)
7. Joseph, B., Jose, K.P.: Analysis of a k-out-of-n reliability system with single server, internal and external service, N-Policy, and multiple server vacations. In: Dudin, A., Nazarov, A., Moiseev, A. (eds.) ITMM WRQ 2023 2023. CCIS, vol. 2163, pp. 3–18. Springer, Cham (2023). https://doi.org/10.1007/978-3-031-65385-8_1
8. Krishnamoorthy, A., Narayanan, V.C., Deepak, T.G.: Reliability of a K-out-of-n system with repair by a service station attending a queue with postponed work. Int. J. Reliab. Qual. Saf. Eng. **14**(04), 379–98 (2007)
9. Krishnamoorthy, A., Sathian, M.K. : Reliability of a k-out-of-n system with repair by a single server extending service to external customers with pre-emption. Reliab. Theory Appl. **11**(2(41)), 61–93 (2016)

10. Liu, B., Wen, Y., Qiu, Q., Shi, H., Chen, J.: Reliability analysis for multi-state systems under K-mixed redundancy strategy considering switching failure. Reliab. Eng. Syst. Saf. **228**, 108814 (2022)
11. Neuts, M.F.: Matrix Geometric Solutions in Stochastic Processes-An Algorithmic Approach. The John Hopkins University Press, Baltimore (1981)
12. Thresiamma, N.J., Jose, K.P.: N-policy for a production inventory system with positive service time. In: Dudin, A., Nazarov, A., Moiseev, A. (eds.) ITMM 2021. CCIS, vol. 1605, pp. 52–66. Springer, Cham (2021). https://doi.org/10.1007/978-3-031-09331-9_5
13. Wang, G., Hu, L., Zhang, T., Wang, Y.: Reliability modeling for a repairable (k1, k2)-out-of-n: G system with phase-type vacation time. Appl. Math. Model. **91**, 311–21 (2021)
14. Wu, W., Tang, Y., Yu, M., Jiang, Y.: Reliability analysis of a k-out-of-n: G repairable system with single vacation. Appl. Math. Model. **38**(24), 6075–6097 (2014)
15. Yang, D.Y., Tsao, C.L.: Reliability and availability analysis of standby systems with working vacations and retrial of failed components. Reliab. Eng. Syst. Saf. **182**, 46–55 (2019)

# On Recursive Marginal and MAP Inference in State Observation Models

Branislav Rudić[1](✉)[iD], Valentin Sturm[1][iD], and Dmitry Efrosinin[2][iD]

[1] Linz Center of Mechatronics GmbH, Altenberger Str. 69, 4040 Linz, Austria
{branislav.rudic,valentin.sturm}@lcm.at
[2] JKU, Institute of Stochastics, Altenberger Str. 69, 4040 Linz, Austria
dmitry.efrosinin@jku.at
https://www.lcm.at/, https://www.jku.at/en/institute-of-stochastics

**Abstract.** Maximum A Posteriori (MAP) inference in state observation models typically covers decoding either marginal MAP state estimates or the joint MAP state sequence estimate. This paper addresses a novel yet fundamental MAP inference method denoted as predecessor decoding. This method recursively decodes the most probable predecessors of a chosen initial state using only the marginal distributions from a forward filtering pass. We elaborate on the motivations, abstract relations and analogues, and in particular, the differences between marginal MAP, joint MAP, and MAP predecessors. We conclude by comparing recent results, where predecessor decoding has been utilized for Gaussian mixture models.

**Keywords:** State Observation Models · Marginal Inference · MAP Inference · Recursive Bayesian Inference · Filtering · Smoothing · Decoding

## 1 Introduction

This paper is an extension of the conference paper [1] originally presented at the 2024 23rd International Conference on Information Technologies and Mathematical Modelling (ITMM'2024). In addition to the content of [1], this work includes a discussion on marginal inference in general state observation models, highlights the relation between the proposed MAP inference method and its marginal inference analogue, and presents a comparison based on simulation results from [2].

State observation models are utilized in various fields such as signal processing, navigation, telecommunications, finance, and more [3]. We attribute this status as state-of-the-art tool in such a broad field of applications to both the great flexibility in modeling different real world problems and the wide variety of available tools for inference and estimation. These inference methods encompass both marginal and Maximum A Posteriori (MAP) inference, applicable to models with either discrete or continuous state spaces [4].

For example, in the context of discrete-state models, specifically Hidden Markov Models (HMM), marginal inference is realized by the forward-backward algorithm, whereas the Viterbi algorithm addresses the joint MAP problem [5]. Unfortunately, grid-based as well as particle-based approaches, such as [6,7], suffer form the curse of dimensionality [8]. In continuous-state models, efficient marginal inference techniques have been discovered presuming Gaussian distributions [3,9]. In such models, the marginal means coincide with the joint MAP estimate [10,11]. Gaussian assumptions can lead to relatively poor results, e.g., in dynamic systems with nonlinear or non-Gaussian characteristics, but especially in systems where the actual posterior distributions might become multimodal.

As a motivating example, consider an application designated to track the positions, linear and angular velocities and accelerations, orientations, and possible other features of a dynamically moving object indirectly observed through measurements affected by multimodal noise. Instead of discretizing or over-simplifying such high-dimensional and non-Gaussian systems to fit into a discrete-state or Gaussian model, more appropriate approximations may be obtained using more general distributions such as convex mixtures. For Gaussian mixture models, a recursive MAP predecessor decoder was recently introduced in [2], which results exhibit joint MAP characteristics, i.e., coherent state estimates consistent with the specified model. This is in stark contrast to traditional inference methods, which generated implausible state trajectories. Our exemplary application would consequently benefit from such coherent results, as improved plausibility inherently increases acceptance among users of any such application.

Given the property, that mixture distributions can be used to approximate any density function to an arbitrary degree, appropriate inference methods hold considerable potential for a range of applications. For mixture models, it has only recently been shown in [12], that an analytic solution exists for recursively inferring the smoothed distributions. The authors believe that the popularity of the non-recursive two-filter approach to smoothing [13] over the analytically intricate recursive smoother [12], is the main reason why its MAP analogue, the recursive decoder, has not yet been discussed.

### 1.1 Contribution and Outline

In this article, the relationship between MAP predecessors and the traditional marginal and MAP inference methods for general state observation models is discussed. In contrast to [2], we emphasize that we do not assume any specific underlying distribution models.

Section 2 covers the fundamentals, scope, graph, and notation related to the underlying model. Marginal inference in state observation models is discussed in Sect. 3, with the focus on recursive smoothing in Sect. 3.1, which is strongly related to the recursive MAP predecessor decoder discussed in Sect. 5, whereas Sect. 4 revises the traditional MAP inference methods. In Sect. 6, we present numeric results of the novel MAP inference method in comparison to traditional methods, to corroborate our findings. Final conclusions are drawn in Sect. 7.

## 2 Model Fundamentals

This section introduces the probabilistic model along with the notation used, establishing the underlying framework for the inference techniques discussed in Sects. 3 through 5. The model comprises two stochastic processes, the dynamic *states* $x_t$ and *observations* $o_t$. A sequence of consecutive random variables, e.g., $x_0, x_1, \ldots, x_T$, is abbreviated as $x_{0:T}$, thereby restricting this paper to discrete-time models. As the name suggests, the $o_{1:T}$ are observable, with each observation $o_t$ causally depending on a corresponding state $x_t$, which remains hidden from the observer. The state is dynamic and can change over time, with each state $x_t$ depending on the previous state $x_{t-1}$ and where the initial state distribution $p(x_0)$ is assumed given. The initial distribution and the dependencies, or rather conditionals

$$p(x_0) \equiv \textit{initialization}, \tag{1}$$
$$p(x_t \mid x_{t-1}) \equiv \textit{transition model}, \tag{2}$$
$$p(o_t \mid x_t) \equiv \textit{observation model}, \tag{3}$$

are the fundamental parameters and have to be specified, either model-based or data-driven, for each possible realization of the variables and for all $t \geq 1$. This work considers probabilistic models over hidden states $x_{0:T}$ and given observations $o_{1:T}$, where the joint distribution factorizes into a product of (1)–(3)

$$p(x_{0:T}, o_{1:T}) = p(x_0) \prod_{t=1}^{T} p(x_t \mid x_{t-1}) p(o_t \mid x_t). \tag{4}$$

Section 5 addresses a novel inference method for this model; to emphasize its significance, it is worth outlining the broad scope of such models. Throughout the cited literature many different names can be found for models satisfying (4), often depending on the context, individual approach or assumptions regarding the involved variables and distributions. For instance,

**Stochastic Dynamical System** [6,9,14,15], often simply referred to as Dynamic System, is used in the context of systems theory;
**Hidden Markov Model** [4,5] traditionally refers to models with discrete-state variables;
**Kalman Filter Model** [10,11] is sometimes used to denote the underlying model of the filtering method discovered by Stratonovich, Kalman, and Bucy, this model is identical to the
**Hidden Gauss-Markov Model** [10,11] indicating that Gaussian distributions are presumed;
**State Space Model** [3,6,7,13] does not presume whether variables are discrete or continuous, or if distributions are Gaussian or non-Gaussian, but can include an input (control) variable additional to output (observation) and state variables;

**Dynamic Bayesian Network** [4] refers to more general probabilistic models which may include a greater number of random variables, respectively an acyclic dynamic network, rather than a single dynamic state variable.

Instead, the authors have adopted the term **State Observation Model** based on [4], which aligns best with the scope of the present paper, since it only presumes the presence of state and observation variables. More importantly, [4] offers a systematic approach to probabilistic graphical models. The probabilistic graph of the state observation model is shown in Fig. 1, where the arrows depict dependencies between variables, white nodes ○ indicate hidden variables, and gray nodes ⬤ indicate that a realization of the random variable is assumed given, either by observation, or in Sect. 5 also by estimation.

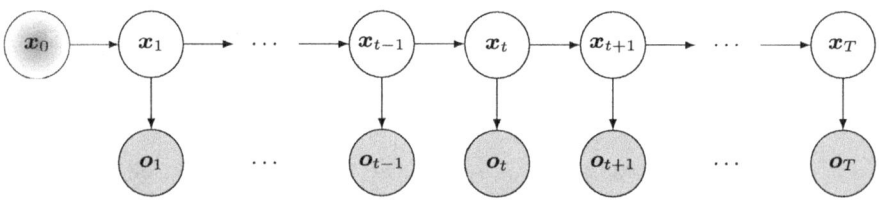

**Fig. 1.** Probabilistic Graph of the State Observation Model.

Distributions, such as $p(\boldsymbol{x}_0)$, are shown as density maps ⊙ inside the respective nodes. In Sects. 3 to 5 the graphical notation is further extended by likelihoods ⊚, MAP estimates ⊙, and dashed arrows indicating how information is passed through the graph in the course of an inference task.

## 3 Marginal Inference in State Observation Models

Marginal inference in state observation models (4) can be subdivided into predicting, filtering and smoothing, which refer to computing the marginal distributions

$$p(\boldsymbol{x}_t \mid \boldsymbol{o}_{1:s}) \equiv \begin{cases} predicting & \Longleftrightarrow s < t \\ filtering & \Longleftrightarrow s = t \\ smoothing & \Longleftrightarrow s > t \end{cases}. \tag{5}$$

These marginals can be inferred either by marginalization of the joint posterior, or recursively based on the factors in (4) thereby exploiting independencies among the random variables of a state observation model. In most cases, only the latter option is feasible. The recursive formulas for the marginals referenced in (5) can for instance also be found in [12], which also includes their derivations.

The filtering recursion consists of an update and a (one-step) prediction step

$$p(\boldsymbol{x}_t \mid \boldsymbol{o}_{1:t}) = \frac{p(\boldsymbol{x}_t \mid \boldsymbol{o}_{1:t-1}) p(\boldsymbol{o}_t \mid \boldsymbol{x}_t)}{p(\boldsymbol{o}_t \mid \boldsymbol{o}_{1:t-1})}, \tag{6}$$

$$p(\boldsymbol{x}_t \mid \boldsymbol{o}_{1:t-1}) = \int p(\boldsymbol{x}_t \mid \boldsymbol{x}_{t-1}) p(\boldsymbol{x}_{t-1} \mid \boldsymbol{o}_{1:t-1}) \, \mathrm{d}\boldsymbol{x}_{t-1}, \tag{7}$$

obtained by applying Bayes' rule and the law of total probability, in (6) and (7), respectively. As already mentioned, filtering refers to both update- (6) and prediction-step (7), where the scalar denominator in the update-step can be omitted $p(\boldsymbol{x}_t \mid \boldsymbol{o}_{1:t}) \propto p(\boldsymbol{x}_t \mid \boldsymbol{o}_{1:t-1}) p(\boldsymbol{o}_t \mid \boldsymbol{x}_t)$. The generic recursive Bayesian filtering formulas (6)–(7) unfold in many different forms, depending on case-specific state space and distribution representation. For instance, in the discrete-state case it becomes the HMM forward algorithm. For continuous-state variables and Gaussian assumptions regarding (1)–(3), Stratonovich, Kalman, and Bucy have discovered (more or less separately around 1960) recursive formulations for filtered mean and covariance matrix. Except for such relatively simple cases, it is deemed impossible to compute closed-form expressions for the marginals [16]. To approximate the filtering marginals in the non-Gaussian case, mixture distributions have been utilized in [14,15] and particle-based methods for instance in [6]. Particle- and grid-based approaches suffer from the curse of dimensionality [8], while mixture approaches have to employ reduction techniques to maintain a feasible number of mixture components [17], which otherwise increases exponentially. However, these challenges are primarily of a practical nature and are specific to the respective filtering approach. With regard to inferring the smoothing marginal $p(\boldsymbol{x}_t \mid \boldsymbol{o}_{1:T})$, complications emerge at an earlier stage and at a more abstract level.

## 3.1 The Smoothing Problem

For smoothing, there are two fundamentally different approaches when inferring the marginal distribution $p(\boldsymbol{x}_t \mid \boldsymbol{o}_{1:T})$. One is based on a recursive or *fixed-lag* decomposition, the other one is based on a non-recursive *two-filter* factorization of $p(\boldsymbol{x}_t \mid \boldsymbol{o}_{1:T})$. Both smoothing approaches apply a form of forward/backward-principle, consisting of two passes through the model graph, one running forward and the other backward in time as illustrated in Fig. 2 and 3.

The recursive approach is obtained by marginalizing the joint distribution $p(\boldsymbol{x}_t, \boldsymbol{x}_{t+1} \mid \boldsymbol{o}_{1:T})$ in (8), applying Bayes' rule on $p(\boldsymbol{x}_t \mid \boldsymbol{x}_{t+1}, \boldsymbol{o}_{1:t})$ in (9), and exploiting conditional independencies

$$p(\boldsymbol{x}_t \mid \boldsymbol{o}_{1:T}) = \int p(\boldsymbol{x}_t, \boldsymbol{x}_{t+1} \mid \boldsymbol{o}_{1:T}) \, \mathrm{d}\boldsymbol{x}_{t+1} \tag{8}$$

$$= \int p(\boldsymbol{x}_t \mid \boldsymbol{x}_{t+1}, \boldsymbol{o}_{1:t}) p(\boldsymbol{x}_{t+1} \mid \boldsymbol{o}_{1:T}) \, \mathrm{d}\boldsymbol{x}_{t+1} \tag{9}$$

$$= p(\boldsymbol{x}_t \mid \boldsymbol{o}_{1:t}) \int \frac{p(\boldsymbol{x}_{t+1} \mid \boldsymbol{x}_t)}{p(\boldsymbol{x}_{t+1} \mid \boldsymbol{o}_{1:t})} p(\boldsymbol{x}_{t+1} \mid \boldsymbol{o}_{1:T}) \, \mathrm{d}\boldsymbol{x}_{t+1}. \tag{10}$$

The filtered $p(\boldsymbol{x}_t \mid \boldsymbol{o}_{1:t})$ and predicted $p(\boldsymbol{x}_{t+1} \mid \boldsymbol{o}_{1:t})$ marginals can be obtained in a forward filtering pass. At $t = T$, the smoothing and filtering marginals are equal, such that the filtered $p(\boldsymbol{x}_T \mid \boldsymbol{o}_{1:T})$ serves as an initialization for the subsequent backward smoothing pass, which is based on the recursive backward correction step (10) for $t < T$. Figure 2 visualizes fixed-lag smoothing with exemplary filtering and smoothing distributions plotted as density maps within the nodes.

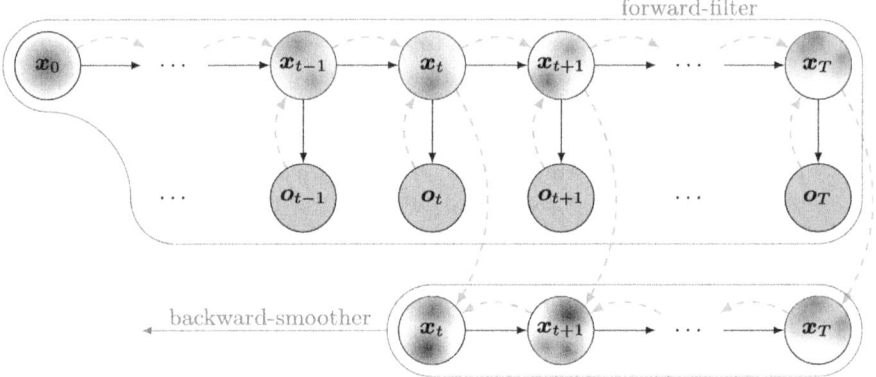

**Fig. 2.** Graphical interpretation of how the marginal distribution $p(\boldsymbol{x}_t \mid \boldsymbol{o}_{1:T})$ is recursively inferred by the forward-filtering/backward-smoothing approach, or *fixed-lag smoothing*.

The problem within this approach is the division with a density in (10), which occurs after applying Bayes' rule on the temporal dependency between $\boldsymbol{x}_t$ and $\boldsymbol{x}_{t+1}$. Note that a division also occurs when applying Bayes' rule on the causal dependency between $\boldsymbol{x}_t$ and $\boldsymbol{o}_t$ in (6), however, since $\boldsymbol{o}_t$ is given, the denominator in this case is a normalizing scalar and can be omitted. We want to underline the importance of this key insight by anticipating that a similar *trick* is exploited in predecessor decoding, see Lemma 1 in Sect. 5.

For the special case where the model is Gaussian, i.e., all factors (1)–(3) are Gaussian distributions, Rauch, Tung, and Striebel (RTS) have discovered recursive formulations for the smoothed mean and covariance matrix in [9] effectively solving (10) for Gaussian models. For mixture models, it has recently been shown in [12] how the division by a mixture density reduces to divisions by the component densities, and how the RTS smoother extends for Gaussian mixture models. However, for non-Gaussian (component) densities the recursive smoothing approach via (10) remains problematic.

To overcome this problem, Kitagawa motivates a two-filter approach in [13] that does not explicitly contain division by a non-Gaussian density. This approach is based on the Bayesian decomposition of the smoothing marginal into

$$p(\boldsymbol{x}_t \mid \boldsymbol{o}_{1:T}) = \frac{p(\boldsymbol{x}_t \mid \boldsymbol{o}_{1:t})\, p(\boldsymbol{o}_{t+1:T} \mid \boldsymbol{x}_t)}{p(\boldsymbol{o}_{t+1:T} \mid \boldsymbol{o}_{1:t})} \tag{11}$$

$$\propto p(\boldsymbol{x}_t \mid \boldsymbol{o}_{1:t})\, p(\boldsymbol{o}_{t+1:T} \mid \boldsymbol{x}_t), \tag{12}$$

resulting in the product of the forward filtered marginal distribution with the *backward information* $p(\boldsymbol{o}_{t+1:T} \mid \boldsymbol{x}_t)$. The latter is not a distribution but a likelihood, which can be recursively filtered backwards in $t$ by

$$p(\boldsymbol{o}_{t+1:T} \mid \boldsymbol{x}_t) = \int p(\boldsymbol{x}_{t+1} \mid \boldsymbol{x}_t)\, p(\boldsymbol{o}_{t+1:T} \mid \boldsymbol{x}_{t+1})\, \mathrm{d}\boldsymbol{x}_{t+1},$$

$$p(\boldsymbol{o}_{t+1:T} \mid \boldsymbol{x}_{t+1}) = p(\boldsymbol{o}_{t+1} \mid \boldsymbol{x}_{t+1})\, p(\boldsymbol{o}_{t+2:T} \mid \boldsymbol{x}_{t+1}).$$

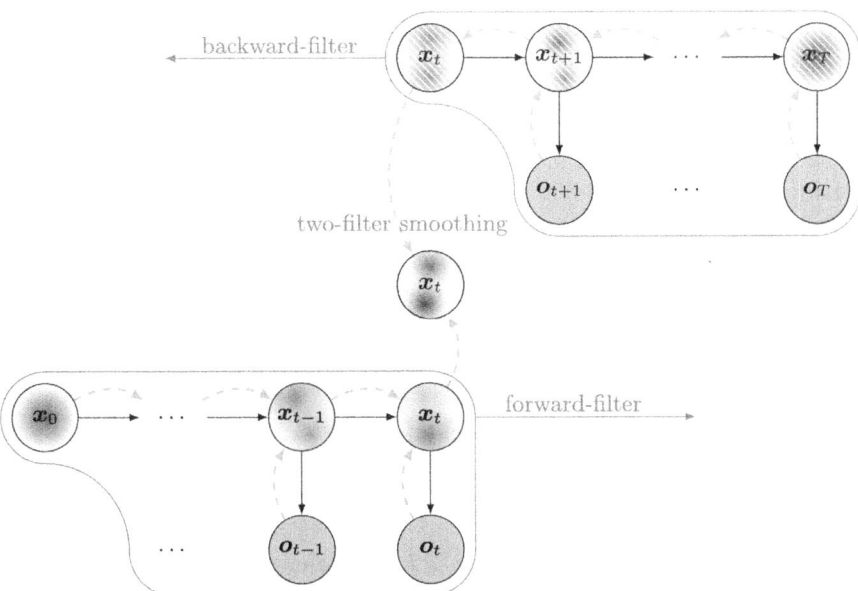

**Fig. 3.** Graphical interpretation of how marginal distribution $p(\boldsymbol{x}_t \mid \boldsymbol{o}_{1:T})$ is inferred by the forward-filtering/backward-filtering approach, or *two-filter smoothing*.

Hence, contrary to the recursive fixed-lag smoother, which implements forward-filtering/backward-smoothing as depicted in Fig. 2, the two-filter smoother performs forward-filtering/backward-filtering, as visualized in Fig. 3. Some authors have proposed that the two-filter smoother is also recursive, given that it factorizes into recursive filters. However, the *recursion* is not in terms of the smoothing marginal $p(\boldsymbol{x}_t \mid \boldsymbol{o}_{1:T})$, as in the fixed-lag case. This can be straightforwardly observed by comparing (11) with (10).

As hinted previously, the major problem of the two-filter approach is that the backward-filter is not performed in terms of distributions but likelihoods, which are not guaranteed to be finitely integrable over the state variable $x$. It could be the case that observations $o$ only capture some aspects of the hidden state, leaving the remaining dimensions with zero knowledge, or in probabilistic terms, with infinite uncertainty. One might consider tracking a system of dynamic objects, such as celestial bodies in the macrocosm or elementary particles in the microcosm. Observing only a subset of the objects does not enable the observer to infer the full state distribution of the whole system. In particular, a single observation of the positions is insufficient for inferring the velocities, accelerations, and so forth, of the objects within the system.

Note that a prior distribution would solve the problem, but because of the direction of time, there can only be a prior from the past $p(x_0)$, not from the future $p(x_{T+1})$. In [13,16], it is suggested to make an educated guess by specifying an artificial distribution for the initialization of the backward-corrector, or to buffer sufficient information $o_{T-m:T}$ such that the delayed backward corrector initialization $p(o_{T-m:T} \mid x_{T-m})$ eventually becomes integrable for some $m$. For Gaussian (mixture) distributions again, a more elegant solution is presented in [18], using the information form of the Gaussian distribution, which allows for zero information (infinite uncertainty) among state space dimensions. Inference in non-Gaussian state space models is frequently approached with particle-based methods, which require similar tricks since particle-based methods can only be used to approximate finite measures [16].

In conclusion, more practical solutions exist for the challenges encountered with the two-filter smoother, making it a more favorable option than the fixed-lag smoother. The authors believe that the popularity of the non-recursive two-filter smoother over the recursive fixed-lag smoother is one of the primary reasons why recursive decoding—the MAP analogue to recursive smoothing—has not been discussed until recently in [1,2].

## 4 Marginal MAP and Joint MAP Inference

MAP inference is traditionally performed either by maximizing the marginals or the joint posterior

$$\dddot{x}_t := \arg\max\nolimits_{x_t} \; p(x_t \mid o_{1:T}) \qquad \forall t, \tag{13}$$

$$\widehat{x}_{0:T} := \arg\max\nolimits_{x_{0:T}} \; p(x_{0:T} \mid o_{1:T}). \tag{14}$$

Figure 4 and 5 illustrate exemplary distribution densities and their marginal and joint MAP assignments within the model graph. Computing $\dddot{x}_t$ and $\widehat{x}_{0:T}$ is also denoted posterior decoding and sequence decoding, respectively. While posterior decoding can be reduced to marginal inference and mode-finding, decoding the joint assignment $\widehat{x}_{0:T}$ presents a greater challenge in state observation models.

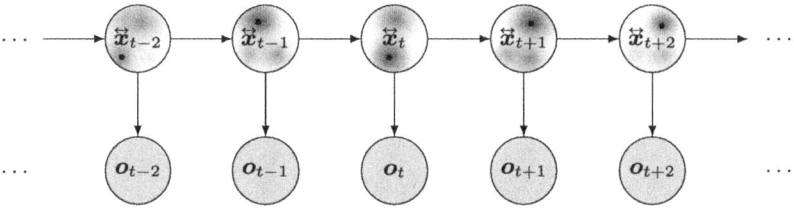

**Fig. 4.** Model graph with incoherent marginal MAP assignments (dots).

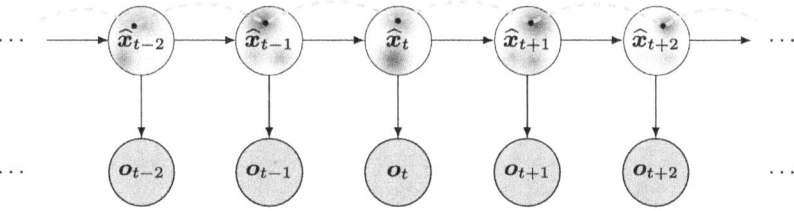

**Fig. 5.** Model graph with coherent joint MAP assignments (connected dots).

Analytic solutions are limited to special cases, such as HMM and linear Gaussian models [7,11,16]. Given the existence of established methods for decoding coherent joint MAP estimates $\hat{x}_{0:T}$ for these relatively simple models, the introduction of an additional coherent decoder would have been superfluous. This is considered to be another significant reason why predecessor decoding has not yet been addressed. However, in the case of more general models, the novel decoder described in the following section should prove to become a valuable addition.

## 5 MAP Predecessors

Contrary to the traditional inference methods from Sect. 4, this section is concerned with the novel MAP predecessor decoder.

**Definition 1.** *The MAP predecessor of a given $x_{t+1}^\star$ w.r.t. $o_{1:t}$ is*

$$\vec{x}_t := \arg\max\nolimits_{x_t} p(x_t \mid x_{t+1}^\star, o_{1:t}). \tag{15}$$

We are interested in inferring the MAP predecessors iteratively, i.e., the given state for which the MAP predecessor is to be inferred is the previously decoded predecessor $x_{t+1}^\star := \vec{x}_{t+1}$. In this manner, and as illustrated in Fig. 6, we recursively obtain a MAP predecessor sequences $\vec{x}_{0:T}$, where only the initial $x_T^\star$ is specified explicitly, although a reasonable choice would be to initialize the decoder in the marginal MAP estimate

$$x_T^\star = \arg\max\nolimits_{x_T} p(x_T \mid o_{1:T}) \tag{16}$$

as is the case for the illustrated examples shown in Fig. 6 and 7c.

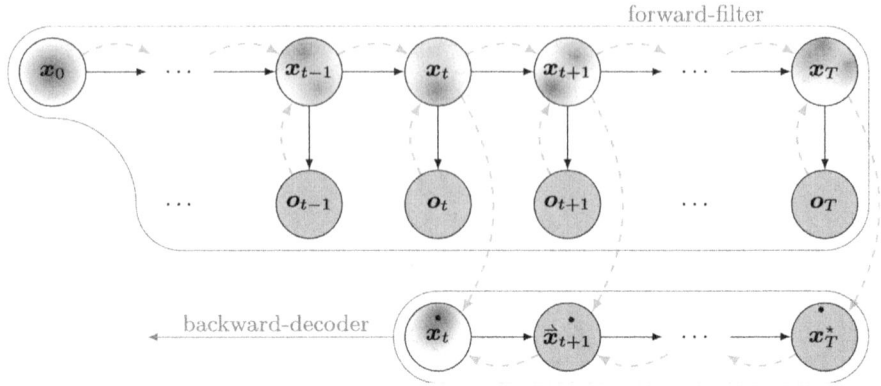

**Fig. 6.** Graphical interpretation of how MAP estimate $\vec{\tilde{x}}_t$, respectively the predecessor distribution $p(x_t \mid \vec{\tilde{x}}_{t+1}, o_{1:t})$ are inferred by the proposed predecessor decoder, respectively by the forward-filtering/backward-decoding decoder.

**Lemma 1.** *The MAP predecessor can be inferred recursively via*

$$\vec{\tilde{x}}_t = \arg\max_{x_t} p(x_t \mid o_{1:t}) p(\vec{\tilde{x}}_{t+1} \mid x_t). \tag{17}$$

*Proof.* By Bayes' theorem the distributions in (15) and (17) are proportional, differing only by scalar factor $p(\vec{\tilde{x}}_{t+1} \mid o_{1:t})$, i.e., $p(x_t \mid \vec{\tilde{x}}_{t+1}, o_{1:t}) p(\vec{\tilde{x}}_{t+1} \mid o_{1:t}) = p(x_t \mid o_{1:t}) p(\vec{\tilde{x}}_{t+1} \mid x_t)$.

By induction on (17), a predecessor sequence $\vec{\tilde{x}}_{0:T}$ is obtained based on the filtering marginals, the most likely transitions, and where the initialization $\vec{\tilde{x}}_T$ can be chosen freely. In an application, the initialization would typically be a point of interest where one aims to identify a probable sequence of preceding states with regard to a given sequence of observations.

Please note, that $\vec{\tilde{x}}_t$ maximizes $p(x_t \mid \vec{\tilde{x}}_{t+1}, o_{1:t}) \propto p(x_t, \vec{\tilde{x}}_{t+1} \mid o_{1:t})$ and by conditional independence (or *Bayes-Ball* [19]) also $p(x_t, \vec{\tilde{x}}_{t+1:T} \mid o_{1:T})$. With regard to the joint posterior $p(x_{0:T} \mid o_{1:T})$, note that

$$\vec{\ddot{x}}_t = \arg\max_{x_t} \int p(x_{0:t-1}, x_t, x_{t+1:T} \mid o_{1:T}) \, dx_{-t}, \tag{18}$$

$$\vec{\tilde{x}}_t = \arg\max_{x_t} \int p(x_{0:t-1}, x_t, \vec{\tilde{x}}_{t+1:T} \mid o_{1:T}) \, dx_{0:t-1}, \tag{19}$$

$$\widehat{x}_t = \arg\max_{x_t} \; p(\widehat{x}_{0:t-1}, x_t, \widehat{x}_{t+1:T} \mid o_{1:T}). \tag{20}$$

The problem with marginal MAP (18) is that $\vec{\ddot{x}}_{0:T}$ can contain implausible or zero transitions, while MAP predecessor decoding (19) yields coherent sequences $\vec{\tilde{x}}_{0:T}$ as demonstrated in [2] and Fig. 7. The joint MAP assignment $\widehat{x}_{0:T}$ in (20) is also coherent but intricate for general models [4,7].

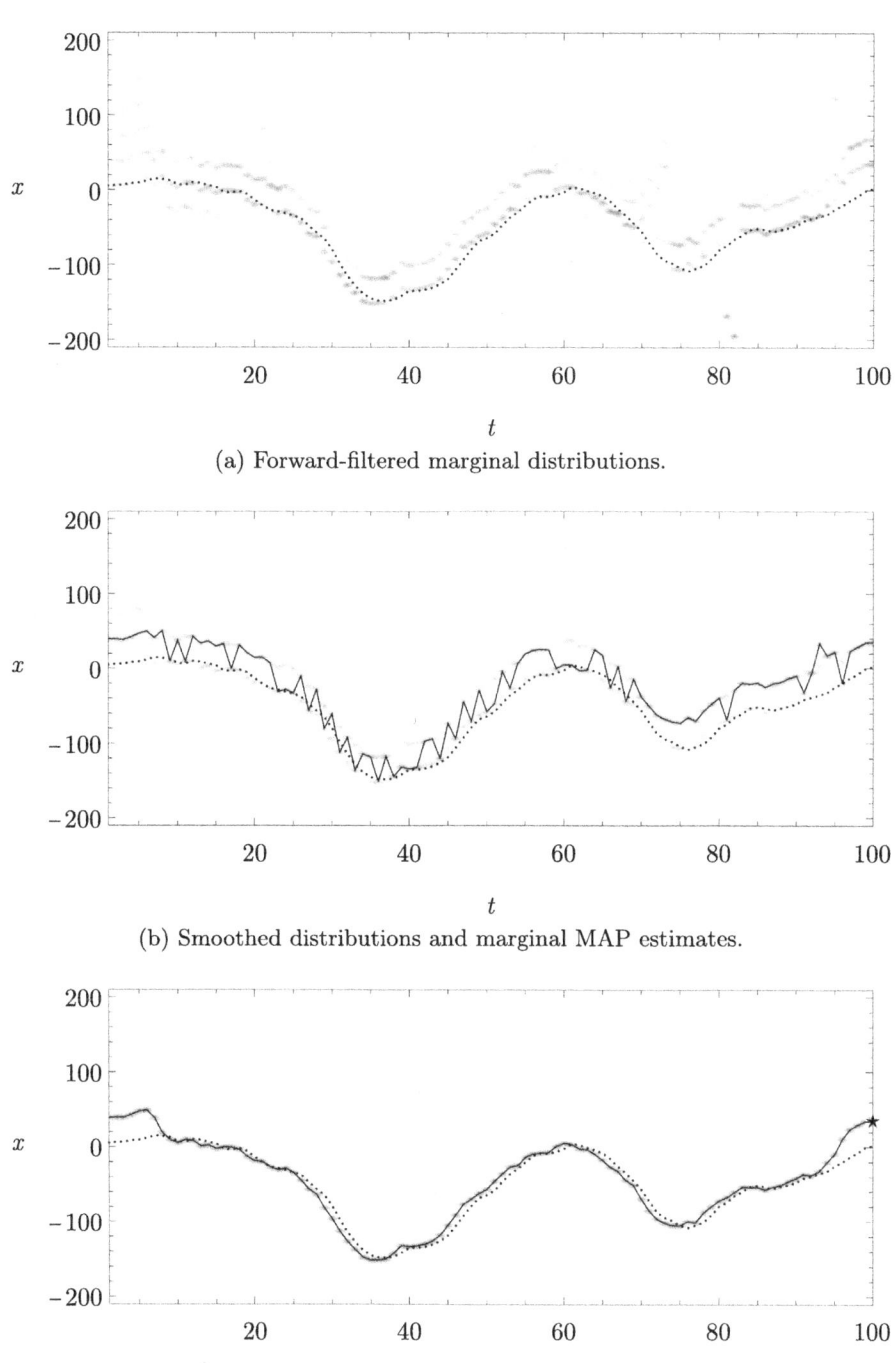

(a) Forward-filtered marginal distributions.

(b) Smoothed distributions and marginal MAP estimates.

(c) MAP predecessors of starred point.

**Fig. 7.** Inference comparison with example results from [2], ground truth is dotted.

## 6 Numeric Results

Figure 7 shows a comparison of inference results from a work in progress [2] on decoding valid trajectories. The shown example is based on a simulated tracking scenario where the ground truth trajectory is affected by non-Gaussian noise, resulting in multimodal Gaussian mixture distributions. Furthermore, [2] fails to address the abstract relations and motivations as discussed in present paper.

## 7 Conclusion

The current article elaborates on a novel MAP inference method, denoted as predecessor decoding, and thoroughly discusses it's properties and provides a distinction and improvement from existing methods used in discrete-time state observation models. The authors identify profound reasons why this fundamental method has not yet been discussed. For one, the marginal inference analogue is problematic and a work-around is preferred; for another, in some relatively simple models the joint MAP can be decoded, thereby spoiling the contributions of MAP predecessors. Similar to marginal MAP, predecessor decoding is performed by finding the modes of marginal distributions, yet the sequence of predecessors is coherent, a characteristic akin to joint MAP. This is achieved by inferring the predecessors recursively where each marginal is conditioned on the previous result. Beyond an initialization, which can be chosen freely, MAP predecessor decoding only requires the distributions encountered in forward filtering, thus its applicability is inherent wherever a forward filter is already in use. The hypothesized coherence of results is validated in an experiment section, demonstrating the strength of our approach.

**Acknowledgments.** This work has been supported by the COMET-K2 Center of the Linz Center of Mechatronics (LCM) funded by the Austrian federal government and the federal state of Upper Austria.

**Disclosure of Interests.** Author Dmitry Efrosinin is a member of the international program committee of the 2024 23rd International Conference on Information Technologies and Mathematical Modelling (ITMM'2024).

## References

1. Rudić, B., Pichler-Scheder, M., Efrosinin, D.: Maximum a posteriori predecessors in state observation models. In: 23rd International Conference on Information Technologies and Mathematical Modelling (ITMM 2024), pp. 391–395. Karshi (2024). to appear
2. Rudić, B., Pichler-Scheder, M., Efrosinin, D.: Valid decoding in Gaussian mixture models. In: Fazekas, I. (ed.) 2024 IEEE 3rd Conference on Information Technology and Data Science (CITDS), pp. 175–180. Debrecen (2024)
3. Särkkä, S.: Bayesian Filtering and Smoothing. Institute of Mathematical Statistics Textbooks. Cambridge University Press, Cambridge (2013)

4. Koller, D., Friedman, N.: Probabilistic Graphical Models: Principles and Techniques. Adaptive Computation and Machine Learning. MIT Press, Cambridge (2009)
5. Rabiner, L.R.: A tutorial on hidden Markov models and selected applications in speech recognition. Proc. IEEE **77**(2), 257–286 (1989)
6. Gustafsson, F.: Particle filter theory and practice with positioning applications. IEEE Aerosp. Electron. Syst. Mag. **25**(7), 53–82 (2010)
7. Godsill, S., Doucet, A., West, M.: Maximum a posteriori sequence estimation using Monte Carlo particle filters. Ann. Inst. Stat. Math. **53**, 82–96 (2001)
8. Daum, F.E., Huang, J.: Curse of dimensionality and particle filters. In: 2003 IEEE Aerospace Conference Proceedings (Cat. No.03TH8652), vol. 4, pp. 1979–1993 (2003)
9. Rauch, H.E., Tung, F., Striebel, C.T.: Maximum likelihood estimates of linear dynamic systems. AIAA J. **3**, 1445–1450 (1965)
10. Ainsleigh, P.L.: Theory of Continuous-State Hidden Markov Models and Hidden Gauss-Markov Models. Technical report 11274, Naval Undersea Warfare Cent., Newport, RI (2001)
11. Ainsleigh, P.L., Kehtarnavaz, N., Streit, R.L.: Hidden Gauss-Markov models for signal classification. IEEE Trans. Signal Process. **50**(6), 1355–1367 (2002)
12. Rudić, B., Sturm, V., Efrosinin, D.: On the analytic solution to recursive smoothing in mixture models. In: 2024 58th Asilomar Conference on Signals, Systems, and Computers, pp. 1–5. Pacific Grove(2024). to appear
13. Kitagawa, G.: The two-filter formula for smoothing and an implementation of the Gaussian-sum smoother. Ann. Inst. Stat. Math. **46**(4), 605–623 (1994)
14. Sorenson, H.W., Alspach, D.L.: Recursive Bayesian estimation using Gaussian sums. Automatica **7**(4), 465–479 (1971)
15. Alspach, D.L., Sorenson, H.W.: Nonlinear Bayesian estimation using Gaussian sum approximations. IEEE Trans. Autom. Control **17**(4), 439–448 (1972)
16. Briers, M., Doucet, A., Maskell, S.: Smoothing algorithms for state-space models. Ann. Inst. Stat. Math. **62**, 61–89 (2009)
17. Ardeshiri, T., Granström, K., Özkan, E., Orguner, U.: Greedy reduction algorithms for mixtures of exponential family. IEEE Signal Process. Lett. **22**(6), 676–680 (2015)
18. Balenzuela, M.P., Dahlin, J., Bartlett, N., Wills, A.G., Renton, C., Ninness, B.: Accurate Gaussian mixture model smoothing using a two-filter approach. In: 2018 IEEE Conference on Decision and Control (CDC), pp. 694–699 (2018)
19. Shachter, R.D.: Bayes-Ball: the rational pastime (for determining irrelevance and requisite information in belief networks and influence diagrams). In: 14th Conference on Uncertainty in Artificial Intelligence (UAI 1998), pp. 480–487. Madison (1998)

# Asymptotic Analysis of Sojourn Time in Retrial Queueing System with Non-persistent Customers and Feedback

Ekaterina Fedorova[✉], Anatoly Nazarov, and Daria Nikolaeva

National Research Tomsk State University, Lenina Avenue, 36, Tomsk, Russia
moiskate@mail.ru

**Abstract.** In the paper, a single-server retrial queueing system M/M/1 with non-persistent customers and feedback is considered. Customers arrive to the system according Poisson stationary process. Service times and delay times have exponential distributions. There is classical policy of retrials (i.e. multiple access). The asymptotic stationary probability distribution of the number of customers in the orbit and sojourn time of a marked customer in the system are derived by the asymptotic analysis method under a long delay condition.

**Keywords:** retrial queue · s-persistent customers · feedback · sojourn time · asymptotic analysis

## 1 Introduction

In the field of robotics and telemedicine, multimodal networks are of great interest. By modality it is called physically recorded elements of communication (human-machine and/or human-human), including both the transmitted data (message) and individual information. The set of multimodal data and their size may vary depending on the task [1]. Because of high load of communication channels in real life, delays in transmission appear. Sometimes, data become not relevant after the delay. It defines the actual direction of science research as preliminary estimation and prediction of time characteristics in networks.

Most of studies of this problem are based on simulation. In the paper, we consider a mathematical model of networks as a retrial queuing system (RQ) with feedback and non-persistence customers.

Retrial queues is a class of queuing theory models [2,3]. RQ is also called as queuing system with repeated calls. In such models, there is a virtual place – orbit, where unserved calls wait perform random delay before next attempt to

---

The research is supported by Russian Science Foundation according to the research project No. 24-21-00454, https://rscf.ru/project/24-21-00454/.

rich a server. In classical retrial queue, a random access protocol in orbit takes place, i.e. any call has access to server at any time moment.

The waiting time distribution is more complicated problem in retrial queueing systems. It was studied by M. Neuts [4], G. Falin [5], A. Gomez-Corral [6], J. Artalejo [7,8], R. Nobel [9], B. Kim [10], etc. [11–13]. In the paper, we apply the method of asymptotic analysis under a long delay developed by Nazarov [14,15].

The rest of the paper is organized as follows. In Sect. 2, the model under study is described. Section 3 is devoted to obtaining the asymptotic probability distribution of the number of customers in the orbit under a long delay condition. In Sect. 4, the analysis of the asymptotic probability distribution of the sojourn time of customers in the system is provided. In Sect. 5, we demonstrate some numerical examples and estimate the accuracy of the asymptotic results. Section 6 consists some conclusions.

## 2 Mathematical Model

In the paper, a retrial queueing system $M/M/1$ is considered. It means that customers arrive to the system according a Poisson stationary process with rates $\lambda$. The system has one server. The service time of each customer is exponential distributed with rate $\mu$. If a server is busy, an arrival customer goes to an orbit, where it waits during exponentially distributed random time with rate $\sigma$. The orbit capacity are not limited. We suppose that there is multiple access retrial policy (classical RQ). From the orbit, a customer try again to get the service. If the server is free, it begins the service, otherwise it returns up to the orbit with probability $s$ or goes away with probability $1 - s$ (non-persistent customers). After the service, the customer may require a repeated service (feedback), so it goes to the orbit with probability $r$ or goes away with probability $1 - r$. The model structure is illustrated on Fig. 1.

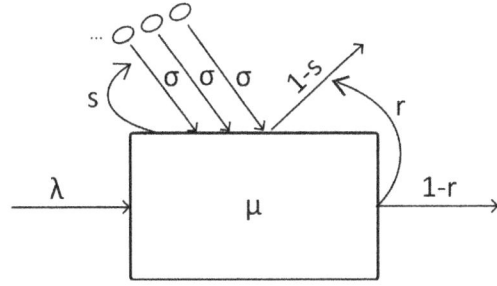

**Fig. 1.** Retrial queue $M/M/1$ with s-persistent customers and feedback

The aim of the study is analysis of stationary characteristics of the discribed model such as the probability distributions of number of customers in the orbit and sojourn time of customers in the system.

## 3 Number of Customers

First of all, let us study a random process of the number of customers in considered model. We denote a random process of the number of customers in the orbit by $i(t)$ and a random process of the server states by $k(t)$: $k(t) = 0$, if the server is free, $k(t) = 1$, if the server is busy.

Process $\{k(t), i(t)\}$ is Markovian, thus we write the following system of Kolmogorov equations for probability distribution $P(k, i, t) = P\{k(t) = k, i(t) = i\}$:

$$\begin{cases} \dfrac{\partial P(0,i,t)}{\partial t} = -(\lambda + i\sigma)P(0,i,t) + \mu P(1,i,t)(1-r) + r\mu P(1,i-1,t), \\ \dfrac{\partial P(1,i,t)}{\partial t} = -(\lambda + \mu + (1-s)i\sigma)P(1,i,t) + \sigma(i+1)P(0,i+1,t) \\ + \lambda P(0,i,t) + \lambda P(1,i-1,t) + \sigma(1-s)(i+1)P(1,i+1,t). \end{cases} \quad (1)$$

In steady state System (1) has the following form

$$\begin{cases} -(\lambda + i\sigma)P(0,i) + \mu P(1,i)(1-r) + r\mu P(1,i-1) = 0, \\ -(\lambda + \mu + (1-s)i\sigma)P(1,i) + \sigma(i+1)P(0,i+1) + \lambda P(0,i) \\ + \lambda P(1,i-1) + \sigma(1-s)(i+1)P(1,i+1) = 0. \end{cases} \quad (2)$$

Let us introduce the partial characteristic functions:

$$H(k, u, t) = \sum_i e^{jui} P(k, i).$$

From Eq. (2), we obtain the following equations for the characteristic functions

$$\begin{cases} -\lambda H(0,u) + j\sigma \dfrac{\partial H(0,u)}{\partial u} + \mu(1-r)H(1,u) + \mu r e^{ju} H(1,u) = 0, \\ -(\lambda + \mu)H(1,u) + j\sigma(1-s)\dfrac{\partial H(1,u)}{\partial u} - j\sigma \dfrac{\partial H(0,u)}{\partial u} e^{-ju} \\ + \lambda H(0,u) + \lambda e^{ju} H(1,u) - j\sigma(1-s)\dfrac{\partial H(1,u)}{\partial u} e^{-ju} = 0. \end{cases} \quad (3)$$

### 3.1 Asymptotic Analysis

For solving System (3), we apply the method of asymptotic analysis under the limit condition of a long delay $\sigma \to 0$ [14].

*The first order asymptotics.* Let us introduce the following notations

$$\sigma = \varepsilon, \quad u = \varepsilon w, \quad H(0, u) = F(0, w, \varepsilon), \quad H(1, u) = F(1, w, \varepsilon). \quad (4)$$

where $\varepsilon$ is an infinitesimal parameter.

We have the following asymptotic equations:

$$\begin{cases} -\lambda F(0,w,\varepsilon) + j\dfrac{\partial F(0,w,\varepsilon)}{\partial w} + \mu(1-r)F(1,w,\varepsilon) \\ + \mu r e^{j\varepsilon w} F(1,w,\varepsilon) = 0, \\ -(\lambda + \mu)F(1,w,\varepsilon) + j(1-s)\dfrac{\partial F(1,w,\varepsilon)}{\partial w} - j\dfrac{\partial F(0,w,\varepsilon)}{\partial w} e^{-j\varepsilon w} \\ + \lambda F(0,w,\varepsilon) + \lambda e^{j\varepsilon w} F(1,w,\varepsilon) - j(1-s)\dfrac{\partial F(1,w,\varepsilon)}{\partial w} e^{-j\varepsilon w} = 0. \end{cases} \quad (5)$$

For deriving an additional equation, let us sum up all equations of (5).

$$e^{-j\varepsilon w}\left(j\frac{\partial F(0,w,\varepsilon)}{\partial w}+j(1-s)\frac{\partial F(1,w,\varepsilon)}{\partial w}\right)+(\mu r+\lambda)F(1,w,\varepsilon)=0. \quad (6)$$

By writing Eqs. (5)–(6) under limit $\varepsilon \to 0$, we obtain

$$\begin{cases} -\lambda F(0,w)+j\dfrac{\partial F(0,w)}{\partial w}+\mu F(1,w)=0, \\ -\mu F(1,w)+\lambda F(0,w)-j\dfrac{\partial F(0,w)}{\partial w}=0, \\ j\dfrac{\partial F(0,w)}{\partial w}+j(1-s)\dfrac{\partial F(1,w)}{\partial w}+(\mu r+\lambda)F(1,w)=0. \end{cases} \quad (7)$$

Let us find a solution in the following form

$$F(k,w)=R_k\exp\{jw\kappa_1\}. \quad (8)$$

By substituting (8) into System (7), we obtain

$$\begin{cases} -\lambda R_0-\kappa_1 R_0+\mu R_1=0, \\ -\kappa_1 R_0-(1-s)\kappa_1 R_1+(\mu r+\lambda)R_1=0. \end{cases} \quad (9)$$

Taking into account normalization condition $R(0)+R(1)=1$, we can derive stationary probabilities of the server states as

$$R_0=\frac{\mu}{\mu+\lambda+\kappa_1}, \quad R_1=\frac{\lambda+\kappa_1}{\mu+\lambda+\kappa_1}. \quad (10)$$

Then from the last equation of System (7), $\kappa_1$ can be obtained:

$$\kappa_1=\frac{(\mu(1-r)-s\lambda)+\sqrt{(\mu(1-r)-s\lambda)^2-4(\mu r+\lambda)\lambda(1-s)}}{2(1-s)}. \quad (11)$$

Turning up to asymptotic notations (21), we obtain that the first order asymptotic characteristic function $H_1(u)=\exp\left\{ju\dfrac{\kappa_1}{\sigma}\right\}$.

*The second order asymptotics.* We suppose that

$$H(0,u)=e^{ju\frac{\kappa_1}{\sigma}}H_2(0,u), \quad H(1,u)=e^{ju\frac{\kappa_1}{\sigma}}H_2(1,u). \quad (12)$$

By substituting (12) into System (3), we obtain the following equations:

$$\begin{cases} -(\lambda+\kappa_1)H_2(0,u)+j\sigma\dfrac{\partial H_2(0,u)}{\partial u}+(\mu+\mu r(e^{ju}-1))H_2(1,u)=0, \\ -((\lambda(1-e^{ju})+\mu)+(\lambda+\kappa_1 e^{-ju})H_2(0,u)-j\sigma\dfrac{\partial H_2(0,u)}{\partial u}e^{-ju} \\ +(1-s)(1-e^{-ju})\kappa_1)H_2(1,u)+j\sigma(1-s)(1-e^{-ju})\dfrac{\partial H_2(1,u)}{\partial u}=0. \end{cases} \quad (13)$$

By summing all equations, we have

$$-\kappa_1 H_2(0,u) + \sigma j \frac{\partial H_2(0,u)}{\partial u} + (e^{ju}(\mu r + \lambda) - (1-s)\kappa_1)H_2(1,u) \\ + j\sigma(1-s)\frac{\partial H_2(1,u)}{\partial u} = 0. \tag{14}$$

Let us introduce the following notations:

$$\sigma = \varepsilon^2,\ u = \varepsilon w,\ H(0,u) = F(0,w,\varepsilon),\ H(1,u) = F(1,w,\varepsilon). \tag{15}$$

Substituting (15) into System (13), we have

$$\begin{cases} -(\lambda+\kappa_1)F(0,w,\varepsilon) + \varepsilon j \dfrac{\partial F(0,w,\varepsilon)}{\partial w} \\ +(\mu(1-r)+\mu r e^{j\varepsilon w})F(1,w,\varepsilon) = 0, \\ -(\lambda+\mu)F(1,w,\varepsilon) - (1-s)(1-e^{-j\varepsilon w})\kappa_1 F(1,w,\varepsilon) \\ +j\varepsilon(1-s)(1-e^{-j\varepsilon w})\dfrac{\partial F(1,w,\varepsilon)}{\partial w} + (\lambda+\kappa_1 e^{-j\varepsilon w})F(0,w,\varepsilon) \\ -j\varepsilon \dfrac{\partial F(0,w,\varepsilon)}{\partial w}e^{-j\varepsilon w} + \lambda e^{j\varepsilon w}F(1,w,\varepsilon) = 0. \end{cases} \tag{16}$$

And from (14), we obtain

$$-\kappa_1 F(0,w,\varepsilon) + \varepsilon j \frac{\partial F(0,w,\varepsilon)}{\partial w} \\ +((\mu r + \lambda)e^{j\varepsilon w} - (1-s)\kappa_1)F(1,w,\varepsilon) + j\varepsilon(1-s)\frac{\partial F(1,w,\varepsilon)}{\partial w} = 0. \tag{17}$$

Let us find the solution in the following form

$$F(k,w,\varepsilon) = \Phi(w)(R_k + j\varepsilon w f_k) + O(\varepsilon^2).$$

By substituting into System (16), we obtain the following equations under the limit $\varepsilon \to 0$:

$$\begin{cases} jw\Phi(w)[-(\lambda+\kappa_1)f_0 + \mu r R_1 + \mu f_1] + j\Phi'(w)R_0 = 0, \\ jw\Phi(w)[-\mu f_1 + (\lambda-(1-s)\kappa_1)R_1 + \\ +(\lambda+\kappa_1)f_0 - \kappa_1 R_0] - j\Phi'(w)R_0 = 0. \end{cases} \tag{18}$$

From the differential equations, we can see that function $\Phi(w)$ has the form: $\Phi(w) = \exp\left\{\dfrac{(jw)^2}{2}\kappa_2\right\}$, where $\kappa_2$ is defined as

$$\kappa_2 = \frac{-(\lambda+\kappa_1)f_0 + \mu r R_1 + \mu f_1}{R_0}. \tag{19}$$

From Eq. (17), we have an equation for $f_k$

$$-\kappa_1 f_0 + (\mu r + \lambda)f_1 + (\mu r + \lambda)R_1 - (1-s)\kappa_1 f_1 - (R_1(1-s) + R_0)\kappa_2 = 0.$$

To uniquely define $f_k$, we need to add an additional equation, e.g. $f_0 + f_1 = 0$.

Turning to asymptotic notations, $H_2(u) = \exp\left\{\dfrac{(ju)^2 \kappa_2}{2\sigma}\right\}$.

In this way, we have proved that the asymptotic characteristic function of number of customers in the orbit has Gaussian form

$$H(u) = \exp\left\{ju\dfrac{\kappa_1}{\sigma} + \dfrac{(ju)^2 \kappa_2}{2\sigma}\right\}.$$

## 4  Sojourn Time

Further, we provide the analysis of sojourn time of customers in the system for the considered model. First of all, let us denote process of remaining time of a marked customer in the system by $T(t)$. Also we need to introduce process $n(t)$ described a place of the marked customer:

$$n(t) = \begin{cases} 0, \text{the marked customer is in the orbit,} \\ 1, \text{the marked customer is on the server.} \end{cases}$$

We denote $P(k(t) = k, n(t) = n, i(t) = i, T(t) < T) = P(k, n, i, T, t)$. Conditional characteristic functions for $T(t)$ can be written in the following form:

$$G(k, n, i, u, t) = M\left\{e^{juT(t)} \mid k(t) = k,\ i(t) = i,\ n(t) = n\right\}.$$

Kolmogorov equations for conditional characteristic functions $G(k, n, i, u, t)$ in steady state have the following form

$$\begin{cases} -(\lambda + i\sigma - ju)G(0,0,i,u) + \lambda G(1,0,i,u) \\ +\sigma(i-1)G(1,0,i-1,u) + \sigma G(1,1,i-1,u) = 0, \\ -(\lambda + \mu - ju + (1-s)i\sigma)G(1,0,i,u) + \mu r\, G(0,0,i+1,u) \\ +\mu(1-r)G(0,0,i,u) + \sigma(i-1)(1-s)G(1,0,i-1,u) \\ +\lambda G(1,0,i+1,u) + \sigma(1-s) = 0, \\ -(\lambda + \mu + i\sigma(1-s) - ju)G(1,1,i,u) + \mu r\, G(0,0,i+1,u) \\ +\sigma i(1-s)G(1,1,i-1,u) + \lambda G(1,1,i+1,u) + \mu(1-r) = 0. \end{cases} \quad (20)$$

For System (20) solving, we will similarly use the method of asymptotic analysis under a long delay condition.

### 4.1  Asymptotic Analysis

We will consider the asymptotic condition of a long delay condition ($\sigma \to 0$) as in previous section.

Let us introduce the following notations

$$\sigma = \varepsilon,\ u = \varepsilon w,\ i\sigma = i\varepsilon = x,\ G(k,n,i,u) = F(k,n,x,w,\varepsilon). \quad (21)$$

where $\varepsilon$ is an infinitesimal parameter.

From System (20), we obtain

$$\begin{cases} -(\lambda + x - j(\varepsilon w))F(0,0,x,w,\varepsilon) + \lambda F(1,0,x,w,\varepsilon) \\ +(x-\varepsilon)F(1,0,x-\varepsilon,w,\varepsilon) + \varepsilon F(1,1,x-\varepsilon,w,\varepsilon) = 0, \\ -(\lambda + \mu + (1-s)x - j(\varepsilon w))F(1,0,x,w,\varepsilon) + \mu r F(0,0,x+\varepsilon,w,\varepsilon) \\ +\mu(1-r)F(0,0,x,w,\varepsilon) + (x-\varepsilon)(1-s)F(1,0,x-\varepsilon,w,\varepsilon) \\ +\lambda F(1,0,x+\varepsilon,w,\varepsilon) + \varepsilon(1-s) = 0, \\ -(\lambda + \mu + x(1-s) - j(\varepsilon w))F(1,1,x,w,\varepsilon) + \mu r F(0,0,x+\varepsilon,w,\varepsilon) \\ +x(1-s)F(1,1,x-\varepsilon,w,\varepsilon) + \lambda F(1,1,x+\varepsilon,w,\varepsilon) + \mu(1-r) = 0. \end{cases} \quad (22)$$

Under limit $\varepsilon \to 0$, we derive

$$F(0,0,x,w) = F(1,0,x,w), \\ F(1,1,x,w) = rF(0,0,x,w) + (1-r). \quad (23)$$

For the first and the second equations of System (22), we apply Taylor series for functions $F(k,n,x \pm \varepsilon, w, \varepsilon)$. After some transformations, we obtain

$$\begin{cases} -(\lambda+x)F(0,0,x,w,\varepsilon) + (\lambda+x)F(1,0,x,w,\varepsilon) = \\ = -j\varepsilon w F(0,0,x,w,\varepsilon) + x\varepsilon \dfrac{\partial F(1,0,x,w,\varepsilon)}{\partial x} + \varepsilon F(1,0,x,w,\varepsilon) \\ +\varepsilon(1-s) - \varepsilon F(1,1,x,w,\varepsilon) + O(\varepsilon^2), \\ \mu F(1,0,x,w,\varepsilon) - \mu F(0,0,x,w,\varepsilon) = (\varepsilon(1-s) + j(\varepsilon w))F(1,0,x,w,\varepsilon) \\ +(\mu r \varepsilon - \varepsilon x(1-s) + \lambda \varepsilon)\dfrac{\partial F(1,0,x,w,\varepsilon)}{\partial x} + \varepsilon(1-s) + O(\varepsilon^2). \end{cases} \quad (24)$$

Multiplying by appropriate factors and equating the equations, we set $\varepsilon \to 0$.

$$-jw\mu F(0,0,x,w) + x\mu \dfrac{\partial F(1,0,x,w)}{\partial x} + \mu F(1,0,x,w) \\ -\mu F(1,1,x,w) = (\lambda+x)((1-s)+jw)F(1,0,x,w) \\ +(\lambda+x)(\mu r - x(1-s) + \lambda)\dfrac{\partial F(1,0,x,w)}{\partial x} + (\lambda+x)(1-s) \quad (25)$$

Taking into account expressions (23), we have

$$F(0,0,x,w) = \dfrac{(\lambda+x)(1-s) + \mu(1-r)}{(\lambda+x)((1-s) + \mu(1-r) - jw(\lambda+x+\mu)}.$$

Under a long delay condition, the steady state mean of process $i(t)$ tends to $\kappa_1/\varepsilon$ defined by (11), so we finally obtain

$$F(0,0,w) = \dfrac{\gamma}{\gamma - jw}, \quad (26)$$

where $\gamma = \dfrac{(\lambda+\kappa_1)(1-s) + \mu(1-r)}{(\lambda+\kappa_1+\mu)} = (1-s)R_1 + (1-r)R_0$.

In this way, the characteristic function of sojourn time of the marked customer in the retrial queue with non-persistence and feedback are asymptotically defined as

$$G(u) = \dfrac{\gamma\sigma}{\gamma\sigma - ju}(1-\gamma) + \gamma. \quad (27)$$

## 5 Numerical Examples

For demonstrating the accuracy of the asymptotic method, we present the comparison of asymptotic distribution $P(i)$ and exact one $D_n(i)$ calculated by a numerical algorithm for different values of the model parameters.

In Fig. 2, a comparison of the distributions is demonstrated for the following parameters
$$\lambda = 3, \quad \mu = 1, \quad \sigma = 0.1, \quad r = 0.1, \quad s = 0.1.$$

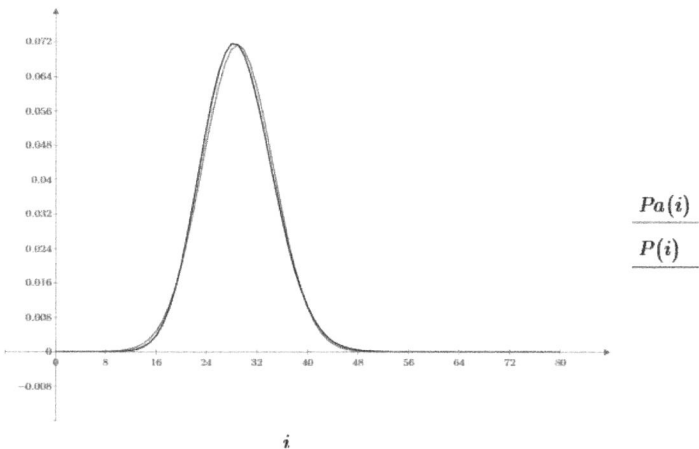

**Fig. 2.** Comparison of asymptotic and exact probability distributions for $\sigma = 0.1$

For the estimation of the asymptotic analysis accuracy, we usually use Kolmogorov distance. The values are presented in Table 1.

**Table 1.** The Kolmogorov distances for different $\sigma$ and $\lambda$

| $\sigma$ | 1 | 0.1 | 0.01 |
|---|---|---|---|
| $\lambda = 1$ | 0.290 | 0.019 | 0.010 |
| $\lambda = 3$ | 0.260 | 0.015 | 0.005 |

In Fig. 3, the asymptotic sojourn time probability distribution is demonstrated for the same parameters.

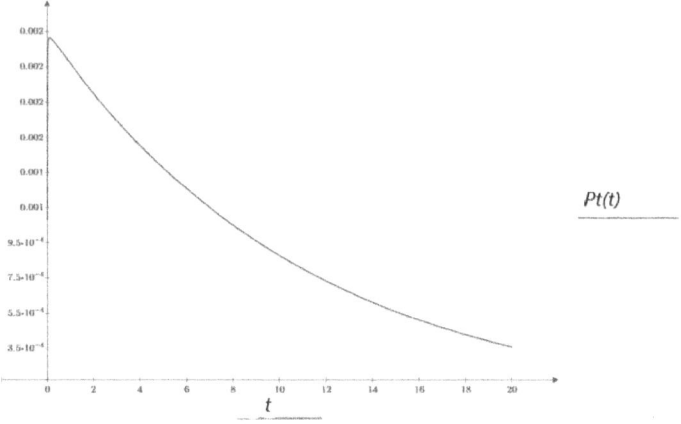

**Fig. 3.** Asymptotic sojourn time probability distribution

## 6 Conclusion

In the study, we have considered the retrial queueing system with non-persistent customers and feedback. The study is conducted by the method asymptotic analysis under a condition of a long delay. We have derived the asymptotic characteristic functions for the probability distributions of number of calls in the orbit and sojourn time of customers in the system. Through numerical analysis, we have concluded that the asymptotic method accuracy increases with decreasing of a retrial rate.

## References

1. Ryndin, A., Pakulova, E., Basov, O., Veselov, G.: Modelling of multi- path transmission system of various priority multimodal information. In: 2020 IEEE 14th International Conference on Application of Information and Communication Technologies (AICT), pp. 1–5 (2020)
2. Artalejo, J.R., Gomez-Corral, A.: Retrial Queueing Systems, p. 267. Springer, Berlin (2008)
3. Falin, G.I., Templeton, J.: Retrial Queues, p. 320. Chapman and Hall, London (1997)
4. Neuts, M.: The joint distribution of the virtual waiting time and the residual busy period for the M/G/1 queue. J. Appl. Probab. **5**, 224–229 (1968)
5. Falin, G., Fricker, C.: On the virtual waiting time in an M/G/1 retrial queue. J. Appl. Probab. **28**(2), 446–460 (1991)
6. Gómez-Corral, A., Ramalhoto, M.: On the waiting time distribution and the busy period of a retrial queue with constant retrial rate. Stochast. Model. Appl. **3**, 37–47 (2000)
7. Artalejo, J.R., Chakravarthy, S.R., Lopez-Herrero, M.J.: The busy period and the waiting time analysis of a MAP/M/c queue with finite retrial group. Stoch. Anal. Appl. **25**(2), 445–469 (2007)

8. Artalejo, J.R., Gómez-Corral, A.: Waiting time analysis of the M/G/1 queue with finite retrial group. Naval Res. Logistics (NRL) **54**(5), 524–529 (2007)
9. Nobel, R., Tijms, H.: Waiting-time probabilities in the M/G/1 retrial queue. Stat. Neerl. **60**(3), 73–78 (2006)
10. Lee, S.W., Kim, B., Kim, J.: Analysis of the waiting time distribution in M/G/1 retrial queues with two way communication. Ann. Oper. Res. **310** (2022)
11. Atencia, I., Galán-García, J.L.: Sojourn times in a queueing system with breakdowns and general retrial times. Mathematics **9**, 2882 (2021)
12. Lin, Y.B., Liu, T.H., Tsai, Y.C., Chang, F.M.: Waiting time control chart for M/G/1 retrial queue. Computation **12**, 191 (2024)
13. Kim, B., Kim, J.: Higher moments of the waiting time distribution in M/G/1 retrial queues. Oper. Res. Lett. **39**(3), 224–228 (2011)
14. Nazarov, A., Sztrik, J., Kvach, A., Tóth, Á.: Asymptotic sojourn time analysis of finite-source M/M/1 retrial queueing system with collisions and server subject to breakdowns and repairs. Ann. Oper. Res. **288**(1), 417–434 (2019). https://doi.org/10.1007/s10479-019-03463-0
15. Nazarov, A., Samorodova, M.: Waiting time asymptotic analysis of a M/M/1 retrial queueing system under two types of limiting condition. In: Dudin, A., Nazarov, A., Moiseev, A. (eds.) Information Technologies and Mathematical Modelling. Queueing Theory and Applications. ITMM 2020. Communications in Computer and Information Science, vol. 1391. Springer, Cham (2021)

# Implementation of a Convolution Algorithm to the Evaluation of Stationary Characteristics of Resource Loss System with Resource-Dependent Service Times

Artem Nazarin[1(✉)] and Eduard Sopin[1,2]

[1] Peoples' Friendship University of Russia (RUDN University), Moscow, Russian Federation
nazaryin_ai@pfur.ru
[2] Institute of Informatics Problems, Federal Research Center Computer Science and Control of Russian Academy of Sciences, Moscow, Russian Federation

**Abstract.** In this paper, a multi-server resource loss system with resource requirement-dependent service times is considered. We develop a convolutional algorithm for the evaluation of the stationary probability distribution and various system characteristics, such as the loss probability and the average volume of occupied resources. In the case study, we calculate these characteristics under the assumption of the Erlang distribution for the service time and several distributions for the resource requirements. We also compare the execution time of the developed convolution algorithm with direct calculations of probabilistic measures.

**Keywords:** Resource loss system · Random requirements · Blocking probability · Normalization constant · Resource-dependent service time

## 1 Introduction

### 1.1 Background

Resource loss systems (ReLS) are widely used in wireless communication systems' performance analysis [1]. One of the main features of such systems is heterogeneous volume of customers, i.e. a customer needs not only a server, but also a random volume of system resources until the departure. Therefore, such systems allow to model the service process of sessions at the base station of a wireless network, in which different volume of resources may be needed to maintain the same bitrate, depending on the quality of the radio channel. Another important element of resource loss systems in application to the analysis of wireless communication network systems is the limited volume of available resources. In [2] these resources are interpreted as customers' or messages' volumes and the

---

The research was funded by the Russian Science Foundation, project no. 22-79-10128, https://rscf.ru/en/project/22-79-10128/.

total volume of customers presented in the system is limited by capacity of the buffer. If the system is full, an arriving customer is lost. However, systems with infinite amount of available resources also have applications [3].

In [4], a multi-server resource loss system with limited amount of resources is introduced with the total amount of resources occupied by all customers taken for tracking. In most previously studied ReLS, it was assumed that the service time does not depend on the volume of resource requirements [5,6], which reflects the features of servicing real-time traffic. However, to take into account the features of elastic traffic transmission in a wireless network, it is necessary to consider ReLS with service time and resource requirements specified by a joint distribution function. In case of real-time traffic, the service time does not depend on resource requirements. It is characterized by the service duration and the data transfer rate. In case of elastic traffic, the service time is not predetermined. Instead, it changes based on the service requirements. In [7], a model of the system with service time that depends on the amount of resources required is considered. This model can be used to analyze data transmission characteristics under conditions of elastic traffic.

Note that the formulas for calculating stationary probabilities and characteristics presented in [4] are too complex for direct calculations due to multiple convolutions of resource requirements' cumulative distribution function. In [8], a convolution algorithm is presented for calculating stationary probabilities and system characteristics based on [4], with a service time independent of the resource requirements.

This paper presents the development of a convolution algorithm for the case with discrete resource requirements and resource requirement-dependent service times. We introduce formulas for calculating the loss probability, the average amount of occupied resources and the average amount of customers. We also provide a numerical analysis example that consists of three parts. In the first part, the characteristics of one distribution with two different initial values are compared. The second part includes the comparison of characteristics for three resource requirements distributions. In both parts the results for independent and resource-dependent service times are considered. Finally, we compare the calculation times for the developed convolution algorithm and direct stationary probabilities calculation.

## 2 Model

### 2.1 Model Description

Consider a multi-server resource loss system with the maximum number of servers $N$ and $R$ resource units, $R < \infty$. Customers arrive according to the Poisson flow with arrival rate $\lambda$. The service times are mutually independent. Let $H_j(x)$ be the conditional cumulative distribution function of the service time provided that $j$ discrete resources are required. The distribution of resource requirements is determined by $\{p_j\}, j = 1, 2, .., R$.

The state of the system at time $t$ is described by the stochastic process $X(t) = (\xi(t), \gamma(t))$. Here $\xi(t)$ is the number of customers in the system at time $t$. Each customer occupies an integer number of $\gamma(t)$ resource units, $\gamma(t) < R$. The newly arrived $i$-th customer requires a random number of resources $r_i \geq 0$. If at the arrival instant $t_i$, $\xi(t_i) = N$ or $(R - \gamma(t_i) < r_i)$, then the customer is lost. At the end of the service, the customer leaves the system and releases the whole volume of occupied resources. If $\xi(t) < N$ or $R - \gamma(t) \geq r_i$ then the customer is accepted for service.

Let $b_i = \int_0^R x d(H_j(x))$ be the conditional average service time, provided that the customer requires $i$ resource units.

Then the average service time $b$ can be found as:

$$b = \sum_{i=0}^{R} b_i p_i. \qquad (1)$$

Denote

$$g_i = \frac{1}{b} b_i p_i. \qquad (2)$$

Then, based on [7], the stationary distribution of the process $X(t)$ can be defined as follows:

$$q_0 = \lim_{t \to \infty} P\{\xi(t) = 0\} = (1 + \sum_{k=1}^{N} \frac{\rho^k}{k!} \sum_{r=0}^{R} g_r^{(k)}), \qquad (3)$$

$$q_k(r) = \lim_{t \to \infty} P\{\xi(t) = k, \gamma(t) = r\} = q_0 \cdot g_r^{(k)} \cdot \frac{\rho^k}{k!}, 1 \leq k \leq N, 0 \leq r \leq R, \qquad (4)$$

where $g_r^{(k)}$—$k$-fold convolution of the distribution $g_r$, $\rho = \lambda b$. The probability of losing a customer can be expressed as:

$$\pi = 1 - q_0 \sum_{k=0}^{N-1} \frac{\rho^k}{k!} \sum_{r=0}^{R} g_r^{(k+1)}. \qquad (5)$$

The average number of occupied resource units is represented by the following expression:

$$B = q_0 \sum_{k=1}^{N} \frac{\rho^k}{k!} \sum_{r=0}^{R} r g_r^{(k)} \qquad (6)$$

The average amount of customers takes the following form:

$$\overline{N} = q_0 \sum_{k=1}^{N} \frac{\rho^k}{k!} \sum_{r=0}^{R} g_r^{(k)} \qquad (7)$$

## 2.2 Convolution Algorithm

Based on the formulas (1)–(7), we can see that the calculations of the stationary distribution and probability characteristics are complicated by the need to calculate multiple convolutions. We introduce convolution algorithm for a model with a service time that depends on the resource requirements. Let us introduce the following functions of integer non-negative arguments:

$$G(n,r) = \sum_{k=0}^{n} \frac{\rho^k}{k!} \sum_{j=0}^{r} g_j^{(k)}. \tag{8}$$

Note that by definition $G(N, R) = q_0^{-1}$ is a normalization constant.

**Theorem 1.** *Functions $G(n,r)$ satisfy the following recurrence relation:*

$$G(n,r) = G(n-1,r) + \frac{\rho}{n} \sum_{j=0}^{r} g_j (G(n-1, r-j) - G(n-2, r-j)), \tag{9}$$

*for $2 \le n \le N$, with the initial values*

$$G(0, r) = 1, \quad 0 \le r \le R, \tag{10}$$

$$G(1, r) = 1 + \rho \sum_{j=0}^{r} g_j, \quad 0 \le r \le R. \tag{11}$$

*Proof.* Considering the difference between $G(n,r)$ and $G(n, r-1)$ and applying the convolution formula, we obtain:

$$G(n,r) - G(n-1,r) = \sum_{k=0}^{n} \frac{\rho^k}{k!} \sum_{j=0}^{r} g_j^{(k)} - \sum_{k=0}^{n-1} \frac{\rho^k}{k!} \sum_{j=0}^{r} g_j^{(k)}$$

$$= \frac{\rho^n}{n!} \sum_{j=0}^{r} \sum_{i=0}^{j} g_j g_{j-i}^{(n-1)} = \frac{\rho^n}{n!} \sum_{i=0}^{r} g_i \sum_{i=j}^{r} g_{j-i}^{(n-1)}$$

$$= \frac{\rho}{n} \sum_{i=0}^{r} g_i \left( \frac{\rho^{n-1}}{(n-1)!} \sum_{i=0}^{r-i} g_{j-i}^{(n-1)} \right)$$

$$= \frac{\rho}{n} \sum_{i=0}^{r} g_i (G(n-1, r-i) - G(n-2, r-i)).$$

**Theorem 2.** *The loss probability $\pi$ can be expressed in terms of functions $G(n,r)$ as follows:*

$$\pi = 1 - G^{-1}(N, R) \sum_{i=0}^{R} g_i G(N-1, R-i). \tag{12}$$

*Proof.* By rewriting formula (5) in terms of functions $G(n,r)$, we get

$$1 - q_0 \sum_{k=0}^{N-1} \frac{\rho^k}{k!} \sum_{j=0}^{R} g_j^{(k+1)} = 1 - G^{-1}(N,R) \sum_{k=0}^{N-1} \frac{\rho^k}{k!} \sum_{j=0}^{R} g_j^{(k+1)} =$$

$$1 - G^{-1}(N,R) \sum_{k=0}^{N-1} \frac{\rho^k}{k!} \sum_{j=0}^{R} \sum_{i=0}^{j} g_i g_{j-i}^{(k)} = 1 - G^{-1}(N,R) \sum_{k=0}^{N-1} \frac{\rho^k}{k!} \sum_{i=0}^{j} g_i \sum_{j=0}^{R} g_{j-i}^{(k)}$$

$$= 1 - G^{-1}(N,R) \sum_{k=0}^{N-1} \frac{\rho^k}{k!} \sum_{i=0}^{j} g_i \sum_{r=0}^{R-i} g_j^{(k)} = 1 - G^{-1}(N,R) \sum_{i=0}^{R} g_i G(N-1, R-i).$$

**Theorem 3.** *The average number of occupied resource units $B$ can be expressed in terms of functions $G(n,r)$ as follows:*

$$B = R - G^{-1}(N,R) \sum_{i=1}^{R} G(N, R-i). \tag{13}$$

*Proof.* Let $\bar{B}$ be the average number of unoccupied resources. Then, by definition:

$$B = R - \bar{B}$$

$$\bar{B} = \sum_{j=0}^{R} (R-j) \sum_{k=0}^{N} q_0 \frac{\rho^k}{k!} g_j^{(k)} = G^{-1}(N,R) \sum_{k=0}^{N} \frac{\rho^k}{k!} \sum_{j=0}^{R} (R-r) g_j^{(k)} =$$

$$G^{-1}(N,R) \sum_{k=0}^{N} \frac{\rho^k}{k!} \sum_{r=0}^{R} \sum_{i=1}^{R-j} g_j^{(k)} = G^{-1}(N,R) \sum_{k=0}^{N} \frac{\rho^k}{k!} \sum_{i=0}^{R} \sum_{j=0}^{R-i} g_j^{(k)} =$$

$$G^{-1}(N,R) \sum_{i=1}^{R} G(N, R-i).$$

**Theorem 4.** *The average amount of customers $\overline{N}$ can be expressed in terms of functions $G(n,r)$ as follows:*

$$\overline{N} = \rho G^{-1}(N,R) \sum_{i=0}^{R} g_i G(N-1, R-i). \tag{14}$$

*Proof.* By definition:

$$\overline{N} = q_0 \sum_{k=1}^{N} \frac{\rho^k}{k!} \sum_{r=0}^{R} g_r^{(k)} = G^{-1}(N,R)\rho \sum_{k=1}^{N} \frac{\rho^{k-1}}{(k-1)!} \sum_{r=0}^{R} \sum_{i=0}^{r} g_{r-i} g_i^{(k-1)} =$$

$$\rho G^{-1}(N,R) \sum_{k=0}^{N-1} \frac{\rho^k}{k!} \sum_{r=0}^{R} \sum_{i=0}^{r} g_i g_{r-i}^{(k)} = \rho G^{-1}(N,R) \sum_{k=0}^{N-1} \frac{\rho^k}{k!} \sum_{i=0}^{R} \sum_{r=i}^{R} g_i g_{r-i}^{(k)} =$$

$$\rho G^{-1}(N,R) \sum_{i=0}^{R} g_i \sum_{k=0}^{N-1} \frac{\rho^k}{k!} \sum_{r=i}^{R} g_{r-i}^{(k)} = \rho G^{-1}(N,R) \sum_{i=0}^{R} g_i \sum_{k=0}^{N-1} \frac{\rho^k}{k!} \sum_{r=0}^{R-i} g_r^{(k)} =$$

$$\rho G^{-1}(N,R) \sum_{i=0}^{R} g_i G(N-1, R-i).$$

As a result, the theorems formulated above determine the convolution algorithm that allow to evaluate the stationary characteristics of a ReLS with resource requirement-dependent service times without calculation of the stationary distribution.

## 3 Case Study

In this section, we consider three cases. In the first case, we analyze the characteristics of one distribution with different distribution parameters and under the condition of service time depending on resource requirements. In the second case, the indicators of three distributions with the same expected value and different variances are compared. The service time is similar. In the third case, we take service time as fixed. Finally, we compare the execution time of the developed convolution algorithm and direct calculations.

We assume that the service time follows the Erlang distribution $\Gamma(k, \theta_j)$ with density $h_j$, where $\theta_j = \frac{j}{R}, j = 1, 2, ..., R$. Take $k = 2$, therefore $b_j = \frac{2R}{j}$. The more resource units are occupied, the shorter the average service time.

Table 1 demonstrates the values of the variables used for the calculations.

**Table 1.** First case parameters

| Variable | Parameter | Value |
|---|---|---|
| $N$ | Number of customers | 100 |
| $R$ | Number of resource units | 100 |
| $p_1$ | Geometric distribution probability parameter | 0.4 |
| $p_2$ | Geometric distribution probability parameter | 0.6 |
| $b_1$ | Average service time for $p_1$ (resource dependent model), s | 153.2476 |
| $b_2$ | Average service time for $p_2$ (resource dependent model), s | 122.172 |
| $\lambda$ | Intensity of arrivals | [0.3,0.65] |

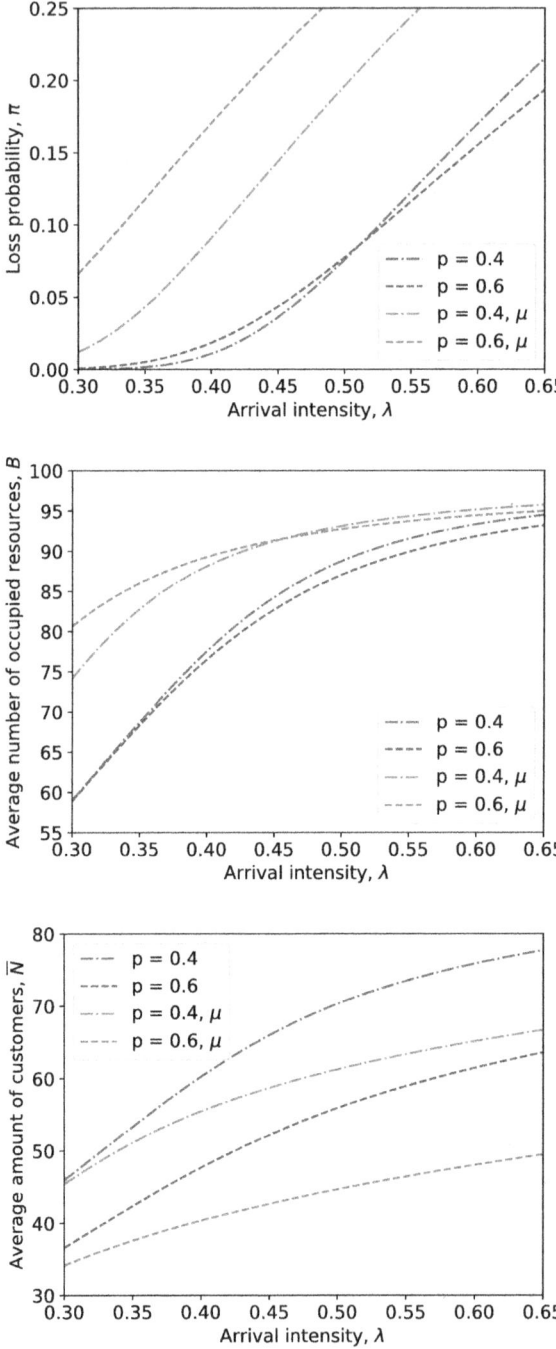

**Fig. 1.** Numerical example results for first case.

First, consider a case where the amount of resource units allocated to the customer follows the truncated geometric distribution. $\lambda = [0.3, 0.65], p_1 = 0.4$ and $p_2 = 0.6$.

The results are depicted in Fig. 1. Initially, the loss probability at $p_1$ is lower than at $p_2$. But with the increase of the arrival intensity, the system takes more "heavy" customers, which, according to the condition, are serviced faster. Thus, at $p = 0.4$, the system receives more customers that do not require many resource units than at $p = 0.6$. Customers that are not resource-demanding but time-demanding fill the system, and when a new one arrives, a loss occurs.

When it comes to the average number of occupied resource units, the distribution with the smaller parameter $p$ occupies a little more. The increase of the parameter $p$ leads to the increase of the variance, which is why more requests with "heavy" resource volumes that are serviced faster arrive. The system is mostly filled with "light" customers. In the case of fixed service time, $p_1$ has a smaller variance but a longer service time. This fact explains that over time, the model with $p_1$ will occupy more resource units on average than the model with $p_2$.

Table 2. Second case parameters

| Variable | Parameter | Value |
|---|---|---|
| $N$ | Number of customers | 100 |
| $R$ | Number of resource units | 100 |
| $m$ | Expected value of distributions | 5 |
| $\sigma^2_{geom}$ | Variance for geometrical distribution | 30 |
| $\sigma^2_{pois}$ | Variance for Poisson distribution | 6 |
| $\sigma^2_{bin}$ | Variance for binomial distribution | 5 |
| $b_{geom}$ | Average service time for geometrical distribution (resource dependent model) | 71.67 |
| $b_{pois}$ | Average service time for Poisson distribution (resource dependent model) | 39.73 |
| $b_{bin}$ | Average service time for binomial distribution (resource dependent model) | 36.346 |

The second case is presented on Fig. 2, and it compares the same characteristics as in the first case, but for three distributions: geometric, Poisson and binomial. The distributions have the same expected values, but different variances. Table 2 presents the parameters of the distributions.

The plot shows that the larger variance of the resource requirements distribution leads to lower loss probability in the models with resource dependent service time. This effect is explained by the fact that in these systems, mainly the most resource-demanding applications are served. Accordingly, with an increase in the dispersion of the distribution, the average volume of lost requests decreases and

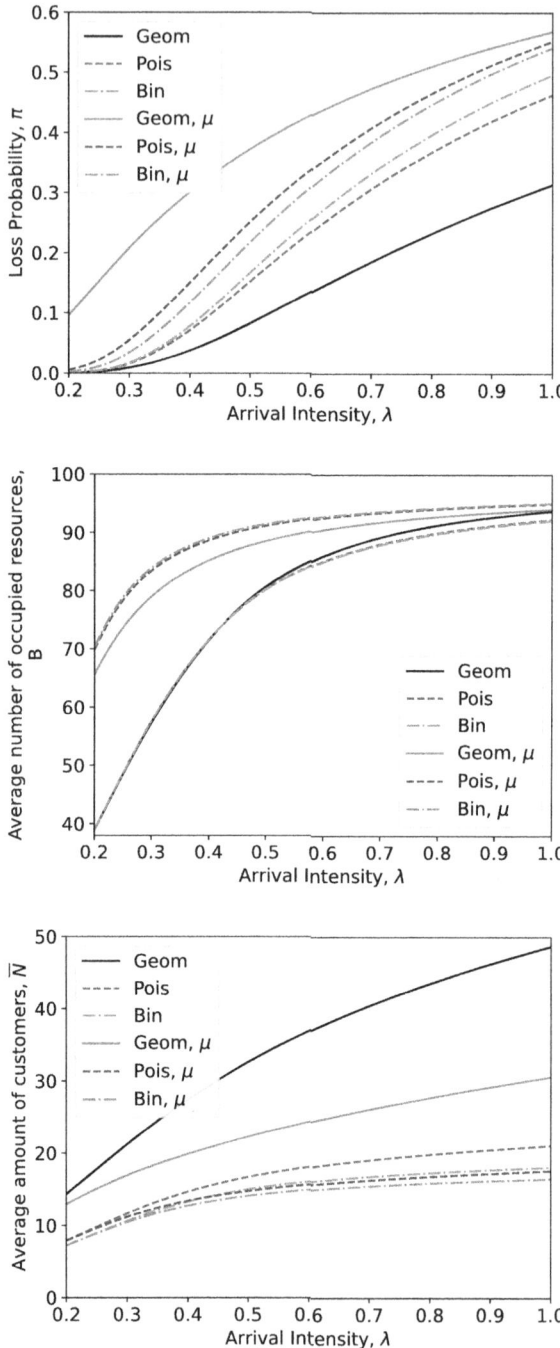

**Fig. 2.** Numerical example results for second case

the average volume of accepted requests increases, which allows a larger number of requests to be accepted for servicing.

However, when time does not depend on resource requirements, the opposite result is obtained. The greater the variance of the distribution, the higher the loss probability, because in this case both "heavy" and "light" customers take the same time to serve and accumulate in the system, which leads to losses of new arrivals.

To sum up, models with service time dependent on arrival demand for resource can service more customers with a lower loss probability.

Finally, in the third case we compare the execution times of direct convolution with the convolutions algorithm. The distribution parameters are the same as in the second case. The results are presented in Table 3.

**Table 3.** Third case parameters

| Parameter | Direct computation, s | Convolution algorithm, s |
|---|---|---|
| $N = R = 100$ | 1.203 | 0.687 |
| $N = R = 300$ | 29.41033 | 16.95827 |
| $N = R = 500$ | 149.87085 | 82.9197 |
| $N = R = 1000$ | 822.02585 | 627.8665 |

## 4 Conclusion

In this paper, we considered a resource loss system with resource-dependent service times. We developed the convolution algorithm to evaluate the system's characteristics. The convolution algorithm can be applied for the evaluation of performance metrics models of wireless networks with elastic traffic. The numerical example showed the dependence of the results on the type of relationship between the distribution of requirements and on the type of service. Moreover, the advantage of the recurrent algorithm over direct calculations was demonstrated.

## References

1. Moltchanov, D., Sopin, E., Begishev, V., Samuylov, A., Koucheryavy, Y., Samouylov, K.: A tutorial on mathematical modeling of 5G/6G millimeter wave and terahertz cellular systems. IEEE Commun. Surv. Tutorials **24**(2), 1072–1116 (2022). https://doi.org/10.1109/COMST.2022.3156207
2. Tikhonenko, O., Ziółkowski, M.: Queueing systems with random volume customers and their performance characteristics. J. Inf. Organ. Sci. **45**(1), 21–38 (2021). https://doi.org/10.31341/jios.45.1.2

3. Lisovskaya, E., Pankratova, E., Moiseeva, S., Pagano, M.: Analysis of a resource-based queue with the parallel service and renewal arrivals. In: Vishnevskiy, V.M., Samouylov, K.E., Kozyrev, D.V. (eds.) DCCN 2020. LNCS, vol. 12563, pp. 335–349. Springer, Cham (2020). https://doi.org/10.1007/978-3-030-66471-8_26
4. Naumov, V.A., Samuilov, K.E., Samuilov, A.K.: On the total amount of resources occupied by serviced customers. Autom. Remote. Control. **77**(8), 1419–1427 (2016). https://doi.org/10.1134/S0005117916080087
5. Begishev, V., et al.: Joint use of guard capacity and multiconnectivity for improved session continuity in millimeter-wave 5G NR systems. IEEE Trans. Veh. Technol. **70**(3), 2657–2672 (2021). https://doi.org/10.1109/TVT.2021.3061906
6. Daraseliya, A., Sopin, E., Moltchanov, D., Koucheryavy, Y., Samouylov, K.: Performance of offloading strategies in collocated deployments of millimeter wave NR-U technology. IEEE Trans. Veh. Technol. **72**(2), 2535–2549 (2021). https://doi.org/10.1109/TVT.2022.3213927
7. Naumov,V., Samuilov, K.: O svyazi resursnyh sistem massovogo obsluzhivaniya s setyami Erlanga. Inform. i eyo primen. **10**(3), 9–14 (2016). https://doi.org/10.14357/19922264160302. in Russian
8. Sopin, E.S., Ageev, K.A., Markova, E.V., Vikhrova, O.G., Gaidamaka, Y.V.: Performance analysis of M2M traffic in LTE network using queuing systems with random resource requirements. Autom. Control. Comput. Sci. **52**(5), 345–353 (2018). https://doi.org/10.3103/S0146411618050127

# Diffusion Approximation for the MAP/GI/1 Retrial Queue with Two-Way Communication

Anatoly Nazarov and Olga Lizyura(✉)

Institute of Applied Mathematics and Computer Science, National Research Tomsk State University, 36 Lenina ave., Tomsk 634050, Russia
oliztsu@mail.ru

**Abstract.** In this paper, we consider a retrial queue with two types of calls: incoming and outgoing calls. The input flow is Markovian arrival process and the service times are arbitrary distributed random variables. In its idle time the server can initiate the service itself and make outgoing calls. We consider the system with multiple types of outgoing calls. We use asymptotic analysis method under the condition of low rate of retrials to study such system and propose the approximation of the stationary probability distribution of the number of calls in the orbit.

**Keywords:** Retrial queue · Markovian arrival process · Two-way communication · Incoming calls · Outgoing calls · Asymptotic-diffusion analysis

## 1 Introduction

Retrial queues are the models of telecommunication and service systems [2,6]. They differ from other queueing models by the following behavior: customers who cannot occupy a server upon arrival join the orbit and retry to access the server after some random delay. Retrial queues with two-way communication are used for modeling blended call centers, where the operator can provide both inbound and outbound calls. In paper [1], the authors describe the impact of using retrial queues as models of call centers. Studies [4,5,7] are devoted to the QoS requirements in telephone services. In [3], the authors provide the investigation of input processes in real call centers.

In literature on retrial queues with two-way communication the models with both Markovian arrival process and arbitrary distribution of service times are not considered due to their complex structure. The main complication is that the analysis requires considering four-dimensional Markov chain. Therefore, we use the asymptotic-diffusion analysis method and build a modification of it for MAP/GI/1 type of retrial queues with two-way communication.

Similar asymptotic results without using diffusion analysis are presented by Sakurai and Phung-Duc in [9]. They study asymptotic behavior of M/G/1 retrial queue with outgoing calls. Nazarov et al. [8] study the slow retrial limit condition for Markovian retrial queues.

The rest of this paper is organized as follows. Section 2 is devoted to the description of MAP/GI/1 retrial queue with multiple types of outgoing calls. Section 3 presents the first step of asymptotic-diffusion analysis of the model, i.e. the derivation of the drift coefficient. Section 4 is devoted to the derivation of the diffusion coefficient. In Sect. 5, we derive the approximation for the stationary probability distribution of the number of calls in the orbit. Section 6 contains some concluding remarks.

## 2 Model Description

We consider single server retrial queue with MAP input. The MAP is driven by Markov chain $m(t)$ with finite set of states $1, 2, ...M$ defined by the generator $\mathbf{Q} = [q_{vm}]$. Let the diagonal matrix $\mathbf{\Lambda}$ denotes of the intensities $\lambda_m$, where $\lambda_m$ is the rate of arrivals in $m$-th state of the MAP. The matrix $\mathbf{D} = [d_{vm}]$ of the probabilities of the arrival at the moment of the underlying Markov chain transition from state $v$ to state $m$. Service times of incoming calls follow an arbitrary distribution with cumulative distribution function $B_1(x)$. If the server is idle at the instant of arrival, then an incoming call occupies it for service.

If the server is busy upon arrival, the incoming call joins the orbit. Incoming calls in orbit make an exponentially distributed delay and reattempt to access the server. The rate of retrials is $\sigma$ for each incoming call in the orbit. After delay, the incoming call behaves the same way as if it first arrived at the system. Thus, it can receive service or make a delay again.

When the server is idle, it makes outgoing calls. We consider the system with multiple types of outgoing calls. The rate of making outgoing calls of type $n$ is equal to $\alpha_n$. Service times of outgoing calls of type $n$ are defined by distribution function $B_n(x)$. We assume $n = 2, 3, \ldots, N$ to simplify derivations.

We denote $i(t)$ as the number of incoming calls in the orbit at instant $t$. Process $k(t)$ reflects the state of the server at instant $t$ as follows:

$$k(t) = \begin{cases} 0, & \text{if the server is idle,} \\ 1, & \text{if an incoming call is in service,} \\ n, & \text{if an outgoing call of type } n \text{ is in service, } n = \overline{2, N}. \end{cases}$$

We also consider the residual time of service $z(t)$, which occurs when the server is busy. Thus, the random process $\{k(t), i(t), m(t), z(t)\}$ with a variable number of components is under investigation.

We denote

$$P\{k(t) = 0, \ i(t) = i, \ m(t) = m\} = P_0(i, m, t),$$

$$P\{k(t) = k, \ i(t) = i, \ m(t) = m, \ z(t) < z\} = P_k(i, m, z, t)$$

and derive the balance equalitions

$$\frac{\partial P_0(i,m,t)}{\partial t} = -\left(\lambda_m + i\sigma + \sum_{n=2}^{N}\alpha_n\right)P_0(i,m,t) + \sum_{k=1}^{N}\frac{\partial P_k(i,m,0,t)}{\partial z}$$

$$+ \sum_{\nu=1}^{M} P_0(i,\nu,t)(1-d_{\nu m})q_{\nu m},$$

$$\frac{\partial P_1(i,m,z,t)}{\partial t} - \frac{\partial P_1(i,m,z,t)}{\partial z} = -\lambda_m P_1(i,m,z,t) - \frac{\partial P_1(i,m,0,t)}{\partial z}$$

$$+ \lambda_m P_1(i-1,m,z,t) + \lambda_m P_0(i,m,t)B_1(z)$$

$$+ (i+1)\sigma P_0(i+1,m,t)B_1(z) + B_1(z)\sum_{\nu=1}^{M}P_0(i,\nu,t)d_{\nu m}q_{\nu m}$$

$$+ \sum_{\nu=1}^{M}\{P_1(i,\nu,z,t)(1-d_{\nu m}) + P_1(i-1,\nu,z,t)d_{\nu m}\}q_{\nu m},$$

$$\frac{\partial P_n(i,m,z,t)}{\partial t} - \frac{\partial P_n(i,m,z,t)}{\partial z} = -\lambda_m P_n(i,m,z,t) - \frac{\partial P_n(i,m,0,t)}{\partial z}$$

$$+ \lambda_m P_n(i-1,m,z,t) + \alpha_n P_0(i,m,t)B_n(z)$$

$$+ \sum_{\nu=1}^{M}\{P_n(i,\nu,z,t)(1-d_{\nu m}) + P_n(i-1,\nu,z,t)d_{\nu m}\}q_{\nu m},\ n=2,\ldots,N.$$

After that, we introduce partial characteristic functions

$$H_0(u,m,t) = \sum_{i=0}^{\infty} e^{jui}P_0(i,m,t),$$

$$H_k(u,m,z,t) = \sum_{i=0}^{\infty} e^{jui}P_k(i,m,z,t),\ k=1,\ldots,N,$$

where $j$ is the imaginary unit. We rewrite the equations as follows:

$$\frac{\partial H_0(u,m,t)}{\partial t} = -\left(\lambda_m + \sum_{n=2}^{N}\alpha_n\right)H_0(u,m,t) + j\sigma\frac{\partial H_0(u,m,t)}{\partial u}$$

$$+ \sum_{k=1}^{N}\frac{\partial H_k(u,m,0,t)}{\partial z} + \sum_{\nu=1}^{M}H_0(u,\nu,t)(1-d_{\nu m})q_{\nu m},$$

$$\frac{\partial H_1(u,m,z,t)}{\partial t} - \frac{\partial H_1(u,m,z,t)}{\partial z} = \lambda_m(e^{ju}-1)H_1(u,m,z,t)$$

$$- \frac{\partial H_1(u,m,0,t)}{\partial z} + \lambda_m H_0(u,m,t)B_1(z) - j\sigma e^{-ju}\frac{\partial H_0(u,m,t)}{\partial u}B_1(z)$$

$$+B_1(z)\sum_{\nu=1}^{M}H_0(u,\nu,t)d_{\nu m}q_{\nu m} + \sum_{\nu=1}^{M}H_1(u,\nu,z,t)(1+(e^{ju}-1)d_{\nu m})q_{\nu m},$$

$$\frac{\partial H_n(u,m,z,t)}{\partial t} - \frac{\partial H_n(u,m,z,t)}{\partial z} = \lambda_m(e^{ju}-1)H_n(u,m,z,t)$$

$$-\frac{\partial H_n(u,m,0,t)}{\partial z} + \alpha_n H_0(u,m,t)B_n(z)$$

$$+\sum_{\nu=1}^{M}H_n(u,\nu,z,t)(1+(e^{ju}-1)d_{\nu m})q_{\nu m}, \quad n=2,\ldots,N. \tag{1}$$

Let us rewrite the equations in matrix form denoting

$$\mathbf{H}_0(u,t) = \{H_0(u,1,t),\ H_0(u,2,t),\ \ldots,\ H_0(u,M,t)\},$$

$$\mathbf{H}_k(u,z,t) = \{H_k(u,1,z,t),\ H_k(u,2,z,t),\ \ldots,\ H_k(u,M,z,t)\},$$

$$\mathbf{A} = \mathbf{\Lambda} + \mathbf{D} \circ \mathbf{Q},$$

and $\mathbf{I}$ as identity matrix. We rewrite system (1) together with the additional equation obtained by summing up the equations of the system and taking the limit by $z \to \infty$

$$\frac{\partial \mathbf{H}_0(u,t)}{\partial t} = \mathbf{H}_0(u,t)\left(\mathbf{Q} - \mathbf{A} - \sum_{n=2}^{N}\alpha_n \mathbf{I}\right) + j\sigma\frac{\partial \mathbf{H}_0(u,t)}{\partial u}$$

$$+\sum_{k=1}^{N}\frac{\partial \mathbf{H}_k(u,0,t)}{\partial z},$$

$$\frac{\partial \mathbf{H}_1(u,z,t)}{\partial t} - \frac{\partial \mathbf{H}_1(u,z,t)}{\partial z} = \mathbf{H}_1(u,z,t)(\mathbf{Q}+(e^{ju}-1)\mathbf{A})$$

$$-\frac{\partial \mathbf{H}_1(u,0,t)}{\partial z} + B_1(z)\mathbf{H}_0(u,t)\mathbf{A} - j\sigma e^{-ju}B_1(z)\frac{\partial \mathbf{H}_0(u,t)}{\partial u},$$

$$\frac{\partial \mathbf{H}_n(u,z,t)}{\partial t} - \frac{\partial \mathbf{H}_n(u,z,t)}{\partial z} = \mathbf{H}_n(u,z,t)(\mathbf{Q}+(e^{ju}-1)\mathbf{A})$$

$$-\frac{\partial \mathbf{H}_n(u,0,t)}{\partial z} + \alpha_n B_n(z)\mathbf{H}_0(u,t), \quad n=2,\ldots,N,$$

$$\frac{\partial \mathbf{H}(u,t)}{\partial t}\mathbf{e} = (e^{ju}-1)\left\{j\sigma e^{-ju}\frac{\partial \mathbf{H}_0(u,t)}{\partial u}\mathbf{e} + (\mathbf{H}(u,t)-\mathbf{H}_0(u,t))\mathbf{A}\mathbf{e}\right\}, \tag{2}$$

where $\mathbf{H}(u,t) = \mathbf{H}_0(u,t) + \sum_{k=1}^{N}\mathbf{H}_k(u,\infty,t)$, $\mathbf{e}$ is the vector of ones.

## 3 First Step of Asymptotic-Diffusion Analysis Method

In system (2), we introduce the following notations:

$$\sigma = \varepsilon, \quad u = w\varepsilon, \quad \tau = t\varepsilon,$$

$$\mathbf{H}_0(u,t) = \mathbf{F}_0(w,\tau,\varepsilon), \quad \mathbf{H}_k(u,z,t) = \mathbf{F}_k(w,z,\tau,\varepsilon), \quad k=1,\ldots,N,$$

and obtain

$$\varepsilon \frac{\partial \mathbf{F}_0(w,\tau,\varepsilon)}{\partial \tau} = \mathbf{F}_0(w,\tau,\varepsilon)\left(\mathbf{Q} - \mathbf{A} - \sum_{n=2}^{N} \alpha_n \mathbf{I}\right) + j\frac{\partial \mathbf{F}_0(w,\tau,\varepsilon)}{\partial w}$$

$$+ \sum_{k=1}^{N} \frac{\partial \mathbf{F}_k(w,0,\tau,\varepsilon)}{\partial z},$$

$$\varepsilon \frac{\partial \mathbf{F}_1(w,z,\tau,\varepsilon)}{\partial \tau} - \frac{\partial \mathbf{F}_1(w,z,\tau,\varepsilon)}{\partial z} = \mathbf{F}_1(w,z,\tau,\varepsilon)(\mathbf{Q} + (e^{jw\varepsilon}-1)\mathbf{A})$$

$$- \frac{\partial \mathbf{F}_1(w,0,\tau,\varepsilon)}{\partial z} + B_1(z)\mathbf{F}_0(w,\tau,\varepsilon)\mathbf{A} - je^{-jw\varepsilon}B_1(z)\frac{\partial \mathbf{F}_0(w,\tau,\varepsilon)}{\partial w},$$

$$\varepsilon \frac{\partial \mathbf{F}_n(w,z,\tau,\varepsilon)}{\partial \tau} - \frac{\partial \mathbf{F}_n(w,z,\tau,\varepsilon)}{\partial z} = \mathbf{F}_n(w,z,\tau,\varepsilon)(\mathbf{Q} + (e^{jw\varepsilon}-1)\mathbf{A})$$

$$- \frac{\partial \mathbf{F}_n(w,0,\tau,\varepsilon)}{\partial z} + \alpha_n B_n(z)\mathbf{F}_0(w,\tau,\varepsilon), \quad n=2,\ldots,N,$$

$$\varepsilon \frac{\partial \mathbf{F}(w,\tau,\varepsilon)}{\partial \tau}\mathbf{e}$$

$$= (e^{jw\varepsilon}-1)\left\{je^{-jw\varepsilon}\frac{\partial \mathbf{F}_0(w,\tau,\varepsilon)}{\partial w}\mathbf{e} + (\mathbf{F}(w,\tau,\varepsilon) - \mathbf{F}_0(w,\tau,\varepsilon))\mathbf{A}\mathbf{e}\right\}. \quad (3)$$

We consider system (3) in the limit by $\varepsilon \to 0$:

$$\mathbf{F}_0(w,\tau)\left(\mathbf{Q} - \mathbf{A} - \sum_{n=2}^{N} \alpha_n \mathbf{I}\right) + j\frac{\partial \mathbf{F}_0(w,\tau)}{\partial w} + \sum_{k=1}^{N} \frac{\partial \mathbf{F}_k(w,0,\tau)}{\partial z} = 0,$$

$$\frac{\partial \mathbf{F}_1(w,z,\tau)}{\partial z} - \frac{\partial \mathbf{F}_1(w,0,\tau)}{\partial z} + \mathbf{F}_1(w,z,\tau)\mathbf{Q} + B_1(z)\mathbf{F}_0(w,\tau)\mathbf{A}$$

$$- jB_1(z)\frac{\partial \mathbf{F}_0(w,\tau)}{\partial w} = 0,$$

$$\frac{\partial \mathbf{F}_n(w,z,\tau)}{\partial z} - \frac{\partial \mathbf{F}_n(w,0,\tau)}{\partial z} + \mathbf{F}_n(w,z,\tau)\mathbf{Q}$$

$$+ \alpha_n B_n(z)\mathbf{F}_0(w,\tau) = 0, \quad n=2,\ldots,N,$$

$$\frac{\partial \mathbf{F}(w,\tau)}{\partial \tau}\mathbf{e} = jw\left\{j\frac{\partial \mathbf{F}_0(w,\tau)}{\partial w}\mathbf{e} + (\mathbf{F}(w,\tau) - \mathbf{F}_0(w,\tau))\mathbf{A}\mathbf{e}\right\}. \quad (4)$$

We will seek the solution of system (4) in the following form:
$$\mathbf{F}_0(w,\tau) = \mathbf{R}_0 e^{jwx(\tau)}, \quad \mathbf{F}_k(w,z,\tau) = \mathbf{R}_k(z)e^{jwx(\tau)}, \quad k=1,\ldots,N.$$

Substituting the solution into the system, we obtain

$$\mathbf{R}_0\left(\mathbf{Q} - \mathbf{A} - \sum_{n=2}^{N}\alpha_n \mathbf{I} - x(\tau)\mathbf{I}\right) + \sum_{k=1}^{N}\mathbf{R}'_k(0) = 0,$$

$$\mathbf{R}'_1(z) - \mathbf{R}'_1(0) + \mathbf{R}_1(z)\mathbf{Q} + B_1(z)\mathbf{R}_0(\mathbf{A} + x(\tau)\mathbf{I}) = 0,$$
$$\mathbf{R}'_n(z) - \mathbf{R}'_n(0) + \mathbf{R}_n(z)\mathbf{Q} + \alpha_n B_n(z)\mathbf{R}_0 = 0, \quad n = 2,\ldots,N,$$
$$x'(\tau) = -x(\tau)\mathbf{R}_0 \mathbf{e} + (\mathbf{R} - \mathbf{R}_0)\mathbf{A}\mathbf{e}. \tag{5}$$

We denote
$$a(x) = -x\mathbf{R}_0(x)\mathbf{e} + (\mathbf{R} - \mathbf{R}_0(x))\mathbf{A}\mathbf{e}. \tag{6}$$

Function $a(x)$ is the drift coefficient of diffusion process approximating the number of incoming calls in the orbit. In further analysis, we omit argument $\tau$ to simplify derivations.

We consider the second equation of system (5) in the limit by $z \to \infty$
$$\mathbf{R}'_1(0) = \mathbf{R}_1 \mathbf{Q} + \mathbf{R}_0(\mathbf{A} + x\mathbf{I}).$$

After that, we substitute obtained vector $\mathbf{R}'_1(0)$ into the second equation of system (5)
$$\mathbf{R}'_1(z) = (\mathbf{R}_1 - \mathbf{R}_1(z))\mathbf{Q} + (1 - B_1(z))\mathbf{R}_0(\mathbf{A} + x\mathbf{I}). \tag{7}$$

Making the same procedure for the third equation of system (5), we obtain equation
$$\mathbf{R}'_n(z) = (\mathbf{R}_n - \mathbf{R}_n(z))\mathbf{Q} + \alpha_n(1 - B_n(z))\mathbf{R}_0, \quad n = 2,\ldots,N. \tag{8}$$

Further, we apply Laplace-Stieltjes transform to equations (7) and (8) as follows:
$$\mathbf{R}^*_1(s)(\mathbf{Q} + s\mathbf{I}) = \mathbf{R}_1 \mathbf{Q} + (1 - B^*_1(s))\mathbf{R}_0(\mathbf{A} + x\mathbf{I}), \tag{9}$$
$$\mathbf{R}^*_n(s)(\mathbf{Q} + s\mathbf{I}) = \mathbf{R}_n \mathbf{Q} + \alpha_n(1 - B^*_n(s))\mathbf{R}_0, \quad n = 2,\ldots,N. \tag{10}$$

Denoting $s_m$ as eigenvalues and $\mathbf{v}_m$ as eigenvectors of matrix $-\mathbf{Q}$, $m = 2,\ldots,M$, we substitute $s = s_m$ into Eqs. (9) and (10) and multiply them by vector $\mathbf{v}_m$:
$$\mathbf{R}^*_1(s_m)(\mathbf{Q} + s_m \mathbf{I})\mathbf{v}_m = \mathbf{R}_1 \mathbf{Q}\mathbf{v}_m + (1 - B^*_1(s_m))\mathbf{R}_0(\mathbf{A} + x\mathbf{I})\mathbf{v}_m,$$
$$\mathbf{R}^*_n(s_k)(\mathbf{Q} + s_m \mathbf{I})\mathbf{v}_m = \mathbf{R}_n \mathbf{Q}\mathbf{v}_m + \alpha_n(1 - B^*_n(s_m))\mathbf{R}_0\mathbf{R}_m, \quad n = 2,\ldots,N.$$

Since the left parts of the equations are zeroes, we obtain
$$\mathbf{R}_1(x)\mathbf{Q}\mathbf{v}_m + (1 - B^*_1(s_m))\mathbf{R}_0(x)(\mathbf{A} + x\mathbf{I})\mathbf{v}_m = 0,$$

$$\mathbf{R}_n(x)\mathbf{Q}\mathbf{v}_m + \alpha_n(1 - B_n^*(s_m))\mathbf{R}_0(x)\mathbf{v}_m = 0, \ n = 2, \ldots, N.$$

The determinant of matrix $\mathbf{Q}$ is zero. Thus, there is always $s_1 = 0$ among the eigenvalues. Thereby, we derive additional equations multiplying equations (7) and (8) by vector $\mathbf{e}$ and taking the limit by $z \to \infty$:

$$\mathbf{R}_1(x)\mathbf{e} = b_1 \mathbf{R}_0(x)(\mathbf{A} + x\mathbf{I})\mathbf{e},$$

$$\mathbf{R}_n(x)\mathbf{e} = \alpha_n b_n \mathbf{R}_0(x)\mathbf{e},$$

where $b_k$ is the mean of corresponding distribution $B_k(x)$, $k = 1, \ldots, N$.

Finally, we add the normalization condition and obtain the system of matrix equations defining vectors $\mathbf{R}_k$, $k = 0, \ldots, N$ as follows:

$$\mathbf{R}_1(x)\mathbf{Q}\mathbf{v}_m + (1 - B_1^*(s_m))\mathbf{R}_0(x)(\mathbf{A} + x\mathbf{I})\mathbf{v}_m = 0,$$

$$\mathbf{R}_1(x)\mathbf{e} = b_1 \mathbf{R}_0(x)(\mathbf{A} + x\mathbf{I})\mathbf{e},$$

$$\mathbf{R}_n(x)\mathbf{Q}\mathbf{v}_m + \alpha_n(1 - B_n^*(s_m))\mathbf{R}_0(x)\mathbf{v}_m = 0, \ n = 2, \ldots, N,$$

$$\mathbf{R}_n(x)\mathbf{e} = \alpha_n b_n \mathbf{R}_0(x)\mathbf{e}, \ n = 2, \ldots, N,$$

$$\sum_{k=0}^{N} \mathbf{R}_k(x) = \mathbf{R}. \tag{11}$$

Here vector $\mathbf{R}$ is the steady state distribution of process $m(t)$, which is solution of the following system:

$$\mathbf{R}\mathbf{Q} = \mathbf{0}, \ \mathbf{R}\mathbf{e} = 1.$$

Having vectors $\mathbf{R}_k(x)$, we can directly calculate drift coefficient $a(x)$. The next step of the analysis is derivation of diffusion coefficient.

## 4  Second Step of Asymptotic-Diffusion Analysis Method

We introduce the following notations in system (2):

$$\mathbf{H}_0(u,t) = e^{j\frac{u}{\sigma}x(\sigma t)}\mathbf{H}_0^{(2)}(u,t), \ \mathbf{H}_k(u,z,t) = e^{j\frac{u}{\sigma}x(\sigma t)}\mathbf{H}_k^{(2)}(u,z,t), \ k = 1, \ldots, N,$$

and obtain

$$\frac{\partial \mathbf{H}_0^{(2)}(u,t)}{\partial t} + jux'(\sigma t)\mathbf{H}_0^{(2)}(u,t) = \mathbf{H}_0^{(2)}(u,t)\left(\mathbf{Q} - \mathbf{A} - \sum_{n=2}^{N}\alpha_n\mathbf{I} - x(\sigma t)\mathbf{I}\right)$$

$$+ j\sigma\frac{\partial \mathbf{H}_0^{(2)}(u,t)}{\partial u} + \sum_{k=1}^{N}\frac{\partial \mathbf{H}_k^{(2)}(u,0,t)}{\partial z},$$

$$\frac{\partial \mathbf{H}_1^{(2)}(u,z,t)}{\partial t} - \frac{\partial \mathbf{H}_1^{(2)}(u,z,t)}{\partial z} + jux'(\sigma t)\mathbf{H}_1^{(2)}(u,z,t)$$

$$= \mathbf{H}_1^{(2)}(u,z,t)(\mathbf{Q} + (e^{ju}-1)\mathbf{A}) - \frac{\partial \mathbf{H}_1^{(2)}(u,0,t)}{\partial z}$$

$$+ B_1(z)\mathbf{H}_0^{(2)}(u,t)(\mathbf{A} + e^{-ju}x(\sigma t)\mathbf{I}) - j\sigma e^{-ju}B_1(z)\frac{\partial \mathbf{H}_0^{(2)}(u,t)}{\partial u},$$

$$\frac{\partial \mathbf{H}_n^{(2)}(u,z,t)}{\partial t} - \frac{\partial \mathbf{H}_n^{(2)}(u,z,t)}{\partial z} + jux'(\sigma t)\mathbf{H}_n^{(2)}(u,z,t)$$

$$= \mathbf{H}_n^{(2)}(u,z,t)(\mathbf{Q} + (e^{ju}-1)\mathbf{A}) - \frac{\partial \mathbf{H}_n^{(2)}(u,0,t)}{\partial z}$$

$$+ \alpha_n B_n(z)\mathbf{H}_0^{(2)}(u,t), \quad n = 2,\dots,N,$$

$$\frac{\partial \mathbf{H}^{(2)}(u,t)}{\partial t}\mathbf{e} + jux'(\sigma t)\mathbf{H}^{(2)}(u,t)\mathbf{e} = (e^{ju}-1)\left\{j\sigma e^{-ju}\frac{\partial \mathbf{H}_0^{(2)}(u,t)}{\partial u}\mathbf{e}\right.$$

$$\left. - e^{-ju}x(\sigma t)\mathbf{H}_0^{(2)}(u,t)\mathbf{e} + (\mathbf{H}^{(2)}(u,t) - \mathbf{H}_0^{(2)}(u,t))\mathbf{A}\mathbf{e}\right\}. \quad (12)$$

After that, we make the following substitutions in system (12):

$$\sigma = \varepsilon^2, \quad u = w\varepsilon, \quad \tau = t\varepsilon^2,$$

$$\mathbf{H}_0^{(2)}(u,t) = \mathbf{F}_0^{(2)}(w,\tau,\varepsilon), \quad \mathbf{H}_k^{(2)}(u,z,t) = \mathbf{F}_k^{(2)}(w,z,\tau,\varepsilon), \quad k=1,\dots,N,$$

and rewrite the system

$$\varepsilon^2 \frac{\partial \mathbf{F}_0^{(2)}(w,\tau,\varepsilon)}{\partial \tau} + jw\varepsilon x'(\tau)\mathbf{F}_0^{(2)}(w,\tau,\varepsilon)$$

$$= \mathbf{F}_0^{(2)}(w,\tau,\varepsilon)\left(\mathbf{Q} - \mathbf{A} - \sum_{n=2}^{N}\alpha_n \mathbf{I} - x(\tau)\mathbf{I}\right)$$

$$+ j\varepsilon\frac{\partial \mathbf{F}_0^{(2)}(w,\tau,\varepsilon)}{\partial w} + \sum_{k=1}^{N}\frac{\partial \mathbf{F}_k^{(2)}(w,0,\tau,\varepsilon)}{\partial z},$$

$$\varepsilon^2 \frac{\partial \mathbf{F}_1^{(2)}(w,z,\tau,\varepsilon)}{\partial \tau} - \frac{\partial \mathbf{F}_1^{(2)}(w,z,\tau,\varepsilon)}{\partial z} + jw\varepsilon x'(\tau)\mathbf{F}_1^{(2)}(w,z,\tau,\varepsilon)$$

$$= \mathbf{F}_1^{(2)}(w,z,\tau,\varepsilon)(\mathbf{Q} + (e^{jw\varepsilon}-1)\mathbf{A}) - \frac{\partial \mathbf{F}_1^{(2)}(w,0,\tau,\varepsilon)}{\partial z}$$

$$+ B_1(z)\mathbf{F}_0^{(2)}(w,\tau,\varepsilon)(\mathbf{A} + e^{-jw\varepsilon}x(\tau)\mathbf{I}) - j\varepsilon e^{-jw\varepsilon}B_1(z)\frac{\partial \mathbf{F}_0^{(2)}(w,\tau,\varepsilon)}{\partial w},$$

$$\varepsilon^2 \frac{\partial \mathbf{F}_n^{(2)}(w,z,\tau,\varepsilon)}{\partial \tau} - \frac{\partial \mathbf{F}_n^{(2)}(w,z,\tau,\varepsilon)}{\partial z} + jw\varepsilon x'(\tau)\mathbf{F}_n^{(2)}(w,z,\tau,\varepsilon)$$

$$= \mathbf{F}_n^{(2)}(w,z,\tau,\varepsilon)(\mathbf{Q} + (e^{jw\varepsilon}-1)\mathbf{A}) - \frac{\partial \mathbf{F}_n^{(2)}(w,0,\tau,\varepsilon)}{\partial z}$$

$$+\alpha_n B_n(z)\mathbf{F}_0^{(2)}(w,\tau,\varepsilon),\ n=2,\ldots,N,$$

$$\varepsilon^2\frac{\partial \mathbf{F}^{(2)}(w,\tau,\varepsilon)}{\partial \tau}\mathbf{e}+jw\varepsilon x'(\tau)\mathbf{F}^{(2)}(w,\tau,\varepsilon)\mathbf{e}=(e^{jw\varepsilon}-1)$$

$$\times\left\{j\varepsilon e^{-jw\varepsilon}\frac{\partial \mathbf{F}_0^{(2)}(w,\tau,\varepsilon)}{\partial w}\mathbf{e}-e^{-jw\varepsilon}x(\tau)\mathbf{F}_0^{(2)}(w,\tau,\varepsilon)\mathbf{e}+\right.$$

$$\left.(\mathbf{F}^{(2)}(w,\tau,\varepsilon)-\mathbf{F}_0^{(2)}(w,\tau,\varepsilon))\mathbf{A}\mathbf{e}\right\}. \tag{13}$$

We consider first three equations of system (13) using Taylor expansions up to $O(\varepsilon^2)$

$$jw\varepsilon x'(\tau)\mathbf{F}_0^{(2)}(w,\tau,\varepsilon)=\mathbf{F}_0^{(2)}(w,\tau,\varepsilon)\left(\mathbf{Q}-\mathbf{A}-\sum_{n=2}^N\alpha_n\mathbf{I}-x(\tau)\mathbf{I}\right)$$

$$+j\varepsilon\frac{\partial \mathbf{F}_0^{(2)}(w,\tau,\varepsilon)}{\partial w}+\sum_{k=1}^N\frac{\partial \mathbf{F}_k^{(2)}(w,0,\tau,\varepsilon)}{\partial z}+O(\varepsilon^2),$$

$$jw\varepsilon x'(\tau)\mathbf{F}_1^{(2)}(w,z,\tau,\varepsilon)=\frac{\partial \mathbf{F}_1^{(2)}(w,z,\tau,\varepsilon)}{\partial z}-\frac{\partial \mathbf{F}_1^{(2)}(w,0,\tau,\varepsilon)}{\partial z}$$

$$+\mathbf{F}_1^{(2)}(w,z,\tau,\varepsilon)(\mathbf{Q}+jw\varepsilon\mathbf{A})+B_1(z)\mathbf{F}_0^{(2)}(w,\tau,\varepsilon)(\mathbf{A}+(1-jw\varepsilon)x(\tau)\mathbf{I})$$

$$-j\varepsilon B_1(z)\frac{\partial \mathbf{F}_0^{(2)}(w,\tau,\varepsilon)}{\partial w}+O(\varepsilon^2),$$

$$jw\varepsilon x'(\tau)\mathbf{F}_n^{(2)}(w,z,\tau,\varepsilon)=\frac{\partial \mathbf{F}_n^{(2)}(w,z,\tau,\varepsilon)}{\partial z}-\frac{\partial \mathbf{F}_n^{(2)}(w,0,\tau,\varepsilon)}{\partial z}$$

$$+\mathbf{F}_n^{(2)}(w,z,\tau,\varepsilon)(\mathbf{Q}+jw\varepsilon\mathbf{A})+\alpha_n B_n(z)\mathbf{F}_0^{(2)}(w,\tau,\varepsilon)+O(\varepsilon^2),\ n=2,\ldots,N.$$

We will seek the solution of the last system in the following form:

$$\mathbf{F}_0^{(2)}(w,\tau,\varepsilon)=\Phi(w,\tau)\{\mathbf{R}_0+jw\varepsilon\mathbf{f}_0\}+O(\varepsilon^2), \tag{14}$$

$$\mathbf{F}_k^{(2)}(w,z,\tau,\varepsilon)=\Phi(w,\tau)\{\mathbf{R}_k(z)+jw\varepsilon\mathbf{f}_k(z)\}+O(\varepsilon^2),\ k=1,\ldots,N, \tag{15}$$

then we have

$$jw\varepsilon x'(\tau)\Phi(w,\tau)\mathbf{R}_0=\Phi(w,\tau)\mathbf{R}_0\left(\mathbf{Q}-\mathbf{A}-\sum_{n=2}^N\alpha_n\mathbf{I}-x(\tau)\mathbf{I}\right)$$

$$+jw\varepsilon\Phi(w,\tau)\mathbf{f}_0\left(\mathbf{Q}-\mathbf{A}-\sum_{n=2}^N\alpha_n\mathbf{I}-x(\tau)\mathbf{I}\right)+j\varepsilon\frac{\partial\Phi(w,\tau)}{\partial w}\mathbf{R}_0$$

$$+\Phi(w,\tau)\sum_{k=1}^N\mathbf{R}'_k(z)+jw\varepsilon\Phi(w,\tau)\sum_{k=1}^N\mathbf{f}'_k(z)+O(\varepsilon^2),$$

$$jw\varepsilon x'(\tau)\Phi(w,\tau)\mathbf{R}_1(z) = \Phi(w,\tau)\mathbf{R}'_1(z) - \Phi(w,\tau)\mathbf{R}'_1(0)$$
$$+jw\varepsilon\Phi(w,\tau)\mathbf{f}'_1(z) - jw\varepsilon\Phi(w,\tau)\mathbf{f}'_1(0) + \Phi(w,\tau)\mathbf{R}_1(z)\mathbf{Q}$$
$$+jw\varepsilon\Phi(w,\tau)\mathbf{f}_1(z)\mathbf{Q} + jw\varepsilon\Phi(w,\tau)\mathbf{R}_1(z)\mathbf{A} + B_1(z)\Phi(w,\tau)\mathbf{R}_0(\mathbf{A}+x(\tau)\mathbf{I})$$
$$+jw\varepsilon B_1(z)\Phi(w,\tau)\mathbf{f}_0(\mathbf{A}+x(\tau)\mathbf{I}) - jw\varepsilon x(\tau)B_1(z)\Phi(w,\tau)\mathbf{R}_0$$
$$-j\varepsilon B_1(z)\frac{\partial\Phi(w,\tau)}{\partial w}\mathbf{R}_0 + O(\varepsilon^2),$$
$$jw\varepsilon x'(\tau)\Phi(w,\tau)\mathbf{R}_n(z) = \Phi(w,\tau)\mathbf{R}'_n(z) - \Phi(w,\tau)\mathbf{R}'_n(0)$$
$$+jw\varepsilon\Phi(w,\tau)\mathbf{f}'_n(z) - jw\varepsilon\Phi(w,\tau)\mathbf{f}'_n(0) + \Phi(w,\tau)\mathbf{R}_n(z)\mathbf{Q}$$
$$+jw\varepsilon\Phi(w,\tau)\mathbf{f}_n(z)\mathbf{Q} + jw\varepsilon\Phi(w,\tau)\mathbf{R}_n(z)\mathbf{A} + \alpha_n B_n(z)\Phi(w,\tau)\mathbf{R}_0$$
$$+jw\varepsilon\alpha_n B_n(z)\Phi(w,\tau)\mathbf{f}_0 + O(\varepsilon^2),\ n=2,\ldots,N.$$

Dividing the equations by $jw\varepsilon$ and taking the limit by $\varepsilon \to 0$, we obtain

$$\mathbf{f}_0\left(\mathbf{Q}-\mathbf{A}-\sum_{n=2}^N \alpha_n\mathbf{I}-x\mathbf{I}\right) + \sum_{k=1}^N \mathbf{f}'_k(z) = a(x)\mathbf{R}_0 - \frac{\partial\Phi(w,\tau)/\partial w}{w\Phi(w,\tau)}\mathbf{R}_0,$$

$$\mathbf{f}'_1(z) - \mathbf{f}'_1(0) + \mathbf{f}_1(z)\mathbf{Q} + B_1(z)\mathbf{f}_0(\mathbf{A}+x\mathbf{I}) = a(x)\mathbf{R}_1(z) - \mathbf{R}_1(z)\mathbf{A}$$
$$+xB_1(z)\mathbf{R}_0 + B_1(z)\frac{\partial\Phi(w,\tau)/\partial w}{w\Phi(w,\tau)}\mathbf{R}_0,$$

$$\mathbf{f}'_n(z) - \mathbf{f}'_n(0) + \mathbf{f}_n(z)\mathbf{Q} + \alpha_n B_n(z)\mathbf{f}_0 = a(x)\mathbf{R}_n(z) - \mathbf{R}_n(z)\mathbf{A},\ n=2,\ldots,N.$$

We present vectors $\mathbf{f}_0$ and $\mathbf{f}_k(z)$ in the following form:

$$\mathbf{f}_0 = C\mathbf{R}_0 + \mathbf{g}_0 - \frac{\partial\Phi(w,\tau)/\partial w}{w\Phi(w,\tau)}\varphi_0, \tag{16}$$

$$\mathbf{f}_k(z) = C\mathbf{R}_k(z) + \mathbf{g}_k(z) - \frac{\partial\Phi(w,\tau)/\partial w}{w\Phi(w,\tau)}\varphi_k(z), \tag{17}$$

and transform the last system in two systems of equations

$$\mathbf{g}_0\left(\mathbf{Q}-\mathbf{A}-\sum_{n=2}^N \alpha_n\mathbf{I}-x\mathbf{I}\right) + \sum_{k=1}^N \mathbf{g}'_k(z) = a(x)\mathbf{R}_0,$$

$$\mathbf{g}'_1(z) - \mathbf{g}'_1(0) + \mathbf{g}_1(z)\mathbf{Q} + B_1(z)\mathbf{g}_0(\mathbf{A}+x\mathbf{I})$$
$$= a(x)\mathbf{R}_1(z) - \mathbf{R}_1(z)\mathbf{A} + xB_1(z)\mathbf{R}_0,$$

$$\mathbf{g}'_n(z)-\mathbf{g}'_n(0)+\mathbf{g}_n(z)\mathbf{Q}+\alpha_n B_n(z)\mathbf{g}_0 = a(x)\mathbf{R}_n(z)-\mathbf{R}_n(z)\mathbf{A},\ n=2,\ldots,N. \tag{18}$$

$$\varphi_0\left(\mathbf{Q}-\mathbf{A}-\sum_{n=2}^N \alpha_n\mathbf{I}-x\mathbf{I}\right) + \sum_{k=1}^N \varphi'_k(z) = \mathbf{R}_0,$$

$$\varphi'_1(z) - \varphi'_1(0) + \varphi_1(z)\mathbf{Q} + B_1(z)\varphi_0(\mathbf{A}+x\mathbf{I}) = -B_1(z)\mathbf{R}_0,$$

$$\varphi'_n(z) - \varphi'_n(0) + \varphi_n(z)\mathbf{Q} + \alpha_n B_n(z)\varphi_0 = 0, \ n = 2,\ldots,N. \quad (19)$$

We note that system (19) is derivative of system (5) by $x$. Thus, we can write the solution of the system as follows:

$$\varphi_0(x) = \mathbf{R}'_0(x), \ \varphi_k(z,x) = \frac{\partial \mathbf{R}_k(z,x)}{\partial x}.$$

Further, we consider the second equation of system (18) in the limit by $z \to \infty$:

$$\mathbf{g}'_1(0) = \mathbf{g}_1\mathbf{Q} + \mathbf{g}_0(\mathbf{A} + x\mathbf{I}) - a(x)\mathbf{R}_1 + \mathbf{R}_1\mathbf{A} - x\mathbf{R}_0.$$

Substituting the obtained formula back into the second equation of system (18), we have

$$\mathbf{g}'_1(z) = (\mathbf{g}_1 - \mathbf{g}_1(z))\mathbf{Q} + \mathbf{g}_0(\mathbf{A} + x\mathbf{I})(1 - B_1(z))$$
$$+ (\mathbf{R}_1 - \mathbf{R}_1(z))(\mathbf{A} - a(x)\mathbf{I}) - x(1 - B_1(z))\mathbf{R}_0. \quad (20)$$

Making the same procedure for the third equation of system (18), we obtain equation

$$\mathbf{g}'_n(z) = (\mathbf{g}_n - \mathbf{g}_n(z))\mathbf{Q} + \alpha_n(1 - B_n(z))\mathbf{g}_0$$
$$+ (\mathbf{R}_n - \mathbf{R}_n(z))(\mathbf{A} - a(x)\mathbf{I}), \ n = 2,\ldots,N. \quad (21)$$

We apply Laplace-Stieltjes transform to Eqs. (20) and (21):

$$\mathbf{g}^*_1(s)(\mathbf{Q} + s\mathbf{I}) = \mathbf{g}_1\mathbf{Q} + \mathbf{g}_0(\mathbf{A} + x\mathbf{I})(1 - B^*_1(s))$$
$$+ (\mathbf{R}_1 - \mathbf{R}^*_1(s))(\mathbf{A} - a(x)\mathbf{I}) - x(1 - B^*_1(s))\mathbf{R}_0,$$
$$\mathbf{g}^*_n(s)(\mathbf{Q} + s\mathbf{I}) = \mathbf{g}_n\mathbf{Q} + \alpha_n(1 - B^*_n(s))\mathbf{g}_0$$
$$+ (\mathbf{R}_n - \mathbf{R}^*_n(s))(\mathbf{A} - a(x)\mathbf{I}), \ n = 2,\ldots,N.$$

Substituting into equations eigenvalues $s_m$ of matrix $-\mathbf{Q}$ and multiplying by corresponding eigenvectors $\mathbf{v}_m$, $m = 2,\ldots,M$, we obtain

$$\mathbf{g}^*_1(s_m)(\mathbf{Q} + s_m\mathbf{I})\mathbf{v}_m = \mathbf{g}_1\mathbf{Q}\mathbf{v}_m + \mathbf{g}_0(\mathbf{A} + x\mathbf{I})(1 - B^*_1(s_m))\mathbf{v}_m$$
$$+ (\mathbf{R}_1 - \mathbf{R}^*_1(s_m))(\mathbf{A} - a(x)\mathbf{I})\mathbf{v}_m - x(1 - B^*_1(s_m))\mathbf{R}_0\mathbf{v}_m,$$
$$\mathbf{g}^*_n(s_m)(\mathbf{Q} + s_m\mathbf{I}) = \mathbf{g}_n\mathbf{Q}\mathbf{v}_m + \alpha_n(1 - B^*_n(s_m))\mathbf{g}_0\mathbf{v}_m$$
$$+ (\mathbf{R}_n - \mathbf{R}^*_n(s_m))(\mathbf{A} - a(x)\mathbf{I})\mathbf{v}_m, \ n = 2,\ldots,N.$$

Since the left parts of equations are zeroes, we can write

$$\mathbf{g}_1\mathbf{Q}\mathbf{v}_m + \mathbf{g}_0(\mathbf{A} + x\mathbf{I})(1 - B^*_1(s_m))\mathbf{v}_m + (\mathbf{R}_1 - \mathbf{R}^*_1(s_m))(\mathbf{A} - a(x)\mathbf{I})\mathbf{v}_m$$
$$- x(1 - B^*_1(s_m))\mathbf{R}_0\mathbf{v}_m = 0,$$
$$\mathbf{g}_n\mathbf{Q}\mathbf{v}_m + \alpha_n(1 - B^*_n(s_m))\mathbf{g}_0\mathbf{v}_m$$
$$+ (\mathbf{R}_n - \mathbf{R}^*_n(s_m))(\mathbf{A} - a(x)\mathbf{I})\mathbf{v}_m = 0, \ n = 2,\ldots,N.$$

We also derive additional equations multiplying equations (20) and (21) by vector **e** and taking the limit by $z \to \infty$:

$$\mathbf{g}_1 \mathbf{e} = b_1(\mathbf{g}_0(\mathbf{A} + x\mathbf{I})\mathbf{e} - x\mathbf{R}_0\mathbf{e}) + \int_0^\infty (\mathbf{R}_1 - \mathbf{R}_1(y))dy(\mathbf{A} - a(x)\mathbf{I})\mathbf{e},$$

$$\mathbf{g}_n \mathbf{e} = \alpha_n b_n \mathbf{g}_0 \mathbf{e} + \int_0^\infty (\mathbf{R}_n - \mathbf{R}_n(y))dy(\mathbf{A} - a(x)\mathbf{I})\mathbf{e}, \quad n = 2, \ldots, N.$$

In order to obtain the values of integrals, we consider the functions

$$\mathbf{G}_n(z) = \int_0^z (\mathbf{R}_n - \mathbf{R}_n(y))dy, \quad n = 1, \ldots, N,$$

and apply Laplace-Stieltjes transform

$$\mathbf{G}_n^*(s) = \int_0^\infty e^{-sz} d\left(\int_0^z (\mathbf{R}_n - \mathbf{R}_n(y))dy\right) = \frac{\mathbf{R}_n - \mathbf{R}_n^*(s)}{s}.$$

Taking the limit by $s \to 0$, we obtain

$$\lim_{s \to 0} \mathbf{G}_n^*(s) = \lim_{s \to 0} \frac{\mathbf{R}_n - \mathbf{R}_n^*(s)}{s} = -\frac{d\mathbf{R}_n^*(s)}{ds}\bigg|_{s=0} = -\frac{d\mathbf{R}_n^*(0)}{ds}.$$

Taking into account that $\mathbf{G}_n^*(0) = \mathbf{G}_n(\infty)$ yields

$$\int_0^\infty (\mathbf{R}_n - \mathbf{R}_n(x))dx = -\frac{d\mathbf{R}_n^*(0)}{ds}.$$

From Eqs. (9) and (10), we obtain the system for values $-\frac{d\mathbf{R}_n^*(0)}{ds}$

$$-\frac{d\mathbf{R}_1^*(0)}{ds}\mathbf{Q} = \mathbf{R}_1 - b_1\mathbf{R}_0(\mathbf{A} + x\mathbf{I}),$$

$$-\frac{d\mathbf{R}_1^*(0)}{ds}\mathbf{e} = \frac{1}{2}b_1^{(2)}\mathbf{R}_0(\mathbf{A} + x\mathbf{I})\mathbf{e},$$

$$-\frac{d\mathbf{R}_n^*(0)}{ds}\mathbf{Q} = \mathbf{R}_n - \alpha_n b_n \mathbf{R}_0, \quad n = 2, \ldots, N,$$

$$-\frac{d\mathbf{R}_n^*(0)}{ds}\mathbf{e} = \frac{1}{2}\alpha_n b_n^{(2)}\mathbf{R}_0\mathbf{e}, \quad n = 2, \ldots, N,$$

where $b_k^{(2)}$ are the second raw moments of distributions $B_k(x)$.

Finally, we write the system for vectors $\mathbf{g}_k(x)$:

$$\mathbf{g}_1(x)\mathbf{Q}\mathbf{v}_m + \mathbf{g}_0(x)(\mathbf{A} + x\mathbf{I})(1 - B_1^*(s_m))\mathbf{v}_m$$

$$+(\mathbf{R}_1(x) - \mathbf{R}_1^*(s_m, x))(\mathbf{A} - a(x)\mathbf{I})\mathbf{v}_m - x(1 - B_1^*(s_m))\mathbf{R}_0(x)\mathbf{v}_m = 0,$$

$$\mathbf{g}_1(x)\mathbf{e} - b_1\mathbf{g}_0(x)(\mathbf{A} + x\mathbf{I})\mathbf{e} = -b_1 x \mathbf{R}_0(x)\mathbf{e} - \frac{d\mathbf{R}_1^*(0)}{ds}(\mathbf{A} - a(x)\mathbf{I})\mathbf{e},$$

$$\mathbf{g}_n(x)\mathbf{Q}\mathbf{v}_m + \alpha_n(1 - B_n^*(s_m))\mathbf{g}_0(x)\mathbf{v}_m$$
$$+(\mathbf{R}_n(x) - \mathbf{R}_n^*(s_m, x))(\mathbf{A} - a(x)\mathbf{I})\mathbf{v}_m = 0, \quad n = 2, \ldots, N,$$

$$\mathbf{g}_n(x)\mathbf{e} - \alpha_n b_n \mathbf{g}_0(x)\mathbf{e} = -\frac{d\mathbf{R}_n^*(0)}{ds}(\mathbf{A} - a(x)\mathbf{I})\mathbf{e}, \quad n = 2, \ldots, N,$$

$$\sum_{k=0}^{N} \mathbf{g}_k(x)\mathbf{e} = 0,$$

where the last equation is additional and $\mathbf{R}_k^*(s_m, x)$, $m = 2, \ldots, M$, $k = 1, \ldots, N$ are the solutions of systems

$$\mathbf{R}_1^*(s_m, x)(\mathbf{Q} + s_m\mathbf{I}) = \mathbf{R}_1(x)\mathbf{Q} + (1 - B_1^*(s_m))\mathbf{R}_0(x)(\mathbf{A} + x\mathbf{I}),$$

$$\mathbf{R}_1^*(s_m, x)\mathbf{v}_m = -B_1^{*\prime}(s_m)\mathbf{R}_0(x)(\mathbf{A} + x\mathbf{I})\mathbf{v}_m,$$

$$\mathbf{R}_n^*(s_m, x)(\mathbf{Q} + s_m\mathbf{I}) = \mathbf{R}_n(x)\mathbf{Q} + \alpha_n(1 - B_n^*(s_m))\mathbf{R}_0(x),$$

$$\mathbf{R}_n^*(s_m, x)\mathbf{v}_m = -\alpha_n B_n^{*\prime}(s_m)\mathbf{R}_0(x)\mathbf{v}_m.$$

After that, we consider the last equation of system (13) using Taylor expansions up to $O(\varepsilon^2)$ and making substitutions (14) and (15)

$$\varepsilon^2 \frac{\partial \Phi(w,\tau)}{\partial \tau} + (jw\varepsilon)^2 a(x)\Phi(w,\tau)\mathbf{f}\mathbf{e} = (jw\varepsilon)^2 \Phi(w,\tau)\{-x\mathbf{f}_0\mathbf{e} + (\mathbf{f} - \mathbf{f}_0)\mathbf{A}\mathbf{e}\}$$

$$+(j\varepsilon)^2 w \frac{\partial \Phi(w,\tau)}{\partial w} \mathbf{R}_0 \mathbf{e} + (jw\varepsilon)^2 x \Phi(w,\tau)\mathbf{R}_0\mathbf{e}$$

$$+\frac{(jw\varepsilon)^2}{2}\Phi(w,\tau)\{-x\mathbf{R}_0\mathbf{e} + (\mathbf{R} - \mathbf{R}_0)\mathbf{A}\mathbf{e}\} + O(\varepsilon^3).$$

Taking the limit by $\varepsilon \to 0$, we obtain

$$\frac{\partial \Phi(w,\tau)}{\partial \tau} + (jw)^2 a(x)\Phi(w,\tau)\mathbf{f}\mathbf{e} = (jw)^2 \Phi(w,\tau)\{-x\mathbf{f}_0\mathbf{e} + (\mathbf{f} - \mathbf{f}_0)\mathbf{A}\mathbf{e}\}$$

$$+\frac{(jw)^2}{2}\Phi(w,\tau)\left\{\frac{2}{w\Phi(w,\tau)}\frac{\partial \Phi(w,\tau)}{\partial w}\mathbf{R}_0\mathbf{e} + x\mathbf{R}_0\mathbf{e} + (\mathbf{R} - \mathbf{R}_0)\mathbf{A}\mathbf{e}\right\}.$$

Making substitutions (16) and (17) yields

$$\frac{\partial \Phi(w,\tau)}{\partial \tau} = (jw)^2 \Phi(w,\tau)\{-x\mathbf{g}_0\mathbf{e} + (\mathbf{g} - \mathbf{g}_0)\mathbf{A}\mathbf{e}$$

$$-\frac{\partial \Phi(w,\tau)/\partial w}{w\Phi(w,\tau)}\{-x\boldsymbol{\varphi}_0\mathbf{e} + (\boldsymbol{\varphi} - \boldsymbol{\varphi}_0)\mathbf{A}\mathbf{e} - \mathbf{R}_0\mathbf{e}\}\Big\}$$

$$+\frac{(jw)^2}{2}\Phi(w,\tau)\{2x\mathbf{R}_0\mathbf{e}+a(x)\}.$$

We take into account that

$$-x\varphi_0\mathbf{e}+(\varphi-\varphi_0)\mathbf{Ae}-\mathbf{R}_0\mathbf{e}=a'(x),$$

and denote

$$b(x)=a(x)+2\left[-x\mathbf{g}_0\mathbf{e}+(\mathbf{g}-\mathbf{g}_0)\mathbf{Ae}+x\mathbf{R}_0\mathbf{e}\right]. \qquad (22)$$

Here $b(x)$ is the diffusion coefficient of diffusion process approximating the number of incoming calls in the orbit.

Thereby, the last equation has the following form:

$$\frac{\partial \Phi(w,\tau)}{\partial \tau}=w\frac{\partial \Phi(w,\tau)}{\partial w}a'(x)+\frac{(jw)^2}{2}\Phi(w,\tau)b(x). \qquad (23)$$

## 5 Diffusion Approximation for the Number of Calls in the Orbit

Applying to equation (23) the inverse Fourier transform, we obtain the Fokker-Planck equation for density function $P(y,\tau)$ of some diffusion process $y(\tau)$

$$\frac{\partial P(y,\tau)}{\partial \tau}=-a'(x)\frac{\partial(yP(y,\tau))}{\partial y}+\frac{b(x)}{2}\frac{\partial^2 P(y,\tau)}{\partial y^2}. \qquad (24)$$

Thus, process $y(\tau)$ is the solution of stochastic differential equation

$$dy(\tau)=a'(x)y(\tau)d\tau+\sqrt{b(x)}d\omega(\tau), \qquad (25)$$

where $\omega(\tau)$ is Wiener process.

We introduce diffusion process $z(\tau)=x(\tau)+\varepsilon y(\tau)$, where $\varepsilon=\sqrt{\sigma}$ and function $x(\tau)$ is the solution of differential equation $dx(\tau)=a(x)d\tau$. Then process $z(\tau)$ is the solution of the following stochastic differential equation:

$$dz(\tau)=[a(x)+\varepsilon a'(x)y(\tau)]d\tau+\varepsilon\sqrt{b(x)}d\omega(\tau).$$

We transform the coefficients of the obtained equations as follows:

$$a(x)+\varepsilon a'(x)y=a(x+\varepsilon y)+O(\varepsilon^2)=a(z)+O(\varepsilon^2),$$

$$\varepsilon\sqrt{b(x)}=\varepsilon\sqrt{b(x+\varepsilon y-\varepsilon y)}=\varepsilon\sqrt{b(z-\varepsilon y)}=\varepsilon\sqrt{b(z)}+O(\varepsilon^2),$$

which yields the stochastic differential equation

$$dz(\tau)=a(z)d\tau+\varepsilon\sqrt{b(z)}d\omega(\tau). \qquad (26)$$

Hence, we can write the Fokker-Planck equation for dencity function $\Pi(z,\tau)$ of process $z(\tau)$

$$\frac{\partial \Pi(z,\tau)}{\partial \tau}=-\frac{\partial(a(z)\Pi(z,\tau))}{\partial z}+\frac{\varepsilon^2}{2}\frac{\partial^2(b(z)\Pi(z,\tau))}{\partial z^2}. \qquad (27)$$

Making reverse substitution $\sigma = \varepsilon^2$ and transform the equation for the stationary distribution of process $z(\tau)$

$$-(a(z)\Pi(z))' + \frac{\sigma}{2}(b(z)\Pi(z))'' = 0.$$

It is easy to see that the solution of the last equation is given by

$$\Pi(z) = \frac{C}{b(z)} \exp\left\{\frac{2}{\sigma} \int_0^z \frac{a(x)}{b(x)} dx\right\},$$

where $C$ is an integration constant.

Based on the obtained distribution, we build the approximation of probability distribution of the number of incoming calls in the orbit

$$PD(i) = \frac{\Pi(i\sigma)}{\sum_{n=0}^{\infty} \Pi(n\sigma)}. \tag{28}$$

## 6 Conclusion

We have considered MAP/GI/1 retrial queue with two-way communication and multiple types of outgoing calls. The state of the system is four-dimensional Markov chain, which we investigated using asymptotic-diffusion approach. We have derived drift and diffusion coefficients of diffusion process approximating the number of incoming calls in the orbit. Based on the dencity function of diffusion process, we have build the approximation (28) of the steady state probability distribution of the number of calls in the orbit.

## References

1. Aguir, S., Karaesmen, F., Akşin, O.Z., Chauvet, F.: The impact of retrials on call center performance. OR Spect. **26**(3), 353–376 (2004)
2. Artalejo, J.R.: A classified bibliography of research on retrial queues: progress in 1990–1999. TOP **7**(2), 187–211 (1999)
3. Avramidis, A., Deslauriers, A., L'Ecuyer, P.: Modeling daily arrivals to a telephone call center. Manage. Sci. **50**(7), 896–908 (2004)
4. Bernett, H.G., Fischer, M.J., Masi, D.: Blended call center performance analysis. IT Prof. **4**(2), 33–38 (2002)
5. Bhulai, S., Koole, G.: A queueing model for call blending in call centers. IEEE Trans. Autom. Control **48**(8), 1434–1438 (2003)
6. Falin, G., Templeton, J.G.C.: Retrial Queues, vol. 75. CRC Press, Boca Raton (1997)
7. Gilmore, A., Moreland, L.: Call centres: how can service quality be managed? Ir. Mark. Rev. **13**(1), 3 (2000)
8. Nazarov, A., Phung-Duc, T., Paul, S.: Slow retrial asymptotics for a single server queue with two-way communication and Markov modulated Poisson input. J. Syst. Sci. Syst. Eng. **28**(2), 181–193 (2019)
9. Sakurai, H., Phung-Duc, T.: Scaling limits for single server retrial queues with two-way communication. Ann. Oper. Res. **247**(1), 229–256 (2016)

# Investigation of M/G/1//N System with Impatient Customers, Unreliable Primary and a Backup Server

Ádám Tóth[✉] and János Sztrik

University of Debrecen, University Square 1, Debrecen 4032, Hungary
{toth.adam,sztrik.janos}@inf.unideb.hu

**Abstract.** This paper investigates a finite-source retrial queueing system characterized by request collisions, primary server unreliability, and the inclusion of a backup server. In cases of collisions, when a new job arrives while the service facility is occupied, both jobs are sent to a virtual waiting area called the orbit. Customers in the orbit make further attempts to access the server after a random interval. During server breakdowns, the customer at the server is transferred to the orbit. The system consists of a backup facility when the primary server is unreachable to process requests while the main service unit is under repair. The novelty of this study lies in the implementation of the impatience of the customers and conducting a sensitivity analysis using various service time distributions for the primary customers. We analyzed two scenarios, presenting key performance metrics through visual representations to emphasize the observed differences.

**Keywords:** Simulation · Queueing system · Finite-source model · Backup server · Collisions · Unreliable operation · Impatience

## 1 Introduction

In the current era of increasing traffic volumes and growing user bases, analyzing communication systems and designing optimal configurations present significant challenges. Information exchange plays a crucial role in all aspects of life, making it essential to develop or adapt mathematical and simulation models for telecommunication systems to meet these evolving demands. Retrial queues are highly effective and well-suited for modeling real-world scenarios commonly found in telecommunications, networking, mobile systems, call centers, and related domains. Numerous scholarly works, such as those referenced in [3] and [4], have extensively investigated various aspects of retrial queuing systems characterized by retrial calls.

In some contexts, researchers often assume that service units are always available; however, operational disruptions or unforeseen events can arise, resulting in the rejection of incoming customers. Devices across various industries are prone to malfunctions, making the presumption of their infallible operation overly optimistic and impractical. Similarly, in wireless communication environments, diverse factors can affect transmission rates, causing interruptions during packet delivery. The inherent instability of retrial queuing systems plays a crucial role in influencing both system operations and performance outcomes. Additionally, halting production entirely is unfeasible, as it may cause delays in order fulfillment. As a result, in such situations, machines or operators with reduced processing capacities may remain active to ensure smoother system operations. Furthermore, the authors explore the feasibility of incorporating a backup server capable of providing services at a reduced rate when the primary server is unavailable. A multitude of recent studies has thoroughly investigated retrial queuing systems with unreliable servers, as evidenced by sources such as [1,6,8] and [13].

Waiting is a ubiquitous phenomenon in various aspects of life, often leading to dissatisfaction due to the time spent in queues. This dissatisfaction may lead to requests leaving the system prematurely without receiving service, a phenomenon known as impatience. Such behavior is observed in diverse domains including healthcare applications, call centers, and telecommunication networks. Impatience has been widely explored in numerous academic studies, identifying several distinct behaviors: balking occurs when customers decide not to enter the system due to long queues; jockeying involves customers switching between queues to expedite service; and reneging refers to customers abandoning the queue after waiting for a specific, often extended, duration. The impatience mechanism is a crucial aspect of the model, as it influences the overall system performance by potentially reducing the number of customers waiting in the system and affecting the service dynamics. Studies examining these behaviors include [7,9] and [12].

In technological settings like Ethernet networks or limited communication sessions, task collisions are highly probable. Asynchronous attempts by multiple entities within the source can cause signal interference, requiring retransmissions to resolve the conflicts. Therefore, it is essential to account for this phenomenon in research focused on devising effective strategies to minimize conflicts and reduce the resulting message delays. Studies discussing findings on collisions can be found in publications such as [10] and [11].

The objective of this study is to perform a sensitivity analysis using various service time distributions for the primary server, in order to evaluate the main performance metrics in scenarios that incorporate the feature of impatience of the customers. When the primary server fails, customer service is transferred to the backup facility. During this period, new customers are directed to the backup unit or to the orbit if the backup unit is busy. Our investigation focuses on the impact of the impatient feature, with results obtained through simulation using Simpack [5]. The simulation program is built using core code components designed to calculate the desired metrics over a variety of input parameters. Visual representations are included to demonstrate how different parameters and distributions influence key performance metrics.

## 2 System Model

We analyze a finite-source retrial queuing system of type $M/G/1//N$, illustrated in Fig. 1, which incorporates an unreliable primary service unit, collision events, and a backup service unit. This model features a finite-source, where each of the $N$ individuals generates requests to the system following an exponential distribution with parameter $\lambda$. Arrival times adhere to an exponential distribution with a mean of $\lambda * N$. When queues are absent, arriving jobs are processed immediately according to one of several distributions—gamma, hypo—exponential, hyper-exponential, Pareto, or lognormal-each defined by unique parameters but maintaining identical mean and variance values ($\eta$).

In cases of server busyness, an arriving customer causes a collision with the customer currently being serviced, resulting in both customers being transferred to the orbit. Jobs in the orbit make additional attempts to access the server after a random time following an exponential distribution with parameter $\sigma$. Moreover, the server experiences random breakdowns, with failure times governed by exponential random variables. The failure time is characterized by parameter $\gamma_0$ when the server is busy and $\gamma_1$ when it is idle.

When the primary service unit fails, repairs commence immediately, with the repair time following an exponential distribution characterized by the parameter $\gamma_2$. If the server fails while busy, the customer is promptly moved to the orbit. During the primary server's unavailability, all customers in the source continue to generate requests, which are then directed to the backup server. The backup server functions at a slower rate, modeled by an exponentially distributed random variable with parameter $\mu$, and is assumed to be fully reliable. It operates only when the primary server is out of service. If the backup server is occupied, incoming requests are redirected to the orbit, and no collisions occur at the backup service unit.

Each primary customer in the system is characterized by an impatience property, which reflects their potential decision to leave the system if not served within a certain time frame. This decision to abandon the system is made after a random time period, which follows an exponential distribution with rate parameter $\tau$.

The model assumes that all random variables are entirely independent in its formulation.

## 3 Simulation Results

### 3.1 First Scenario

We employed a statistical module class featuring an advanced analysis tool to estimate the mean and variance of observed variables using the batch mean method. This approach aggregates $n$ consecutive observations from a steady-state simulation to produce a sequence of nearly independent samples. Renowned for its reliability, the batch mean method is widely used to construct confidence

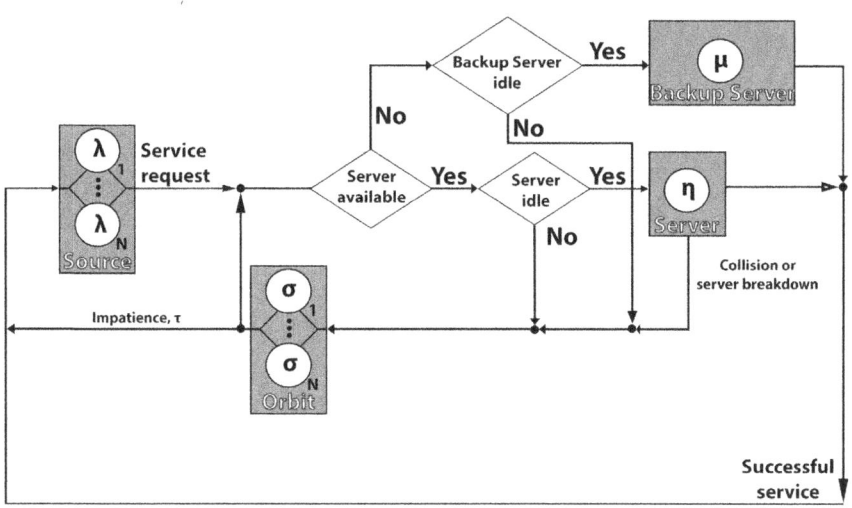

**Fig. 1.** System model

intervals for the steady-state mean of a process. To achieve approximate independence among sample averages, sufficiently large batch sizes are required. Further details on the batch mean method can be found in [2]. Our simulations were conducted with a 99.9% confidence level, terminating once the relative half-width of the confidence interval was reduced to 0.00001.

**Table 1.** Numerical values of model parameters

| N | $\gamma_0$ | $\gamma_1$ | $\gamma_2$ | $\sigma$ | $\mu$ | $\tau$ |
|---|---|---|---|---|---|---|
| 100 | 0.1 | 0.1 | 1 | 0.05 | 0.1 | 0.01 |

In this section, we aimed to determine service time parameters for each distribution to ensure they have identical mean values and variances. Four different distributions were examined to assess their influence on performance metrics. The hyper-exponential distribution was selected specifically to achieve a squared coefficient of variation greater than one. Table 2 outlines the input parameters for the various distributions, while Table 1 lists the values of other relevant parameters.

Figure 2 depicts the correlation between arrival intensity and the mean response time of customers who are successfully served. Successfully served customers refer to those who remain in the system until receiving service, unaffected by impatience. Among the distributions, the Pareto distribution exhibits the highest mean response time, while the distinctions between the remaining distributions become more evident. Of particular note, the gamma distribution yields the shortest mean response time.

**Table 2.** Parameters of service time of primary customers

| Distribution | Gamma | Hyper-exponential | Pareto | Lognormal |
|---|---|---|---|---|
| Parameters | $\alpha = 0.011$ $\beta = 0.011$ | $p = 0.494$ $\lambda_1 = 0.989$ $\lambda_2 = 1.011$ | $\alpha = 2.005$ $k = 0.501$ | $m = -2.257$ $\sigma = 2.125$ |
| Mean | 1 | | | |
| Variance | 90.25 | | | |
| Squared coefficient of variation | 90.25 | | | |

An intriguing observation is that as arrival intensity grows, the mean response time initially increases but then begins to decline after surpassing a certain threshold. This behavior is a hallmark of retrial queuing systems with a finite source and typically emerges under specific parameter settings.

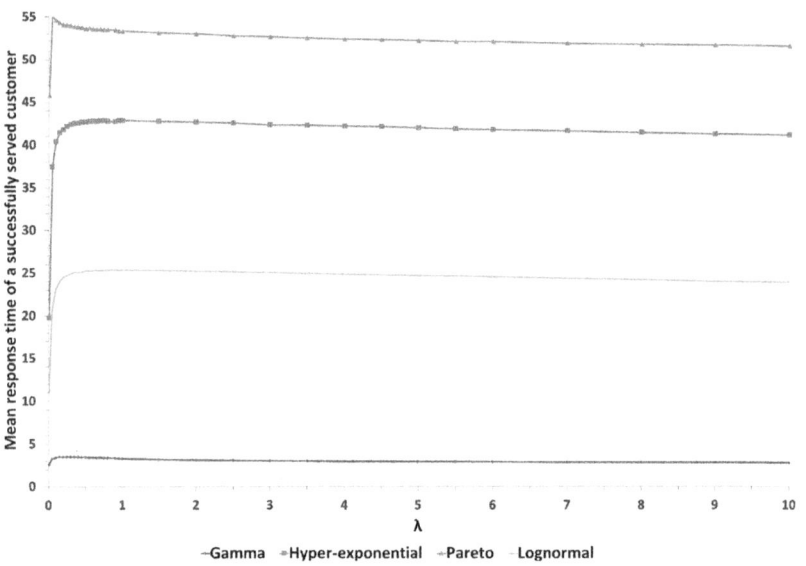

**Fig. 2.** Mean response time vs. arrival intensity

Figure 3 shows the mean response time of those customers who are served by the backup service unit in relation to the arrival rate of incoming customers. Inspecting closely the obtained results the same tendency develops what we observed in the previous figure. The Pareto distribution exhibits the highest values, while the gamma distribution produces the lowest values. Because the service rate of the backup unit is smaller than the rate of the primary service unit the spent time in the system of the customer served by the backup unit is higher on average.

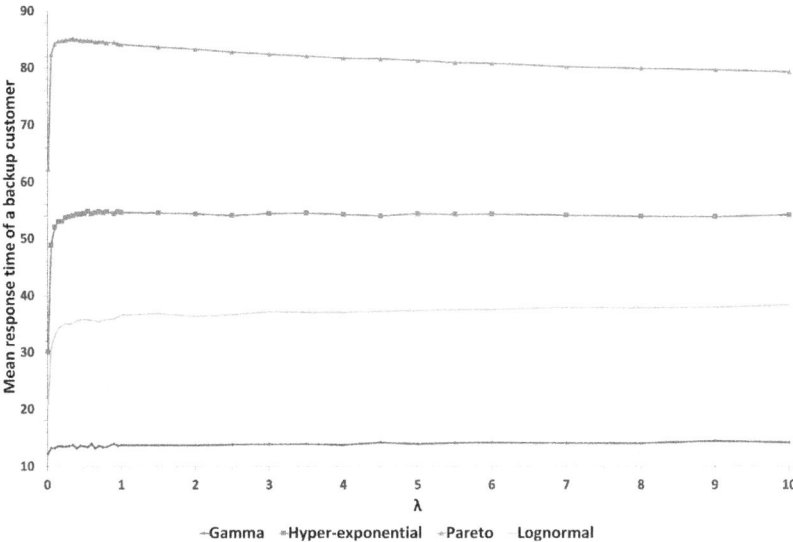

**Fig. 3.** Mean response time of a backup customer vs. arrival intensity

Figure 4 depicts the utilization of the backup service unit as a function of arrival intensity, comparing the different distributions. Unlike the significant differences observed in the previous figures, the results here are relatively close to each other. A closer look reveals that the backup service unit is utilized about 50% of the time, meaning it is occupied by customers for half of the total simulation period. As the arrival intensity increases, the utilization of the backup service unit also rises. However, once the arrival intensity reaches a certain threshold (around 1 in this case), the utilization plateaus.

Figure 5 illustrates the mean number of customer retrials as the arrival intensity increases. There are notable differences between the distributions used, with service times following a Pareto distribution showing particularly high retrial rates. In contrast, with a gamma distribution, requests typically do not attempt to reengage with the service unit, while other distributions exhibit a significantly higher number of collisions. The results also clearly show that after reaching a certain arrival intensity, the number of retrials stabilizes and does not continue to increase.

### 3.2 Second Scenario

Building on the results from the previous section, we next aimed to explore how changing the service time parameters would influence performance metrics. This time, we carefully chose parameters so that the squared coefficient of variation remained below one. Since the squared coefficient of variation for a hypo-exponential distribution is consistently less than one, we substituted the hyper-exponential distribution with its hypo-exponential counterpart. Distributions

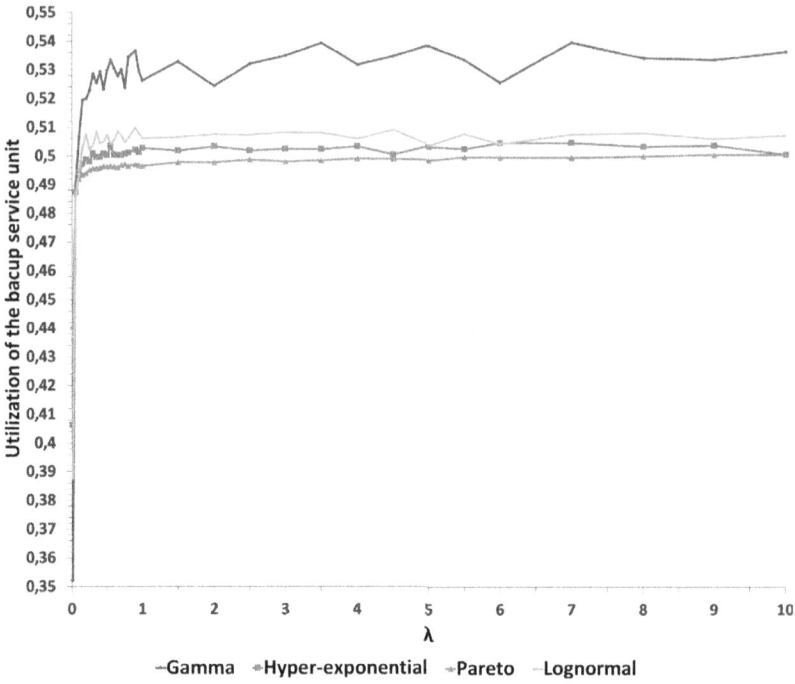

**Fig. 4.** The utilization of the primary service unit vs. arrival intensity

with squared coefficient of variation values below one tend to be more regular, meaning that values are clustered more closely around the mean, which is often the case for hypo-exponential distributions. In queuing theory, squared coefficient of variation below one typically leads to less fluctuation in waiting times, contributing to more predictable performance metrics. With these updated service time parameters, we revisited the same performance figures to assess the impact of these adjustments, as shown in Table 3. All other parameters were kept constant, as specified in Table 1.

**Table 3.** Parameters of service time of primary customers

| Distribution | Gamma | Hypo-exponential | Pareto | Lognormal |
|---|---|---|---|---|
| Parameters | $\alpha = 1.522$ | $\mu_1 = 4.5454$ | $\alpha = 2.588$ | $m = -0.252$ |
| | $\beta = 1.522$ | $\mu_2 = 1.282$ | $k = 0.614$ | $\sigma = 0.71$ |
| Mean | 1 | | | |
| Variance | 0.6568 | | | |
| Squared coefficient of variation | 0.6568 | | | |

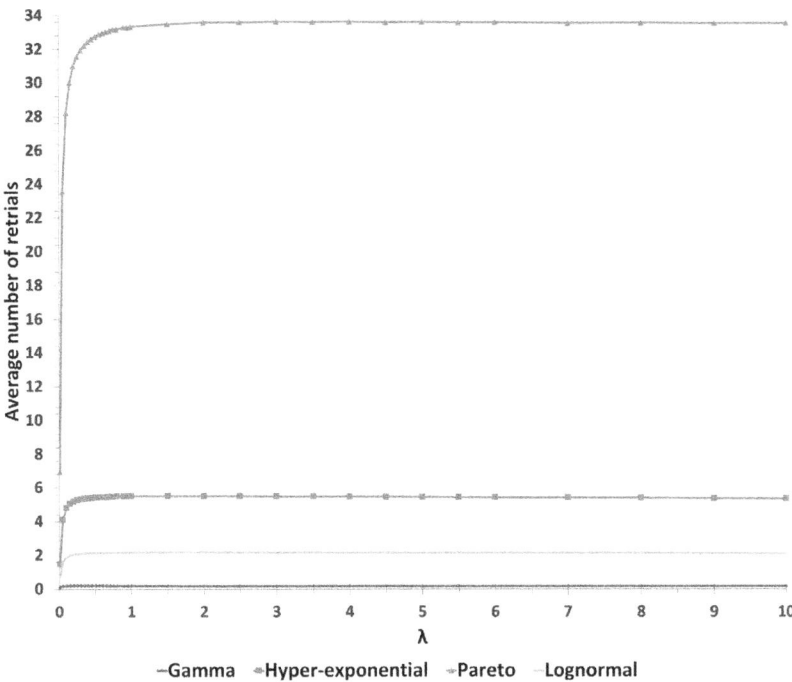

**Fig. 5.** The mean number of retrials vs. arrival intensity

To highlight the differences between the two scenarios, we begin by examining the mean response time of a customer, as shown in Fig. 6. The curves are noticeably closer to each other, with less pronounced differences, except for the Pareto distribution, which still produces significantly higher values compared to the other distributions. Similar to Fig. 2, the mean response time reaches a maximum, a common occurrence in retrial queuing systems with a finite customer pool. The same pattern is observed: after reaching a certain arrival intensity, the mean response time peaks and then gradually decreases as arrival intensity continues to rise.

Figure 7 illustrates the mean response time for customers served by the backup service unit. Similar to the previous scenario, the results follow the same trend observed in the prior figure, with the resulting curves notably close to each other and displaying minimal differences. Since the service rate of the backup unit is lower than that of the primary unit, customers served by the backup unit spend, on average, more time in the system.

Figure 8 illustrates the utilization of the backup service unit as a function of arrival intensity across different distributions. Unlike the substantial differences observed in previous figures, the results here are fairly similar. Closer inspection reveals that the backup service unit's utilization is around 50%, indicating it is

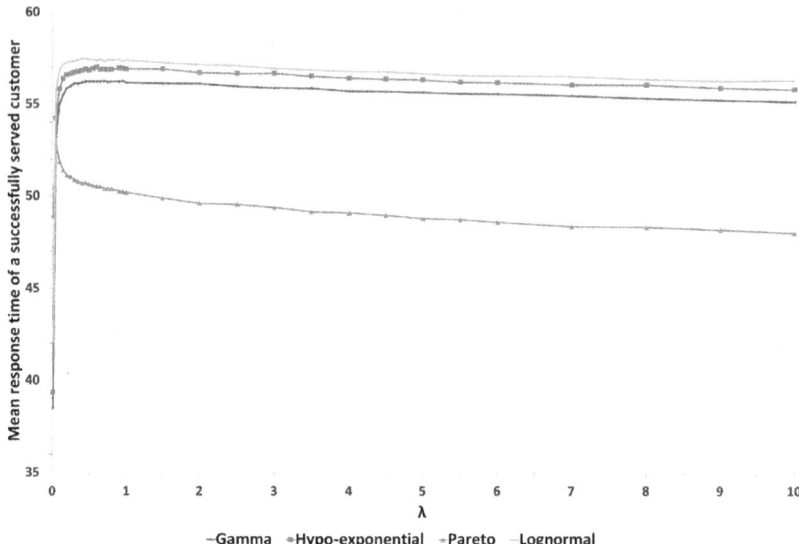

**Fig. 6.** Mean response time vs. arrival intensity

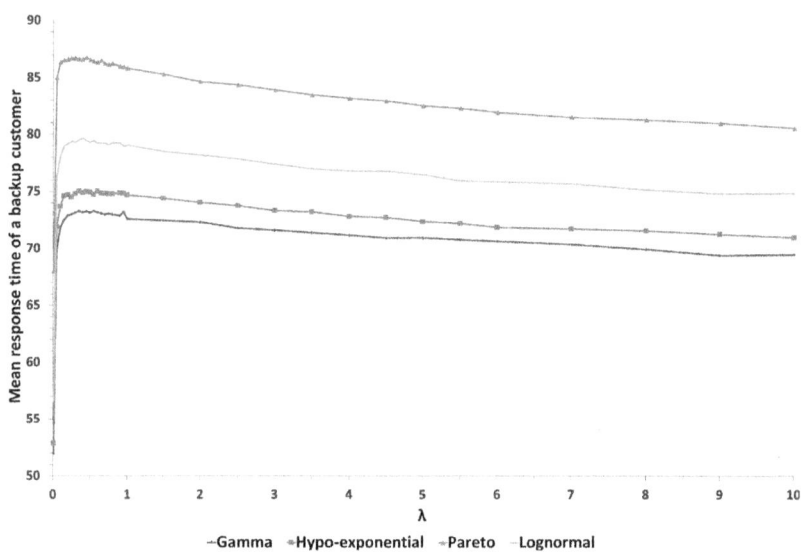

**Fig. 7.** Mean response time of a backup customer vs. arrival intensity

occupied by customers for about half of the total simulation time. A similar trend is evident: as arrival intensity increases, the backup service unit's utilization also rises. Once the arrival intensity reaches a specific threshold (around 1 in this instance), the utilization stabilizes.

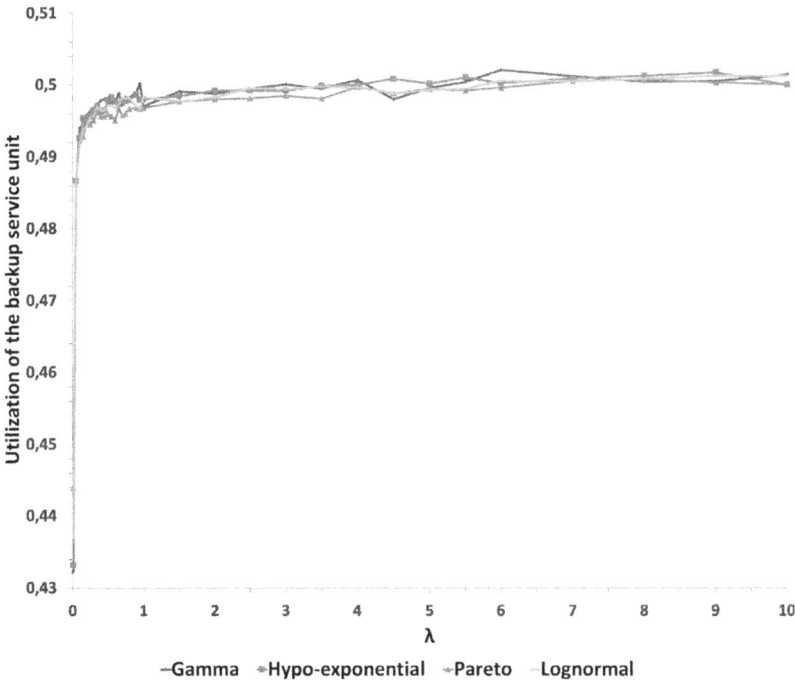

**Fig. 8.** The utilization of the primary service unit vs. arrival intensity

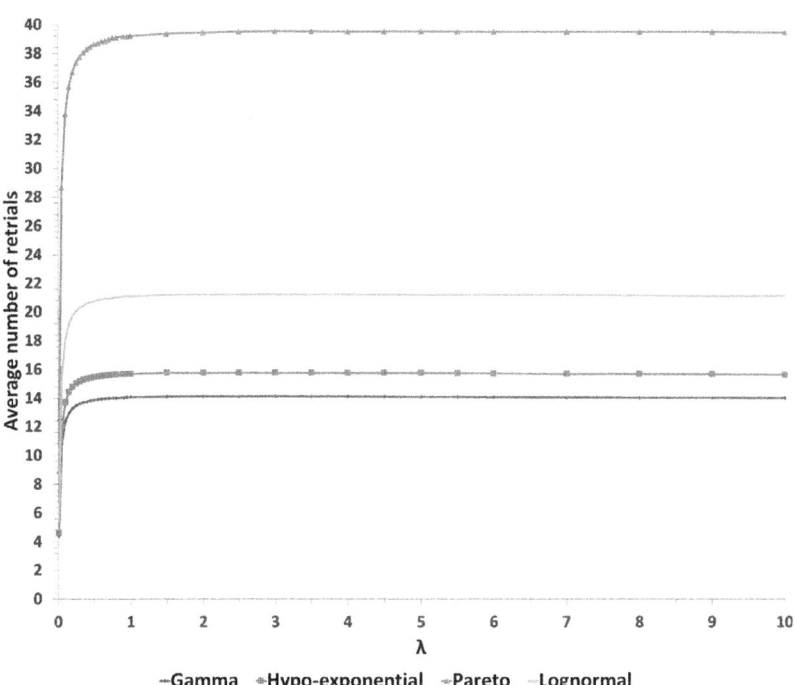

**Fig. 9.** The mean number of retrials vs. arrival intensity

Figure 9, which illustrates the average number of retries, shows a trend similar to the previous scenario. The highest retrial rates are seen with Pareto-distributed service times, while the lowest occur with gamma-distributed service times, although the differences are considerably smaller. Another noteworthy observation, upon closer inspection of the figure, is that the number of retries is higher for all distributions in the scenario using the previous parameter settings.

## 4 Conclusion

We simulated a retrial queuing system modeled as $M/G/1//N$, featuring an unreliable primary server and a backup service unit. The program enabled a sensitivity analysis of key performance metrics, such as the mean response time of successfully served customers. The results revealed significant differences in performance measures when the squared coefficient of variation exceeded one, highlighting the influence of the chosen distribution, while only minor deviations were observed when it was below one. The curves also demonstrated how customer impatience contributes to reducing the average response time for primary customers. Future research will focus on examining the effects of server blocking, incorporating two-way communication, exploring alternative impatience behaviors in different models, and performing sensitivity analyses on other variables, including failure rates.

## References

1. Chakravarthy, S.R., Shruti, Kulshrestha, R.: A queueing model with server breakdowns, repairs, vacations, and backup server. Oper. Res. Perspect. **7**, 100131 (2020). https://doi.org/10.1016/j.orp.2019.100131, https://www.sciencedirect.com/science/article/pii/S2214716019302076
2. Chen, E.J., Kelton, W.D.: A procedure for generating batch-means confidence intervals for simulation: checking independence and normality. SIMULATION **83**(10), 683–694 (2007)
3. Dragieva, V.I.: Number of retrials in a finite source retrial queue with unreliable server. Asia-Pac. J. Oper. Res. **31**(2), 23 (2014). https://doi.org/10.1142/S0217595914400053
4. Fiems, D., Phung-Duc, T.: Light-traffic analysis of random access systems without collisions. Ann. Oper. Res. **277**(2), 311–327 (2017). https://doi.org/10.1007/s10479-017-2636-7
5. Fishwick, P.A.: SimPack: getting started with simulation programming in C and C++. In: In 1992 Winter Simulation Conference, pp. 154–162 (1992)
6. Gharbi, N., Nemmouchi, B., Mokdad, L., Ben-Othman, J.: The impact of breakdowns disciplines and repeated attempts on performances of small cell networks. J. Comput. Sci. **5**(4), 633–644 (2014)
7. Gupta, N.: Article: a view of queue analysis with customer behaviour and priorities. In: IJCA Proceedings on National Workshop-Cum-Conference on Recent Trends in Mathematics and Computing 2011, RTMC, no. 4 (2012)

8. Klimenok, V., Dudin, A., Semenova, O.: Unreliable retrial queueing system with a backup server. In: Vishnevskiy, V.M., Samouylov, K.E., Kozyrev, D.V. (eds.) DCCN 2021. LNCS, vol. 13144, pp. 308–322. Springer, Cham (2021). https://doi.org/10.1007/978-3-030-92507-9_25
9. Kumar, R., Jain, N., Som, B.: Optimization of an M/M/1/N feedback queue with retention of reneged customers. Oper. Res. Decis. **24**, 45–58 (2014). https://doi.org/10.5277/ord140303
10. Kvach, A., Nazarov, A.: Sojourn time analysis of finite source Markov retrial queuing system with collision, chap. 8, pp. 64–72. Springer, Cham (2015)
11. Nazarov, A., Kvach, A., Yampolsky, V.: Asymptotic analysis of closed markov retrial queuing system with collision, chap. 1, pp. 334–341. Springer, Cham (2014)
12. Panda, G., Goswami, V., Datta Banik, A., Guha, D.: Equilibrium balking strategies in renewal input queue with bernoulli-schedule controlled vacation and vacation interruption. J. Industr. Manage. Optim. **12**, 851–878 (2015). https://doi.org/10.3934/jimo.2016.12.851
13. Satheesh R.K., Praba S.K.: A multi-server with backup system employs decision strategies to enhance its service. Res. Square 1–31 (2023). https://doi.org/10.21203/rs.3.rs-2498761/v1

# G-Network with Rewards as a Cluster System Model

Tatiana Rusilko[✉] and Dmitry Salnikov

Yanka Kupala State University of Grodno, 22 Ozheshko St, 230023 Grodno, Belarus
tatiana.rusilko@gmail.com

**Abstract.** The main objective of this paper is study and analytical modeling of a cluster system. A closed G-network with positive and negative requests is used as a mathematical model of the cluster system. Negative requests are associated with negative impact of malicious code and errors on model nodes. In addition, in the G-network there is a sequence of rewards or earnings that is generated in the process of transmitting requests between cluster nodes. The model is studied in the asymptotic case of a large number of requests being processed. The purpose of this paper is to calculate the expected total reward of the cluster system model in the asymptotic case. An ordinary differential equation for the expected reward that the cluster model will earn in a time $t$ if it starts in a given initial state is derived. The presented technique allows to analyse the cluster system efficiency with mathematically specified accuracy.

**Keywords:** Queueing network · G-network · Network with rewards · Asymptotic analysis · Cluster system · Mathematical modelling

## 1 Introduction

Systems designed for parallel data processing are widely used. These systems are based on processing incoming tasks. These tasks arrive at the system nodes and require resources for processing. A cluster is a set of connected computing systems (computers) working together as a unified computing resource.

Queueing networks are one of the effective tools for modeling such objects. G-networks are generalized queueing networks with several types of requests, in particular, positive and negative requests [1]. Their applications include modeling computing systems and networks, evaluating their performance, modeling neural networks, and more [2,3]. As a model of a cluster system, we will use an exponential G-network with closed structure. In this case, the G-network state vector $\mathbf{k}(t)$ with elements representing the number of customers at each node is a continuous-time Markov process on the finite state space.

A Markov process serves as a mathematical model for studying complex systems. R. Howard studied the continuous-time Markov process under the assumption that it receives (earns) a reward of $R_{ij}$ conventional units (c.u.) when the system makes a transition from state $i$ to state $j$, $i \neq j$; the Markov process earns a reward at the rate of $R_i$ c.u. per unit time during all the time that it occupies state $i$ [4]. In Howard's interpretation, the Markov process generates a sequence of rewards as it makes transitions from state to state and is called the "Markov process with rewards" [4]. The reward is thus a random process with a probability distribution governed by the probabilistic relations of the Markov process. In [1], the problem of finding the expected total earnings $V_i(t)$ that the system will earn in a time $t$ if it starts in the state $i$ was solved. Note the term "expected" and notation $V_i(t)$ are used in the sense of "prospective" or relating to the future time interval $t$ if the initial state of the process is $i$.

The concept of Markov processes with rewards was used by M. Matalytski to define exponential queueing networks with rewards, known as HM(Howard – Matalytski)-networks [5,6]. HM-networks, along with servicing and transmitting requests, generate earnings. Earnings can be measured by any physical quantity relevant to the problem, such as energy levels, units of production, monetary units, etc.

The purpose of this paper is the mathematical modelling and efficiency analysis of the cluster system using a closed exponential G-network with rewards. An asymptotic analysis of the cluster model is carried out, which implies an approximation method of the queueing network study under the assumption of a large but limited number of requests [7,8]. We are interested in the total expected earnings of the cluster model if it will operate for a time $t$ with a given initial condition.

## 2 Model Description. Formulation of the Problem

A cluster system, or computer cluster, is a collection of interconnected computers that work collaboratively to perform tasks and applications as a single cohesive unit. This system is designed to enhance performance, reliability, and scalability by leveraging the combined power and resources of its individual components. Cluster systems are often employed in fields requiring intensive computational power, such as meteorology, genetics, astrophysics, and big data analytics. A cluster system operates by distributing computational tasks across multiple interconnected units. Each unit contributes its own processing power, memory, and storage to the overall system, allowing it to handle large-scale computations and data-intensive applications effectively. The fundamental objective is to achieve higher efficiency and throughput compared to a single computer system. One of the key elements of cluster systems is the concept of parallel processing. This involves dividing a large computational task into smaller, more manageable chunks, which are then processed simultaneously by different nodes within the cluster. This method not only accelerates the processing speed but also ensures that resources are utilized optimally. Cluster systems are also highly scalable. As computational demands grow, additional units can be added to the cluster to increase its overall capacity and processing power.

The main components of a cluster system are compute nodes and interconnection network. The individual computers or servers that make up the cluster are referred to as compute nodes, worker nodes, or simply nodes. Each node operates independently but communicates and coordinates with other nodes to share the computational workload. Nodes can vary in terms of processing power, memory capacity, and storage capabilities. The interconnection network is also referred to as the interconnect fabric, communication network, or network. The network is the communication backbone of the cluster, facilitating data transfer and message passing between nodes. High-speed interconnects, such as InfiniBand or Ethernet, are commonly used to ensure low-latency and high-bandwidth communication. In the network, information is transmitted via individual data packets, with data transmission speeds, processing, and packet routing having their own limitations. The cluster system can transmit useful data, outdated or corrupted requests, as well as malicious software (malware).

Let's conceptualize a cluster system model using a queueing network. First of all, it is necessary to establish a correspondence between the components of the cluster system and the queueing network. Compute nodes are represented as queueing systems. Each node has a queue containing incoming tasks. After a task is processed by the node, it may either exit the system if completed, or be routed to another node for further processing. The routing mechanism is modelled using transition probabilities. The interconnection network can be represented as the pathways between compute nodes. It shows how queueing systems are connected into a queueing network. Since there are several types of data in a cluster system, it makes sense to use a generalized queueing network with several types of requests – G-network. Additionally, it is assumed that receiving and processing useful information generates some earnings, while executing outdated tasks and malicious code leads to expenses. Thus, the problem of mathematical modelling of such a cluster system can be solved using a G-network with rewards.

As a model of a cluster system, we will use a closed exponential G-network, consisting of a finite number of nodes $S_0$, $S_1$, ..., $S_n$. We set $K$ is the total number of requests circulating in the G-network. The request in the G-network corresponds to a data packet being processed by the cluster system. The node $S_0$ is an IS-node (Infinite Server) of $K$ identical exponential servers and it plays the role of a fictitious request source. The node $S_0$ generates a Poisson flow of customers with rate of $\lambda_0 k_0$, where $\lambda_0$ is the flow parameter, $k_0$ is the number of customers in the node $S_0$. This arrival flow is divided into a flow of positive and negative requests. Regular requests belong to the positive class, while corrupted, outdated, or malware-containing requests belong to the negative class. And the probability of a positive request arriving at short time interval $\Delta t$ is $\lambda_0 k_0 p_{0i}^+ \Delta t + o(\Delta t)$, the arrival probability of a negative request is $\lambda_0 k_0 p_{0i}^- \Delta t + o(\Delta t)$, $i = \overline{1,n}$, and $\sum_{i=1}^{n} \left( p_{0i}^+ + p_{0i}^- \right) = 1$.

The nodes of the cluster system are modeled by the G-network nodes $S_1$, $S_2$, ..., $S_n$. Each node $S_i$ is a queueing system with $m_i$ servers and unlimited waiting area for positive requests. The probability of completing the positive request service at a node $S_i$ during a short time interval $\Delta t$ is $\mu_i \min(m_i, k_i) \Delta t + o(\Delta t)$, where $k_i$ is the number of requests in the node $S_i$; the probability of completing the service two or more requests is $o(\Delta t)$. The completion of the request service at different nodes in a short time interval $o(\Delta t)$ are mutually independent events.

Requests are served according to the FIFO rule. When a request has completed service in the node $S_i$ it is instantly transferred to the node $S_j$ as positive with probability $p_{ij}^+$ or negative with probability $p_{ij}^-$, otherwise it is transferred to the IS-node $S_0$ with probability $p_{i0}^+ = 1 - \sum_{j=1}^{n}(p_{ij}^+ + p_{ij}^-)$, where $i \neq j$ and $i,j = \overline{1,n}$.
Negative requests arriving at a node are not served by the node servers; therefore, they are considered as signals. Let's assume that a negative request arriving at a node $S_i$, $i = \overline{1,n}$, removes one positive request located at the same node and both of them are immediately transferred to the IS-node $S_0$ as positive requests. It is possible to study other behavior strategies of negative requests.

The state of the cluster model under study is described by a $n$-dimensional continuous-time Markov process on finite state space

$$\mathbf{k}(t) = (k_1(t), k_2(t), \ldots, k_n(t)),$$

where $k_i(t)$ is the number of requests in the node $S_i$ at the moment $t$, $0 \leq k_i(t) \leq K$, $i = \overline{1,n}$, $t \in [0, +\infty)$, and $k_0(t) = K - \sum_{i=1}^{n} k_i(t)$. Let us assume that the cluster model is in the state $\mathbf{k}$ if at some time $t$ components $k_i(t) = k$, $i = \overline{1,n}$, form a vector $\mathbf{k} = (k_1, k_2, \ldots, k_n)$.

Let us suppose that the cluster sytem earns $R_{ij}^+$ c.u. when a positive request makes a transition from the node $S_i$ to the node $S_j$ and it earns $R_{ij}^-$ c.u. when a negative request makes the same transition, $i \neq j$, $i,j = \overline{0,n}$. We call $R_{ij}^+$ and $R_{ij}^-$ the "reward" associated with the transition of a positive and negative request, respectively, from $S_i$ to $S_j$. Suppose further that the cluster model receives a reward at the rate of $R(\mathbf{k})$ c.u. per unit time during all the time that it occupies the state $\mathbf{k}$. The question of interest is: what will be the expected total earnings of the cluster model in a time $t$, $t \in T$, if it is currently in the state $\mathbf{k}$.

## 3 Asymptotic Analysis of the Cluster Model with Rewards

The primary objective of asymptotic methods in queueing theory is to study service processes in queueing systems and networks by deriving appropriate approximations under specific critical (limiting) assumption. The scientific researches of Tomsk State University are widely known in the field of asymptotic methods [9–12].

In this paper, the cluster model is studied in the asymptotic case of a large number of requests $K$. The passage to the limit from a Markov chain $\mathbf{k}(t)$ to a continuous-state Markov process $\boldsymbol{\xi}(t)$ is performed. The mathematical approach used in this paper is based on a discrete model of a continuous Markov process [13].

Let $V(\mathbf{k}, t)$ be the expected total cluster model reward that the model will receive in a time $t$ if it starts in the initial state $\mathbf{k}$. Notation $v(\mathbf{x}, t)$ is used for reward density in the case of asymptotic approximation, $\mathbf{x}$ is the start state, $t$ is remaining time. The concept of reward density is discussed below.

To proceed further we need the following notations: $\mathbf{I}_i$ is a $n$-vector with zero components excluding $i$-th, that equals to 1, $\theta(x)$ is the Heaviside step function, defined as 0 for $x \leq 0$. Let us assume that at the initial moment of time, Markov process $\mathbf{k}(t)$, which describes the state of the cluster model, is in the state $\mathbf{k}$. Based on these assumptions, we examine all potential transitions of $\mathbf{k}(t)$ from the given initial state $\mathbf{k}$ at $t = 0$, along with the corresponding network earnings in a short time interval $\Delta t$. This allows us to relate the expected total reward in a time $t + \Delta t$, $V(\mathbf{k}, t + \Delta t)$, to $V(\mathbf{k}, t)$ by an equation.

- A transition from the state $\mathbf{k}$ to the state $\mathbf{k} + \mathbf{I}_j - \mathbf{I}_i$ with probability

$$\mu_i \min(m_i, k_i) p_{ij}^+ \Delta t + o(\Delta t),$$

indicating that the request was served at the node $S_i$ and joined the node $S_j$ as positive, $i, j = \overline{1, n}$. In this scenario, the model would receive the reward $R_{ij}^+$ plus the expected total reward $V(\mathbf{k} + \mathbf{I}_j - \mathbf{I}_i, t)$ to be made if it starts in the state $\mathbf{k} + \mathbf{I}_j - \mathbf{I}_i$ with the remaining time $t$, $V(\mathbf{k} + \mathbf{I}_j - \mathbf{I}_i, t)$.
- A transition from the state $\mathbf{k}$ to the state $\mathbf{k} - \mathbf{I}_j - \mathbf{I}_i$ with probability

$$\mu_i \min(m_i, k_i) p_{ij}^- \Delta t + o(\Delta t),$$

indicating that the request was served at $S_i$ and transmitted to $S_j$ as negative, $i, j = \overline{1, n}$. In this case, the model reward is $R_{ij}^-$ c.u. plus the reward $V(\mathbf{k} - \mathbf{I}_j - \mathbf{I}_i, t)$ that the network would receive for the remaining time $t$ if the initial state was $\mathbf{k} - \mathbf{I}_j - \mathbf{I}_i$.
- The process transitions from the state $\mathbf{k}$ to the state $\mathbf{k} - \mathbf{I}_i$ in three scenarios. First, with a probability of

$$\mu_i \min(m_i, k_i) p_{i0}^+ \Delta t + o(\Delta t),$$

indicating that a customer was served at $S_i$ and transmitted to the IS-node $S_0$, $i = \overline{1, n}$. The model earns $R_{i0}^+$ c.u. from this transition. Secondly, with the probability of

$$\lambda_0 \left( K - \sum_{i=1}^{n} k_i \right) p_{0i}^- \Delta t + o(\Delta t),$$

which occurs when a negative customer arrives to $S_i$ from the IS-node $S_0$, $i = \overline{1, n}$. The model earns $R_{0i}^-$ c.u. in this case. Thirdly, with the probability of

$$\mu_i \min(m_i, k_i) p_{ij}^- (1 - \theta(k_j)) \Delta t + o(\Delta t),$$

when a positive customer is transmitted as a negative from $S_i$ to an empty node $S_j$, $i, j = \overline{1, n}$. The model earns $R_{ij}^-$ c.u. in this scenario. In each of these three cases, the expected reward $V(\mathbf{k} - \mathbf{I}_i, t)$ is added to the model earnings.
- A transition from the state $\mathbf{k}$ to the state $\mathbf{k} + \mathbf{I}_i$ with probability

$$\lambda_0 \left( K - \sum_{i=1}^{n} k_i \right) p_{0i}^+ \Delta t + o(\Delta t),$$

which happens when a positive customer arrives to $S_i$ from the IS-node $S_0$, $i = \overline{1,n}$. The model reward is $R_{0i}^+$ c.u. plus the expected total reward $V(\mathbf{k}+\mathbf{I}_i, t)$ that the model would earn in the remaining time $t$ if it started in the state $\mathbf{k}+\mathbf{I}_i$.

- During time $\Delta t$ the model remains in the state $\mathbf{k}$ with probability

$$1 - \left[\sum_{i,j=1}^{n} \mu_i \min(m_i, k_i)\left(1 + p_{ij}^-(1-\theta(k_j))\right) + \sum_{i=1}^{n} \lambda_0 \left(K - \sum_{i=1}^{n} k_i\right)\right] \Delta t + o(\Delta t),$$

which results in a reward $R(\mathbf{k})\Delta t$ plus the expected reward that the cluster model will earn in the remaining time $t$, $V(\mathbf{k}, t)$. The probabilities of other transitions and rewards are considered to be of order $o(\Delta t)$.

With the aforementioned details, let us sum up the product of probabilities and rewards across all possible options. Dividing both sides of the resulting equation by $\Delta t$ and taking the limit as $\Delta t \to 0$, we derive the following set of difference-differential equations that completely define $V(\mathbf{k}, t)$:

$$\frac{\partial V(\mathbf{k}, t)}{\partial t} = \sum_{i,j=1}^{n} \mu_i \min(m_i, k_i) p_{ij}^+ \left(V(\mathbf{k}+\mathbf{I}_j-\mathbf{I}_i, t) - V(\mathbf{k}, t)\right)$$

$$+ \sum_{i,j=1}^{n} \mu_i \min(m_i, k_i) p_{ij}^- \left(V(\mathbf{k}-\mathbf{I}_j-\mathbf{I}_i, t) - V(\mathbf{k}, t)\right)$$

$$+ \sum_{i=1}^{n} \mu_i \min(m_i, k_i) p_{i0} \left(V(\mathbf{k}-\mathbf{I}_i, t) - V(\mathbf{k}, t)\right)$$

$$+ \sum_{i=1}^{n} \lambda_0 \left(K - \sum_{i=1}^{n} k_i\right) p_{0i}^- \left(V(\mathbf{k}-\mathbf{I}_i, t) - V(\mathbf{k}, t)\right)$$

$$+ \sum_{i,j=1}^{n} \mu_i \min(m_i, k_i) p_{ij}^-(1-\theta(k_j)) \left(V(\mathbf{k}-\mathbf{I}_i, t) - V(\mathbf{k}, t)\right)$$

$$+ \sum_{i=1}^{n} \lambda_0 \left(K - \sum_{i=1}^{n} k_i\right) p_{0i}^+ \left(V(\mathbf{k}+\mathbf{I}_i, t) - V(\mathbf{k}, t)\right) + Q(\mathbf{k}), \text{ where}$$

$$Q(\mathbf{k}) = R(\mathbf{k}) + \sum_{i,j=1}^{n} \left(\mu_i \min(m_i, k_i)\left(p_{ij}^+ R_{ij}^+ + p_{ij}^- R_{ij}^-(2-\theta(k_j))\right)\right)$$

$$+ \sum_{i=1}^{n} \mu_i \min(m_i, k_i) p_{i0}^+ R_{i0}^+ + \sum_{i=1}^{n} \lambda_0 \left(K - \sum_{i=1}^{n} k_i\right) \left(p_{0i}^- R_{0i}^- + p_{0i}^+ R_{0i}^+\right). \quad (1)$$

According to R. Howard, let us define a quantity $Q(\mathbf{k})$ as the "earning rate" of the cluster model [4]. This earning rate is a combination of reward rates and transition rewards.

The set of equations (1) cannot be solved for large $K$ and $n$. Cluster systems typically handle a large number of requests. Therefore, we can apply an approximation and study the limiting behavior of the random process $\mathbf{k}(t)$ in the asymptotic case of a large number of requests $K$, $K \gg 1$. We proceed to the limit from the Markov chain $\mathbf{k}(t)$ to the continuous Markov process

$$\boldsymbol{\xi}(t) = \frac{\mathbf{k}(t)}{K} = \left(\frac{k_1(t)}{K}, \frac{k_2(t)}{K}, \ldots, \frac{k_n(t)}{K}\right),$$

as $K$ tends to be large. The phase space of the vector $\boldsymbol{\xi}(t)$ is

$$X = \left\{\mathbf{x} = (x_1, x_2, \ldots, x_n) : x_i \geq 0, i = \overline{1,n}, \sum_{i=1}^{n} x_i \leq 1\right\}.$$

The increment of $\xi_i(t)$ in the short time $\Delta t \to 0$ is $\Delta x_i = \varepsilon = 1/K$, and as $K \to \infty$, $\Delta x_i \to 0$. Consequently, the process $\xi_i(t)$ tends to be continuous as $K \to \infty$ ($\varepsilon \to 0$), making the vector $\boldsymbol{\xi}(t)$ a continuous-state Markov process on $X$. In this asymptotic case, the total reward of the cluster model is a continuously changing process that depends on the initial state $\mathbf{x}$, $\mathbf{x} \in X$, and the upcoming time $t$.

In physics, the mass density (volumetric mass density or specific mass) is a substance's mass per unit of volume. Mathematically, it is expressed as the ratio of mass to volume:

$$\rho = \lim_{\varepsilon \to 0} \frac{m(x_1 \leq \xi_1 < x_1 + \varepsilon, x_2 \leq \xi_2 < x_2 + \varepsilon, \ldots, x_n \leq \xi_n < x_n + \varepsilon)}{\varepsilon^n}.$$

In probability theory, probability density is the probability per unit volume. By analogy, let us introduce the concept of "reward density" which refers to the reward that the network earns per unit of state space in time $t$ based on the initial state at the point $\mathbf{x}$:

$$v(\mathbf{x}, t) = \lim_{\varepsilon \to 0} \frac{V(x_1 \leq \xi_1 < x_1 + \varepsilon, x_2 \leq \xi_2 < x_2 + \varepsilon, \ldots, x_n \leq \xi_n < x_n + \varepsilon, t)}{\varepsilon^n}.$$

Realizing the passage to the limit as $K \to \infty$, a $n$-dimensional lattice with vertices at discrete points $\frac{\mathbf{k}}{K} = \left(\frac{k_1}{K}, \frac{k_2}{K}, \ldots, \frac{k_n}{K}\right)$ transforms into set of points $\mathbf{x} = (x_1, x_2, \ldots, x_n) \in X$, the "point density" increases with $K$. It is essential to take into account the continuous change in reward on $X$. Let us approximate the reward as $V(\mathbf{x}K, t) = v(\mathbf{x}, t)\varepsilon^n$ when $x_i \leq \xi_i < x_i + \varepsilon$, $i = \overline{1,n}$. Similarly, considering reward rates and transition rewards, $R(\mathbf{x})$ and $R_{ij}$, we introduce parameters of earning rate per unit of state space $X$, $r(\mathbf{x})$ and $r_{ij}$. Thus, $r(\mathbf{x})\varepsilon^n$ and $r_{ij}\varepsilon^n$ are earning parameters in case of small change in $\frac{\mathbf{k}}{K}$. Let $\mathbf{e}_i = \mathbf{I}_i \varepsilon$.

Therefore, in the asymptotic case being analyzed, Eq. (1) can be expressed as the following partial differential equation:

$$\frac{\partial v(\mathbf{x},t)}{\partial t} = K \sum_{i,j=1}^{n} \mu_i \min(\varepsilon m_i, x_i) p_{ij}^+ \left( v(\mathbf{x}+\mathbf{e}_j-\mathbf{e}_i,t) - v(\mathbf{x},t) \right)$$

$$+ K \sum_{i,j=1}^{n} \mu_i \min(\varepsilon m_i, x_i) p_{ij}^- \left( v(\mathbf{x}-\mathbf{e}_j-\mathbf{e}_i,t) - v(\mathbf{x},t) \right)$$

$$+ K \sum_{i=1}^{n} \mu_i \min(\varepsilon m_i, x_i) p_{i0} \left( v(\mathbf{x}-\mathbf{e}_i,t) - v(\mathbf{x},t) \right) \quad (2)$$

$$+ K \sum_{i=1}^{n} \lambda_0 \left(1 - \sum_{i=1}^{n} x_i\right) p_{0i}^- \left( v(\mathbf{x}-\mathbf{e}_i,t) - v(\mathbf{x},t) \right)$$

$$+ K \sum_{i,j=1}^{n} \mu_i \min(\varepsilon m_i, x_i) p_{ij}^- (1-\theta(x_j)) \left( v(\mathbf{x}-\mathbf{e}_i,t) - v(\mathbf{x},t) \right)$$

$$+ K \sum_{i=1}^{n} \lambda_0 \left(1 - \sum_{i=1}^{n} x_i\right) p_{0i}^+ \left( v(\mathbf{x}+\mathbf{e}_i,t) - v(\mathbf{x},t) \right) + q(\mathbf{x}),$$

$$q(\mathbf{x}) = r(\mathbf{x}) + K \left( \sum_{i,j=1}^{n} \mu_i \min(l_i, x_i) \left( p_{ij}^+ r_{ij}^+ + p_{ij}^- r_{ij}^- (2-\theta(x_j)) \right) \right.$$

$$\left. + \sum_{i=1}^{n} \mu_i \min(l_i, x_i) p_{i0}^+ r_{i0}^+ + \sum_{i=1}^{n} \lambda_0 \left(1 - \sum_{i=1}^{n} x_i\right) \left( p_{0i}^+ r_{0i}^+ + p_{0i}^- r_{0i}^- \right) \right),$$

where $q(\mathbf{x})$ is earning rate on state space $X$.

If $v(\mathbf{x},t)$ is a function that is twice continuously differentiable with respect to $x_i$, then we can apply the Taylor series to second order about the point $\mathbf{x}$ for functions $v(\mathbf{x}+\mathbf{e}_j-\mathbf{e}_i,t)$, $v(\mathbf{x}-\mathbf{e}_i,t)$, and $v(\mathbf{x}+\mathbf{e}_i,t)$, $i,j=\overline{1,n}$. Equation (2) becomes:

$$\frac{\partial v(\mathbf{x},t)}{\partial t} = \sum_{i,j=1}^{n} \mu_i \min(\varepsilon m_i, x_i) p_{ij}^+ \left( \left( \frac{\partial v(\mathbf{x},t)}{\partial x_j} - \frac{\partial v(\mathbf{x},t)}{\partial x_i} \right) \right.$$

$$\left. + \frac{\varepsilon}{2} \left( \frac{\partial^2 v(\mathbf{x},t)}{\partial x_j^2} - 2\frac{\partial^2 v(\mathbf{x},t)}{\partial x_j \partial x_i} + \frac{\partial^2 v(\mathbf{x},t)}{\partial x_i^2} \right) \right)$$

$$+ \sum_{i,j=1}^{n} \mu_i \min(\varepsilon m_i, x_i) p_{ij}^- \left( -\left( \frac{\partial v(\mathbf{x},t)}{\partial x_j} + \frac{\partial v(\mathbf{x},t)}{\partial x_i} \right) \right.$$

$$\left. + \frac{\varepsilon}{2} \left( \frac{\partial^2 v(\mathbf{x},t)}{\partial x_j^2} + 2\frac{\partial^2 v(\mathbf{x},t)}{\partial x_j \partial x_i} + \frac{\partial^2 v(\mathbf{x},t)}{\partial x_i^2} \right) \right)$$

$$+ \sum_{i=1}^{n} \mu_i \min(\varepsilon m_i, x_i) p_{i0}^+ \left( -\frac{\partial v(\mathbf{x},t)}{\partial x_i} + \frac{\varepsilon}{2} \frac{\partial^2 v(\mathbf{x},t)}{\partial x_i^2} \right)$$

$$+ \sum_{i=1}^{n} \lambda_0 \left(1 - \sum_{i=1}^{n} x_i\right) p_{0i}^{-} \left(-\frac{\partial v(\mathbf{x},t)}{\partial x_i} + \frac{\varepsilon}{2} \frac{\partial^2 v(\mathbf{x},t)}{\partial x_i^2}\right)$$

$$+ \sum_{i,j=1}^{n} \mu_i \min(\varepsilon m_i, x_i) p_{ij}^{-}(1 - \theta(x_j)) \left(-\frac{\partial v(\mathbf{x},t)}{\partial x_i} + \frac{\varepsilon}{2} \frac{\partial^2 v(\mathbf{x},t)}{\partial x_i^2}\right)$$

$$+ \sum_{i=1}^{n} \lambda_0 \left(1 - \sum_{i=1}^{n} x_i\right) p_{0i}^{+} \left(\frac{\partial v(\mathbf{x},t)}{\partial x_i} + \frac{\varepsilon}{2} \frac{\partial^2 v(\mathbf{x},t)}{\partial x_i^2}\right) + q(\mathbf{x}) + o(\varepsilon^2).$$

Having grouped first-order and second-order partial derivatives of a function $v(\mathbf{x},t)$ in the resulting equation, we get the compact mathematical expression up to $o(\varepsilon^2) = o(1/K^2)$:

$$\frac{\partial v(\mathbf{x},t)}{\partial t} = -\sum_{i=1}^{n} A_i(\mathbf{x},t) \frac{\partial v(\mathbf{x},t)}{\partial x_i} + \frac{\varepsilon}{2} \sum_{i,j=1}^{n} B_{ij}(\mathbf{x},t) \frac{\partial^2 v(\mathbf{x},t)}{\partial x_i \partial x_j} + q(\mathbf{x}), \quad (3)$$

$$A_i(\mathbf{x},t) = \sum_{j=1}^{n} \mu_j \min(\varepsilon m_j, x_j)(p_{ji}^{-} - p_{ji}^{+} + \delta_{ji})$$

$$+ \mu_i \min(\varepsilon m_i, x_i) \sum_{j=1}^{n} p_{ij}^{-}(1 - \theta(x_j)) - \lambda_0 \left(1 - \sum_{i=1}^{n} x_i\right) \left(p_{0i}^{+} - p_{0i}^{-}\right),$$

$$B_{ii}(\mathbf{x},t) = \sum_{j=1}^{n} \mu_j \min(\varepsilon m_j, x_j)(p_{ji}^{+} + p_{ji}^{-} + \delta_{ji})$$

$$+ \mu_i \min(\varepsilon m_i, x_i) \sum_{j=1}^{n} p_{ij}^{-}(1 - \theta(x_j)) + \lambda_0 \left(1 - \sum_{i=1}^{n} x_i\right) \left(p_{0i}^{+} - p_{0i}^{-}\right),$$

$$B_{ij}(\mathbf{x},t) = \mu_i \min(\varepsilon m_i, x_i) \left(p_{ij}^{-} - p_{ij}^{+}\right), i \neq j,$$

where $\delta_{ji}$ is the Kronecker delta.

The Eq. (3) is the generalized multidimensional Kolmogorov backward equation. It differs from the well-known multidimensional Kolmogorov backward equation only by the inclusion of the earning component $q(\mathbf{x})$.

## 4 Mathematical Model of the Expected Reward

Solving Eq. (3) is a complex problem. Therefore, a series of subsequent approximations is required. It is evident that the diffusion coefficients $B_{ij}(\mathbf{x},t)$ of an Eq. (3) are of order $\varepsilon$. Therefore the term $\frac{\varepsilon}{2} \sum_{i,j=1}^{n} B_{ij}(\mathbf{x},t) \frac{\partial^2 v(\mathbf{x},t)}{\partial x_i \partial x_j}$ on the right-hand side of (3) is $O(\varepsilon^2)$. Thus, up to terms of order $O(\varepsilon^2)$, the reward density is given by the equation:

$$\frac{\partial v(\mathbf{x},t)}{\partial t} = -\sum_{i=1}^{n} A_i(\mathbf{x},t) \frac{\partial v(\mathbf{x},t)}{\partial x_i} + q(\mathbf{x}). \quad (4)$$

Integrating the density $v(\mathbf{x}, t)$ within a $n$-dimensional region $D$, $D \subseteq X$, we obtain the expected total reward that the model will earn in time $t$ given the initial state $\mathbf{x} \in D$:

$$V_D(t) = \int\int\ldots\int_D v(\mathbf{x}, t)d\mathbf{x}.$$

Applying this transformation to both sides of Eq. (4), and utilizing the rules of integration, the Leibniz integral rule, and the linearity of the coefficients $A_i(\mathbf{x}, t)$ in $\mathbf{x}$, along with the boundary condition $A(\mathbf{x},t)v(\mathbf{x},t) = 0$, $\mathbf{x} \in \Gamma(D)$, where $\Gamma(D)$ represents the reflecting boundary of the region $D$ [11], we derive a first order ordinary linear differential equation for the expected reward $V_D(t)$:

$$\frac{d}{dt}V_D(t) = \sum_{i=1}^{n}\frac{\partial A_i(\mathbf{x},t)}{\partial x_i} \cdot V_D(t) + \int\int\ldots\int_D q(\mathbf{x})d\mathbf{x}. \quad (5)$$

The Eq. (5) is a mathematical model of the expected total reward of the cluster system that this system will earn in time $t$ if its initial state belongs to set $D$. Now we have a linear differential equation that completely defines $V_D(t)$ given the known $V_D(0)$.

By analogy with a center of gravity of a material body $D$ in physics, we can determine the equilibrium point of the expected reward, when $\mathbf{x} \in D$:

$$E_i^D(t) = \frac{1}{V_D(t)}\int\int\ldots\int_D x_i v(\mathbf{x},t)d\mathbf{x}, \quad i = \overline{1, n}.$$

## 5 Numerical Example

Let's construct a specific model of a cluster system. Assume that the control node is connected to four compute nodes via an interconnect network. Requests arrive from the external environment to four queueing nodes $S_i, i = \overline{1, n}$. After processing, they return in the opposite direction. The exponential G-network is used as a mathematical model.

Let's set the service parameters: total number of requests $K = 10\,000$, arrival rate $\lambda_0 = \frac{1}{K}$, processing rates $\mu_1 = 0.015$, $\mu_2 = 0.005$, $\mu_3 = 0.009$, $\mu_4 = 0.01$. Non-zero elements of the probability matrix are $p_{01}^+ = 0.2$, $p_{01}^- = 0.05$, $p_{02}^+ = 0.4$, $p_{02}^- = 0.1$, $p_{03}^+ = 0.1$, $p_{04}^+ = 0.15$, $p_{10} = 1$, $p_{20} = 1$, $p_{30} = 1$, $p_{40} = 1$. Let's set the parameters defining the G-network's reward: $r_{01}^+ = 0.3$, $r_{01}^- = -3$, $r_{02}^+ = 0.05$, $r_{02}^- = -1$, $r_{03}^+ = 0.1$, $r_{03}^- = -1$, $r_{04}^+ = 0.1$, $r_{04}^- = -1$, $r_{10}^+ = 0.1$, $r_{20}^+ = 2$, $r_{30}^+ = 0.1$, $r_{40}^+ = 0.1$, and in other cases $r_{ij} = 0$. We will find the expected reward of the cluster model when the initial state $x \in D$, where $D = \{x = (x_1, x_2, x_3, x_4) : 2\varepsilon < x_i \leq 200\varepsilon, i = \overline{1,4}\}$.

Let's use the Eq. (5) with the parameters set above. Figure 1 shows the numerical solution of (5) with the condition $V_D(0) = 0$.

Thus, assuming that the initial state is given, $\mathbf{x} \in D$, it is possible to estimate the expected reward of the cluster system in c.u. as a function of remaining time.

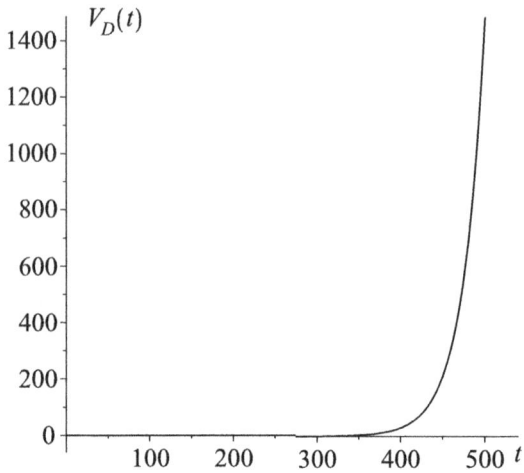

**Fig. 1.** Earnings of the cluster system with remaining time

From Fig. 1 it is clear that despite the presence of errors and malicious software, the overall operation of the cluster system has positive earnings dynamics in the long term.

## 6 Conclusions

This paper presents an analytical model of a cluster system. The exponential queueing G-network with rewards was used as a stochastic cluster system model. The sequence of rewards or incomes is generated by the cluster model as the results of transmitting requests between nodes. Real cluster systems typically handle a large number of requests. In this regard, the model was studied in the asymptotic case of a large number of requests. Ultimately, the results of mathematical modeling can be used to predict the expected total reward of the cluster system as a function of remaining time $t$, when the start state of the network is known. The presented technique allows to analyse the cluster system efficiency with mathematically specified accuracy.

**Acknowledgments.** The research was supported by the State Program of Scientific Research of the Republic of Belarus "Convergence-2025" (sub-program "Mathematical models and methods", assignment 1.6.01.06.

**Disclosure of Interests.** The authors have no competing interests to declare that are relevant to the content of this article.

# References

1. Gelenbe, E.: Product form queueing networks with negative and positive customers. J. Appl. Probab. **28**, 656–663 (1991). https://doi.org/10.2307/3214499
2. Caglayan, M.: G-networks and their applications to machine learning, energy packet networks and routing: introduction to the special issue. Probab. Eng. Inf. Sci. **31**(4), 381–395 (2017). https://doi.org/10.1017/S0269964817000171
3. Rusilko, T.: The G-network as a stochastic data network model. J. Belarusian State Univ. Math. Inf. **2**, 45–54 (2023). https://doi.org/10.33581/2520-6508-2023-2-45-54
4. Howard, R.: Dynamic Programming and Markov Processes. Massachusetts Institute of Technology, Cambridge (1960)
5. Matalytsky, M., Pankov, A.: Probabilistic analysis of income in banking networks. VESTNIK BSU Ser. 1. Phys. Math. Comput. Sci. **9**, 79–92 (2004)
6. Matalytski, M.: Finding expected revenues in G-network with multiple classes of positive and negative customers. Probab. Eng. Inf. Sci. **33**(1), 105–120 (2019). https://doi.org/10.1017/S0269964818000013
7. Rusilko, T.: Asymptotic analysis of a closed G-network of unreliable nodes. J. Appl. Math. Comput. Mech. **21**(2), 91–102 (2022). https://doi.org/10.17512/jamcm.2022.2.08
8. Rusilko, T., Salnikov, D.: Asymptotic analysis of a closed G-network with rewards. Tomsk State Univ. J. Control Comput. Sci. **68**, 38–47 (2024)
9. Nazarov, A., Pavlova, E.: Study of SMO type MMPP|M|N with feedback by the method of asymptotic analysis. Tomsk State Univ. J. Control Comput. Sci. **58**, 47–57 (2022)
10. Moiseeva, S., Bushkova, S., Pankratova, E., et al.: Asymptotic analysis of resource heterogeneous QS (MMPP + 2M)$^{(2,\nu)}$/GI(2)/$\infty$ under equivalently increasing service time. Autom. Remote. Control. **83**(8), 1213–1227 (2022). https://doi.org/10.1134/S0005117922080057
11. Moiseev, A., Nazarov, A., Paul, S.: Asymptotic diffusion analysis of multi-server retrial queue with hyper-exponential service. Mathematics **8**(4), 531 (2020). https://doi.org/10.3390/math8040531
12. Danilyuk, E., Plekhanov, A., Moiseeva, A., Sztrik, J.: Asymptotic diffusion analysis of retrial queueing system M/M/1 with impatient customers, collisions and unreliable servers. Axioms **11**(12), 699 (2022). https://doi.org/10.3390/axioms11120699
13. Tikhonov, V., Mironov, M.: Markov processes. Sovetskoe radio, Moscow (1977)

# On Application of Karamata Slowly Varying Functions in the Theory of Noncritical Markov Branching Systems

Azam A. Imomov[1,2](✉) and Zuhriddin A. Nazarov[2]

[1] Karshi State University, Karshi, Uzbekistan
imomov_azam@mail.ru
[2] Romanovskiy Institute of Mathematics, Tashkent, Uzbekistan

**Abstract.** The paper discusses the continuous-time Markov branching system. We deal only with the noncritical case. The primary task of this paper is to extend to the continuous case of our recent result, which explicitly calculates the famous constant in the theory of subcritical Galton-Watson branching systems, announced by Kolmogorov in 1938. We demonstrate the convergence rate to the Kolmogorov constant for the continuous-time Markov branching system. This result contributes to determining the speed of approximation rate in several classical limit theorems of the theory of Markov branching systems.

**Keywords:** Markov Branching System · Markov chain · Extinction time · Kolmogorov constant · Basic Lemma · Invariant Distribution · Limit Theorems · Convergence rate

## 1 Background and Purpose

The branching stochastic models describe an evolution of the population size of the reproductive individuals' system. These models most clearly illustrate numerous stochastic phenomena occurring both in nature and in human activity. The simple Galton-Watson model originally evolved as a family survival model in the second half of the 19th century, today has numerous generalizations and modifications; see [1–3] and [12].

The advanced current standing of the modern theory of stochastic branching systems has been achieved largely due to its compatibility with the theory of slowly varying functions. The conception of slow variation or, more general, regular variation was first initiated by the famous Serbian mathematician Jovan Karamata in the early 30s of the XX century; see, for instance, [4,9] and [20]. The integration of these two theories began with the publication of the excellent work of Zolotarev [24].

Further effective applications of slowly varying conception in various models of the theory of stochastic branching systems were demonstrated in works [8, 15–19, 23], in which classical results were improved and deeper properties of the random systems under consideration were discovered.

Recall that real-valued, positive and measurable function $\ell(x)$ is said to be slowly varying (SV) at infinity, in the sense of Karamata, if $\ell(\lambda x)/\ell(x) \to 1$ as $x \to \infty$ for each $\lambda > 0$. The SV property can be defined at any finite point by shifting the origin of the function to this point. In what follows, we use the symbol $\mathcal{SV}_\alpha$ to denote the class of Karamata SV-functions at point $\alpha$. The Representation Theorem [20] asserts that each function $\ell(\cdot) \in \mathcal{SV}_\infty$ may be written in the form

$$\ell(x) = c(x) \exp\left(\int_a^x \frac{\varepsilon(u)}{u}\,du\right) \qquad (1)$$

for some $a > 0$, where $c(x)$ has a finite positive limit, $c(x) \to c > 0$, the function $\varepsilon(x)$ is a continuous called the SV index, such that $\varepsilon(x) \to 0$ as $x \to \infty$. In a special case, $\ell(\cdot)$ is said to be normalized if $c(x) \equiv const$. Any SV-function is associated with an appropriate regular varying function. A function $R(x)$ is called regularly varying at infinity with index $\delta$, if it can be expressed as $R(x) = x^\delta \ell(x)$ for some $\ell(\cdot) \in \mathcal{SV}_\infty$. Then $R(\lambda x)/R(x) \to \lambda^\delta$ as $x \to \infty$ for each $\lambda > 0$; see also, [4] and [20] for more information.

In this report, we are interested in the application of SV-functions to enhance and generalize some classical limit results in the theory of stochastic continuous-time Markov Branching (MB) system. The states of the system are determined by the number of individuals and its growth occurs on a continuous-time axis $\mathcal{T} := [0, +\infty)$ according to the following random mechanism. The individuals undergo transformations in accordance with the branching rates $\{a_k, k \in \mathbb{N}_0\}$, where $\mathbb{N}_0 = \{0\} \cup \mathbb{N}$ and $\mathbb{N}$ is a set of natural numbers. Denoting by $Z(t)$ the population size at time $t \in \mathcal{T}$, we have a homogeneous-continuous-time Markov chain. The appropriate q-matrix $\mathbb{Q} = \{q_{ij}\}$ of the chain $\{Z(t), t \in \mathcal{T}\}$ is given as follows:

$$q_{ij} = \begin{cases} ia_{j-i+1} & \text{if } j \geq i \geq 0, \\ ia_0 & \text{if } j = i - 1, \\ 0 & \text{otherwise}. \end{cases} \qquad (2)$$

This in essence regulates the further evolution of the system; see [14].

Consider transition probabilities of the Markov chain $\{Z(t), t \in \mathcal{T}\}$

$$P_{ij}(t) := \mathbb{P}\Big\{Z(\tau + t) = j \mid Z(\tau) = i\Big\} \quad \text{for any} \quad \tau \in \mathcal{T}.$$

They are $i$-fold convolution of the distribution $\mathsf{p}_j(t) := P_{1j}(t)$. Define generating functions (GFs)

$$f(s) = \sum_{j \in \mathbb{N}_0} a_j s^j \quad \text{and} \quad F(t; s) := \sum_{j \in \mathbb{N}_0} \mathsf{p}_j(t) s^j$$

for $s \in [0, 1)$. An important fact is that the GFs $F(t; s)$ and $f(s)$ are related by the Kolmogorov backward equation $\partial F(t;s)/\partial t = f(F(t;s))$ which is solvable under the boundary condition $F(0; s) = s$; see [22].

If $m := \sum_{j \in \mathbb{N}} j a_j$ converges, then $m = f'(1-)$ is the average intensity of the branching rate in the system, which determines an asymptotic classification of the system trajectories. The MB system is divided into the subcritical, critical, and supercritical types, depending on $m < 0$, $m = 0$ and $m > 0$, respectively.

Letting $q$ be the extinction probability of the system initiated by the one founder-individual, it is the smallest root of the equation $f(s) = 0$ for $s \in [0, 1]$. This probability also satisfies the fixed-point equation $F(t; s) = s$ for all $t \in \mathfrak{T}$. We note, that always $q > 0$ being that $a_0 > 0$ by default. In what follows we consider the noncritical system, i.e. $m \neq 0$.

Let $\mathbb{P}\{Z(t) > 0\}$ be the survival probability at time $t$ of the system initiated by the single founder-individual. In the subcritical situation, one of the classical problems is to study the asymptote of this probability. This problem was first processed by Sevastyanov [21] in 1951. He proved that if $f''(1-) < \infty$, then the following asymptotic formula is valid:

$$\mathbb{P}\{Z(t) > 0\} \sim \mathcal{K} e^{mt} \quad \text{as} \quad t \to \infty, \tag{3}$$

where $\mathcal{K}$ is some constant depending on a form of $f(s)$. In the discrete-time case, the limiting coefficient $\mathcal{K}$ is the famous Kolmogorov constant in the theory of subcritical Galton-Watson branching systems, which was announced by Kolmogorov [13] in 1938. In the recent paper [5], the constant in the Galton-Watson case was calculated explicitly depending on the second moment of the offspring law. By analogy with the Galton-Watson case, we refer to $\mathcal{K}$ in (3) as the Kolmogorov constant.

In this report, we propose an explicit calculation analysis of the constant $\mathcal{K}$ in the asymptotic formula (3) and its extending to the supercritical case, depending on the moments of the system's branching rates under consideration.

## 2 Basic Assumption and Main Results

Return to the MB system $\{Z(t), t \in \mathfrak{T}\}$. Let

$$\mathrm{U}_q[0, 1) := \{[0, q) \cup (q, 1)\}$$

be a unit interval with a punctured point $q$. Taking $R(t; s) := q - F(t; s)$, we note that in the subcritical case $q = 1$ and $Q(t) := R(t; 0) = \mathbb{P}\{Z(t) > 0\}$ is the survival probability at time $t$ of the system initiated by the one founder-individual. In this case, bypassing the second moment condition $f''(1-) < \infty$, Zolotarev [24] proved that the existence of the integral

$$I(m) = -\int_0^1 \frac{f(u) + m(1-u)}{(1-u)f(u)} du \tag{4}$$

is necessary and sufficient for the validity of the asymptotic representation of type (3) and the constant $\mathcal{K}$ is related to $I(m)$ by the equality $I(m) = \ln \mathcal{K}$. For convenience, we wrote the integral in (4) in a slightly different form. As shown in [22], that the convergence of integral $I(m)$ is equivalent to that

$$\sum_{k \in \mathbb{N}} a_k k \ln k < \infty. \tag{5}$$

Zolotarev's result motivates us to think about extending the task he has processed, to the noncritical situation. Define

$$\mathcal{A}_q(t; s) := \frac{R(t; s)}{\beta^t}$$

for $s \in U_q[0, 1)$, where $\beta = \exp\{f'(q)\}$. Let $\mathcal{H} := \min\{t : Z(t) = 0\}$ be an extinction time of the genealogical tree of one founder-individual. The classical extinction theorem implies that $\mathbb{P}\{t < \mathcal{H} < \infty | Z(t) = j\} = q^j$ for any $t \in \mathcal{T}$. Taking this into account and using the total probability formula, we obtain $\mathbb{P}\{t < \mathcal{H} < \infty\} = q - F(t; 0)$. So, $Q(t) = R(t; 0) = \mathbb{P}\{t < \mathcal{H} < \infty\}$ is the bridled-survival probability of the genealogical tree of one founder-individual in the noncritical situation. Let

$$\mathcal{A}_q(t) := \mathcal{A}_q(t; 0) = \frac{Q(t)}{\beta^t}.$$

In [10] it was proved that

$$\ln \frac{\mathcal{A}_q(t; s)}{q - s} \longrightarrow \int_s^q \frac{f(u) - f'(q)(u - q)}{f(u)(u - q)} du \text{ as } t \to \infty. \tag{6}$$

In this case, the convergence of the integral in (6) is also equivalent to Zolotarev's condition (5). To verify this, we define the transformed MB system $\{Z_q(t)\}$, generated by the Harris-Sevastyanov transformation $f_q(s) = f(qs)/q$, provided that $q \neq 0$ for the supercritical case. It is easy to see that $\{Z_q(t)\}$ is a subcritical system initiated by one founder-individual, the branching rates of which are

$$\varphi_k = a_k q^{k-1} \quad \text{and} \quad f_q(s) = \sum_{k \in \mathbb{N}} \varphi_k s^k.$$

Therefore the average intensity of the branching rate is $f'_q(1-) = f'(q) < 0$. Next, letting $F_q(t; s) := \mathbb{E} s^{Z_q(t)}$, it immediately follows from the Kolmogorov backward equation that $F_q(t; s) = F(t; qs)/q$. Then $\mathbb{E} Z_q(t) = \beta^t$. Now, using standard transformations, we write the integral in (6) in the form

$$\int_1^s \frac{f_q(u) + f'_q(1)(1 - u)}{(1 - u) f_q(u)} du \tag{7}$$

for $s \in [0, 1)$. Since $q \leq 1$ it follows from (5) that $\sum_{k \in \mathbb{N}} \varphi_k k \ln k < \infty$. Then, in accordance with Zolotarev's conclusion, integral (7) converges. So, Zolotarev's condition (5) is sufficient for converging the integral on the right-hand side of

(6). Thus, since $R(t; qs) = qR_q(t; s)$ it follows that $Q(t) = q\mathbb{P}\{Z_q(t) > 0\}$ and hence, according to formula (3), we have

$$\frac{1}{\mathcal{A}_q(t)} = \frac{\mathbb{E}Z_q(t)}{\mathbb{P}\{t < \mathcal{H} < \infty\}} = \frac{1}{q}\mathbb{E}\left[Z_q(t) \mid Z_q(t) > 0\right] \longrightarrow \frac{1}{q\mathcal{K}} =: \frac{1}{\mathcal{K}_q}$$

as $t \to \infty$. So $\mathcal{K}_q$ can be interpreted as the coefficient of asymptotic equivalence between the expectation $\mathbb{E}Z_q(t) = \beta^t$ of the population size and the survival probability $\mathbb{P}\{Z_q(t) > 0\}$ of the transformed subcritical MB system generated by the Harris-Sevastyanov transformation $f_q(s)$. By analogy with the discrete-time case, we refer to $\mathcal{K}_q$ as the extended Kolmogorov constant; see [5]. Further, since $f_q(s)$ is a power series, in order for Zolotarev's condition (5) to be satisfied, it is necessary and sufficient that the integrand in (7) for some $\nu > 0$ has the form

$$\frac{f_q(s) + f_q'(1)(1-s)}{(1-s)f_q(s)} = \mathcal{O}\left(1-s\right)^{\nu-1} \qquad \text{as} \quad s \uparrow 1. \tag{8}$$

This necessity proposes considering the case when the Harris-Sevastyanov transformation $f_q(s)$ to be admitted the following representation:

$$f_q(s) = |f_q'(1)|(1-s) + (1-s)^{1+\nu}\mathcal{L}\left(\frac{1}{1-s}\right), \tag{9}$$

where $\mathcal{L}(\cdot) \in \mathcal{SV}_\infty$ and we choose that $\nu \in (0, 1)$. This implies that $f''(1) = \infty$ if $\nu < 1$. The following moment assumption is related to the condition (9):

$$m \neq 0 \quad \text{and} \quad \sum_{k \in \mathbb{N}} k^{1+\nu}a_k < \infty \quad \text{for} \quad \nu \in (0, 1). \qquad [A]$$

We state our main result in the following theorem.

**Theorem 1.** *Let the condition (9) with [A] be satisfied. Then*

$$\mathcal{A}_q(t) = \frac{q}{1 + q\Delta(t)}, \tag{10}$$

*where*

$$\Delta(t) = \frac{1}{|\ln \beta| \nu q}L_\Delta(t), \tag{11}$$

*and* $L_\Delta(\cdot) \in \mathcal{SV}_\infty$ *such that* $L_\Delta(t)/\mathcal{L}(\beta^{-t}) \to 1$ *as* $t \to \infty$.

We will devote the rest of the paper to proving Theorem 1 and, in the Appendix section, we will present some of its applications.

## 3 Preliminaries

First, we will prove the following property of SV-functions, which will be important in proving our results.

**Lemma 1.** *Let* $L(\cdot) \in \mathcal{SV}_\infty$. *Then*

$$\int_0^t L(u)\alpha^u du = \begin{cases} L(t)t(1+o(1)) & \text{if } \alpha = 1 \quad \text{(i)} \\ L(t)\dfrac{1-\alpha^t}{|\ln \alpha|}(1+o(1)) & \text{if } \alpha \in (0,1) \quad \text{(ii)} \end{cases} \qquad (12)$$

*as* $t \to \infty$.

*Proof.* Making a change of variables $u = \lambda t$, we find that

$$I(t) := \int_0^t L(u)\alpha^u du = t \int_0^1 L(\lambda t)\alpha^{\lambda t} d\lambda$$
$$= tL(t)\left[\int_0^1 \alpha^{\lambda t} d\lambda + \int_0^1 \left(\frac{L(\lambda t)}{L(t)} - 1\right) \alpha^{\lambda t} d\lambda\right].$$

The first integral in the last step is $(1 - \alpha^t)/t|\ln \alpha|$ for $\alpha \in (0,1)$ and is equal to 1 if $\alpha = 1$. The expression in round brackets of the second integrand is bounded by Potter's Theorem [4, Ch I, §5.4, Theorem 1.5.6] and tends to 0 as $t \to \infty$ uniformly in $\lambda \in (0,1]$. Hence

$$I(t) = L(t)\frac{1-\alpha^t}{|\ln \alpha|}(1+o(1)) \qquad \text{as} \quad t \to \infty$$

for $\alpha \in (0,1)$ and $I(t) = L(t)t(1+o(1))$ as $t \to \infty$ for $\alpha = 1$.
The assertions (12) follows.

Further we observe the asymptotic properties of the function $\mathcal{A}_q(t;s)$ as $t \to \infty$ for $s \in U_q[0,1)$. Indeed, after finding the function $\mathcal{A}_q(s) = \lim_{t \to \infty} \mathcal{A}_q(t;s)$, we obtain the Kolmogorov constant as $\mathcal{K}_q = \mathcal{A}_q(0)$.

**Lemma 2.** *Without any assumptions*

$$\mathcal{A}_q(t;s) \longrightarrow q-s \qquad \text{as} \quad t \to \infty \qquad (13)$$

*for all* $s \in U_q[0,1)$.

*Proof.* The mean value theorem implies

$$f(s) = f'(\xi(s))(s-q)$$

for all $s \in U_q[0,1)$, where $\xi(s) = q - (q-s)\theta$ and $\theta \in (0,1)$. Using this relation, we rewrite the Kolmogorov backward equation $\partial F/\partial t = f(F)$ as follows:

$$\frac{\partial R(t;s)}{\partial t} = f'(\xi(t;s))R(t;s), \qquad (14)$$

where $\xi(t;s) = q - \theta R(t;s)$. Since $R(t;s) > 0$ for $s \in [0,q)$, it follows that $\xi(t;s) < q$. It is known that the function $f(s)$ is concave, so $f'(s)$ increases monotonically. Then, after integrating over $[0,t]$ in (14), we obtain a uniform upper bound $R(t;s) < q\beta^t$. Therefore

$$q - \beta^t \le \xi(t;s) < q \qquad \text{for} \quad s \in [0,q). \tag{15}$$

Simultaneously with this, we observe that $\xi(t;s) > q$ for $s \in (q,1)$. Then once again, integration of equality (14) over $[0,t]$ entails a uniform lower bound $R(t;s) > (q-1)\beta^t$, which implies

$$q < \xi(t;s) < q + \beta^t \qquad \text{for} \quad s \in (q,1). \tag{16}$$

Denoting $q_\pm(t) := q \pm \beta^t$, we can combine relations (15) and (16) into the following inequalities:

$$q_-(t) \le \xi(t;s) < q_+(t) \qquad \text{for all} \quad s \in \mathrm{U}_q[0,1). \tag{17}$$

Now we integrate the relation (14) over $[0,t]$ and obtain

$$\ln \frac{R(t;s)}{q-s} = \int_0^t f'\big(\xi(u;s)\big)du.$$

The combination of last equality with relations (14) and (17), and the concavity property of the function $f'(s)$ leads to the inequalities

$$\int_0^t m_{q-}(u)du \le \ln \frac{R(t;s)}{q-s} < \int_0^t m_{q+}(u)du, \tag{18}$$

where $m_{q\pm}(t) = f'_q\big(q_\pm(t)\big)$. Undoubtedly

$$m_{q\pm}(t) \to m_q = f'(q) \qquad \text{as} \quad t \to \infty$$

and this implies that $m_{q\pm}(\cdot) \in \mathcal{SV}_\infty$. Then assertion (12(i)) entails that

$$\int_0^t m_{q\pm}(u)du \sim f'(q)t \qquad \text{as} \quad t \to \infty.$$

Therefore, it follows from (18) that

$$\ln \frac{R(t;s)}{q-s} = f'(q)t\big(1 + o(1)\big) \qquad \text{as} \quad t \to \infty.$$

The assertion (13) readily follows.

Lemma 2 suggests that we look for the function

$$\Delta(t;s) := \frac{\beta^t}{R(t;s)} - \frac{1}{q-s}$$

for $s \in \mathrm{U}_q[0,1)$.

**Lemma 3.** *If the condition (9) with* **(i)** *is satisfied, then*

$$\Delta(t;s) = \frac{1}{(q-s)^{1-\nu}} \frac{1}{|\ln \beta| \nu q^{\nu}} L_{\Delta}(t) \qquad (19)$$

*for all* $s \in U_q[0,1)$, *where* $L_{\Delta}(\cdot) \in \mathcal{SV}_{\infty}$ *and* $L_{\Delta}(t)/\mathcal{L}(\beta^{-t}) \to 1$ *as* $t \to \infty$.

*Proof.* Replacing $s$ by $F_q(t;s)$ in (9) we write:

$$f_q(F_q(t;s)) = |f'(q)|R_q(t;s) + R_q^{1+\nu}(t;s)\mathcal{L}\left(\frac{1}{R_q(t;s)}\right), \qquad (20)$$

where $R_q(t;s) = 1 - F_q(t;s)$. It is easy to check that the Kolmogorov backward equation holds for this function. Then we have

$$\frac{dF_q(t;s)}{dt} = f_q(F_q(t;s)). \qquad (21)$$

Combining (20) and (21) we have

$$\frac{dR_q(t;s)}{dt} = f'(q)R_q(t;s) - R_q^{1+\nu}(t;s)\mathcal{L}\left(\frac{1}{R_q(t;s)}\right). \qquad (22)$$

Letting

$$\omega(t;s) := \frac{1}{R_q(t;s)} \quad \text{and} \quad B_{\mathcal{L}}(t;s) := \omega^{1-\nu}(t;s)\mathcal{L}(\omega(t;s))$$

we rewrite (22) as the following differential equation:

$$\frac{\partial \omega(t;s)}{\partial t} + f'(q)\omega(t;s) = B_{\mathcal{L}}(t;s). \qquad (23)$$

Further we are looking for the function $\omega(t;s)$ that satisfies Eq. (23) as the product of two functions of the variables $t$ and $s$, so that

$$\omega(t;s) = u(t;s)v(t;s),$$

where the unknown functions $u(t;s)$ and $v(t;s)$ are required to be found. Then Eq. (23) becomes

$$\frac{\partial u(t;s)}{\partial t} v(t;s) + u(t;s)\left(\frac{\partial v(t;s)}{\partial t} + f'(q)v(t;s)\right) = B_{\mathcal{L}}(t;s). \qquad (24)$$

At the same time, we presume that the expression $B_{\mathcal{L}}(t;s)$ on the right-hand side of (24) is known. In fact, it can be estimated asymptotically using Lemma 2.

In what follows, the symbol $\varepsilon > 0$ everywhere stands for an arbitrarily small positive number, which may differ from one place to another.

Let

$$E_{q\pm}(t) := \exp\left\{\int_0^t f'(q_{\pm}(u))du\right\}.$$

Then, it follows from (18) that

$$(1-s)E_{q-}(t) \leq R_q(t;s) < (1-s)E_{q+}(t). \tag{25}$$

Since $f'(q_{\pm}(t))$ are SV at infinity and $q_{\pm}(t) \to q$ as $t \to \infty$, the assertion (12(i)) implies that

$$\left|E_{q\pm}(t) - \beta^t\right| \to 0 \quad \text{as} \quad t \to \infty.$$

Therefore it follows from (25) that for $\varepsilon > 0$, there exists a number $t_0 = t_0(\varepsilon) \in \mathcal{T}$ such that for all $t > t_0$ the following relations hold:

$$\frac{1}{1-s} - \varepsilon < w(t;s)\beta^t \leq \frac{1}{1-s} + \varepsilon \tag{26}$$

for $t$ sufficiently large. Then

$$\frac{1}{(1-s)^{1-\nu}\beta^{(1-\nu)t}} - \varepsilon < w^{1-\nu}(t;s) \leq \frac{1}{(1-s)^{1-\nu}\beta^{(1-\nu)t}} + \varepsilon \tag{27}$$

for arbitrarily (obviously different from the previous) small $\varepsilon > 0$. Therefore, given our notation, we transform relations (27) into the form

$$\frac{1}{\beta^{(1-\nu)t}} - \varepsilon < \frac{B_{\mathcal{L}}(t;s)(1-s)^{1-\nu}}{\mathcal{L}(w(t;s))} \leq \frac{1}{\beta^{(1-\nu)t}} + \varepsilon \tag{28}$$

for arbitrarily small $\varepsilon > 0$. Estimations (26) imply that $(1-s)w(t;s) \sim \beta^{-t}$ as $t \to \infty$. Then there exists $\mathcal{L}_\beta(\cdot) \in \mathcal{SV}_\infty$ such that

$$\frac{\mathcal{L}_\beta(\beta^{-t})}{\mathcal{L}(w(t;s))} \to 1 \quad \text{as} \quad t \to \infty$$

uniformly in $s \in U_q[0,1)$. Therefore, we can further transform (28) as follows:

$$\frac{\mathcal{L}_\beta(\beta^{-t})}{(1-s)^{1-\nu}\beta^{(1-\nu)t}} - \varepsilon < B_{\mathcal{L}}(t;s) \leq \frac{\mathcal{L}_\beta(\beta^{-t})}{(1-s)^{1-\nu}\beta^{(1-\nu)t}} + \varepsilon \tag{29}$$

for $t$ sufficiently large.

Returning to the Eq. (24), we look at its left side. Choose the function $v(t;s)$ such that the multiplier of $u(t;s)$ in the last equation becomes equal to 0. Then we have

$$\begin{cases} \dfrac{\partial v(t;s)}{\partial t} = -f'(q)v(t;s) \\ \dfrac{\partial u(t;s)}{\partial t}v(t;s) = B_{\mathcal{L}}(t;s). \end{cases} \tag{30}$$

Integration the first equation over $[0,t]$ gives

$$v(t;s) = \frac{1}{\beta^t}v(0;s). \tag{31}$$

Substituting this function into the second equation of (30) and subsequent integration over $[0, t]$ leads to the equation

$$u(t; s) = u(0; s) + \frac{1}{v(0; s)} \int_0^t B_{\mathcal{L}}(u; s) \beta^u du. \qquad (32)$$

Substituting the found functions (31) and (32) into $\omega = uv$, we obtain

$$\beta^t \omega(t; s) - \omega(0; s) = \int_0^t B_{\mathcal{L}}(u; s) \beta^u du. \qquad (33)$$

According to (29), the difference

$$\left| B_{\mathcal{L}}(t; s) - \frac{\mathcal{L}_\beta(\beta^{-t})}{(1-s)^{1-\nu} \beta^{(1-\nu)t}} \right|$$

can be made infinitesimal by choosing $\varepsilon > 0$ desirably small. Hence, the integrand term $B_{\mathcal{L}}(t; s)$ in (33) can be majorized by the function

$$\frac{\mathcal{L}_\beta(\beta^{-t})}{(1-s)^{1-\nu} \beta^{(1-\nu)t}}$$

for $t$ sufficiently large and for all $s \in U_q[0, 1)$. Thus, denoting $\ell_\beta(t) := \mathcal{L}_\beta(\beta^{-t})$, we have

$$\int_0^t B_{\mathcal{L}}(u; s) \beta^u du = \frac{1}{(1-s)^{1-\nu}} \int_0^t \beta^{\nu u} \ell_\beta(u) du \qquad (34)$$

for $t$ sufficiently large and for all $s \in U_q[0, 1)$, where $\ell_\beta(\cdot) \in \mathcal{SV}_\infty$. The integral on the right-hand side of (34) can be estimated using statement (12(ii)). We have

$$\int_0^t \beta^{\nu u} \ell_\beta(u) du = \frac{1}{\nu |\ln \beta|} \mathcal{L}_\beta(\beta^{-t})(1 + o(1)) \quad \text{as} \quad t \to \infty. \qquad (35)$$

Now, considering relations (33)–(35) together, we conclude that

$$\frac{\beta^t}{R_q(t; s)} - \frac{1}{1-s} = \frac{1}{\nu |\ln \beta|} \frac{1}{(1-s)^{1-\nu}} L_\Delta(t), \qquad (36)$$

where $L_\Delta(t)/\mathcal{L}_\beta(\beta^{-t}) \to 1$ as $t \to \infty$. Since $R(t; qs) = qR_q(t; s)$, formula (36) directly leads to relation (19).

Lemma 3 is completely proved.

## 4 Proof of Theorem 1

In our designation, we write

$$\mathcal{A}_q(t; s) = \frac{q}{1 + q\Delta(t; s)}, \qquad (37)$$

where as proven in Lemma 3, that

$$\Delta(t;s) = \frac{1}{(q-s)^{1-\nu}} \frac{1}{|\ln \beta| \nu q^{\nu}} L_{\Delta}(t) \qquad (38)$$

for all $s \in U_q[0,1)$, where $L_{\Delta}(\cdot) \in \mathcal{SV}_{\infty}$ and $L_{\Delta}(t)/\mathcal{L}(\beta^{-t}) \to 1$ as $t \to \infty$. Now we put $s = 0$, then (37) and (38) produce (10) and (11) respectively

Thus the theorem is proved completely. □

## Appendix

In this final section, we present some consequences of Theorem 1.

In subcritical case representation (9) becomes

$$f(s) = |m|(1-s) + (1-s)^{1+\nu} \mathcal{L}\left(\frac{1}{1-s}\right), \qquad (39)$$

where $\mathcal{L}(\cdot) \in \mathcal{SV}_{\infty}$ and we choose that $\nu \in (0,1)$. Then Theorem 1 implies the following formula for the survival probability at time $t$ of the system initiated by the one founder-individual:

$$\mathbb{P}\{Z(t) > 0\} = \frac{1}{1 + \Delta(t)} m^t, \qquad (40)$$

where

$$\Delta(t) = \frac{1}{|m|\nu} \mathcal{L}\left(e^{|m|t}\right).$$

Of course, this is an improvement of the Kolmogorov-type theorem proved by Sevastyanov [21], for subcritical processes with an infinite variance of the branching rate intensity. The improvement here is that if $2b := f''(1) < \infty$, then $\nu = 1$ in condition (39). Therefore, $\mathcal{L}(u) \to b$ as $u \to \infty$. Thus, the coefficient-multiplier of $m^t$ on the right-hand side of formula (40) approaches

$$\mathcal{K} := \frac{1}{1 + \frac{b}{|m|}},$$

and this is an explicit form of the continuous-time analogue of the famous Kolmogorov constant.

Now let $m \neq 0$. Consider a new population growth system called the Markov Q-process. Denoting $W(t)$ be the population size at time $t \in \mathcal{T}$ in this system, we have continuous-homogeneous-time irreducible Markov chain with the state space $\mathcal{E} \subset \mathbb{N}$. Its transition probabilities $\mathcal{Q}_{ij}(t)$ for all $i, j \in \mathcal{E}$ are defined by GF

$$w_i(t;s) := \sum_{j \in \mathcal{E}} \mathcal{Q}_{ij}(t) s^j = w(t;s) \left[\frac{F(t;qs)}{q}\right]^{i-1}, \qquad (41)$$

where $w(t;s) := w_1(t;s)$ has a form of

$$w(t;s) = \frac{s}{\beta^t} \left.\frac{\partial F(t;x)}{\partial x}\right|_{x=qs} \quad \text{for any} \quad t \in \mathcal{T}; \qquad (42)$$

see [11]. Detailed information on the properties of the process $\{W(t), t \in \mathcal{T}\}$ and the related results can also be found for instance, in [6,7] and [17].

The backward and forward Kolmogorov equations yield that

$$\left.\frac{\partial F(t;x)}{\partial x}\right|_{x=qs} = \frac{1}{f(qs)} f\left(F(t;qs)\right) \quad \text{for any} \quad t \in \mathcal{T}. \qquad (43)$$

Since $f(s) \sim |f'(q)|(q-s)$ as $s \to q$, relations (42) and (43) entail

$$w(t;s) = s\frac{1}{\beta^t}\frac{|f'(q)|}{f(qs)} R(t;s)\left(1+o(1)\right) \quad \text{as} \quad t \to \infty \qquad (44)$$

for all $s \in [0,1)$. Since $\lim_{t\to\infty} F(t;s) = q$ uniformly in $s \in [0,r]$ for any fixed $r < 1$ (see [21]), it follows from (41) that $w_i(t;s)/w(t;s) \to \infty$ as $t \to \infty$. On the other hand $\lim_{s\downarrow 0}\left[w_i(t;s)/s\right] = \mathcal{Q}_{i1}(t)$. Then combining relations (41) and (44), we have the following analogue of the statement (3):

$$\frac{\mathcal{Q}_{i1}(t)\beta^t}{\mathbb{P}\{t < \mathcal{H} < \infty\}} \to \frac{|\ln\beta|}{a_0} \quad \text{as} \quad t \to \infty.$$

Further improvement of this result is possible due to condition (9).

# References

1. Asmussen, S., Hering, H.: Branching Processes. Birkhäuser, Boston (1983)
2. Athreya, K.B., Ney, P.E.: Branching Processes. Springer (1972)
3. Harris, T.E.: The Theory of Branching Processes. Springer (1963)
4. Bingham, N.H., Goldie, C.M., Teugels J. L.: Regular Variation. Cambridge (1987)
5. Imomov, A.A., Murtazaev, M.: On the Kolmogorov constant explicit form in the theory of discrete-time stochastic branching systems. J. Appl. Probab. **61**(3), 927–941 (2024)
6. Imomov, A.A., Nazarov, Z.A.: Central limit theorem and law of large numbers analogues for the total progeny in the Q-processes. Contemp. Math. **5**(3), 2751–2769 (2024). https://doi.org/10.37256/cm.5320242839
7. Imomov, A.A., Nazarov, Z.A., Moiseeva, S.P.: On estimation of structural parameters in Q-process. Commun. Comput. Inform. Sci. **2163**, 241–254 (2024). https://doi.org/10.1007/978-3-031-65385-8-18
8. Imomov, A.A., Tukhtaev, E.E.: On asymptotic structure of critical Galton-Watson branching processes allowing immigration with infinite variance. Stoch. Model. **39**(1), 118–140 (2023)
9. Imomov, A., Meyliyev, A.: On the application of slowly varying functions with remainder in the theory of Markov branching processes with mean one and infinite variance. Ukr. Math. J. **73**(8), 1225–1237 (2022). https://doi.org/10.1007/s11253-022-01988-5

10. Imomov, A.A.: On conditioned limit structure of the Markov branching process without finite second moment. Malaysian J. Math. Sci. **11**(3), 393–422 (2017)
11. Imomov, A.A.: On Markov continuous time analogue of Q-processes. Theory Prob. Math. Stat. **84**, 57–64 (2012)
12. Jagers, P.: Branching Processes with Biological applications. Wiley/Pitman Press, UK (1975)
13. Kolmogorov, A.N.: K resheniyu odnoy biologicheskoy zadachi. Rep. SRI Math. Mech. Tomsk Univ. **2**, 7–12, 1938. (Russian)
14. Li, J., Cheng, L., Li, L.: Long time behaviour for markovian branching-immigration systems. Discrete Event Dyn. Syst. **31**, 37–57 (2021)
15. Nagaev, S.V., Wachtel, V.: The critical Galton-Watson process without further power moments. J Appl. Prob. **44**(3), 753–769 (2007)
16. Pakes, A.G.: Critical Markov branching process limit theorems allowing infinite variance. Adv. Appl. Prob. **42**, 460–488 (2010)
17. Pakes, A.G.: Revisiting conditional limit theorems for the mortal simple branching process. Bernoulli **5**(6), 969–998 (1999)
18. Pakes, A.G.: Limit theorems for the simple branching process allowing immigration, I. The case of finite offspring mean. Adv. Appl. Prob. **11**, 31–62 (1979)
19. Pakes, A.G.: Some results for non-supercritical Galton-Watson process with immigration. Math. Biosci. **24**, 71–92 (1975)
20. Seneta, E.: Regularly Varying Functions. Springer, Berlin (1985). Translated Russian, Nauka, Moscow
21. Sevastyanov, B.A.: The theory of branching stochastic process. Uspekhi Mat. Nauk. **6**(46), 47–99 (1951)
22. Sevastyanov, B.A.: Branching Processes. Nauka, Moscow (Russian)
23. Wanga, J., Wanga, X., Li, J.: Asymptotic behavior of supercritical branching processes. Stat. Probab. Lett. **195**, 109782 (2023)
24. Zolotarev, V.M.: More exact statements of several theorems in the theory of branching processes. Theory Prob. Appl. **2**, 245–253 (1957)

# Asymptotic Analysis of a Multiserver Retrial Queue with Disasters

Natalya Meloshnikova and Ekaterina Fedorova(✉)

National Research Tomsk State University, Lenina Avenue, 36, Tomsk, Russia
moiskate@mail.ru

**Abstract.** In the paper, a multiserver retrial queueing system with disasters is considered. We have two arrival Poisson processes – "positive" and "negative" calls. Service time is exponentially distributed. If all servers are inaccessible, an arrival positive call goes into an orbit, where it performs an exponential random delay. Disasters occur at moments of negative call arrivals, which destroy all positive calls in the system (in servers and the orbit). In the paper, the model is studied by the method of asymptotic analysis under the condition of a high rate of the positive arrival process. It is proved that the asymptotic stationary probability distribution of the number of calls in the orbit is exponential. Some numerical results are presented. The modified asymptotics is proposed based on the estimation of "zero" state probability.

**Keywords:** retrial queue · negative calls · asymptotic analysis · high arrival rate · disasters

## 1 Introduction

In classical queueing theory, models can be categorized into systems with queues, systems with losses, and queuing networks, but retrial queueing systems are of particular interest. Retrial Queueing Systems (RQ systems) are mathematical models with repeated calls applying to analyze and optimize various telecommunications systems, mobile networks, and call-centers, etc. [1,2]. The main difference of these models is that an unserviced arrival call is sent to a virtual place – an orbit and makes a random delay.

Examples of real systems include also cloud computing technologies, (like as IaaS, PaaS and SaaS) [3]. Cloud technologies represent a way of providing IT resources over the internet. When a large number of users try to connect, those who do not immediately receive the service are sent to a virtual "orbit", where they wait for a random time before reconnecting to the required technology.

The foundational works in the study of RQ systems are written in [1] and [2], focusing on the full description and mathematical analysis of RQ systems and also some numerical methods.

In the paper, we consider the retrial model with "disasters" caused by negative calls. E. Gelenbe [4] introduced the concept of negative calls, which have a detrimental impact on the system (called G-system and G-networks). In real systems, negative calls may represent viruses, hacker attacks, system breakdowns, and so on. Other researches make further studies of the effects of negative calls, which can delete normal calls or change the system's state. The most extensive biography is in [5].

Disasters in RQ are studied by [6,7]. Shin [6] study a multiserver RQ with four types of arrivals (positive, disasters, negative for deleting customers in orbit and negative for deleting customers in service area). There, all arrivals occur according to a Markovian arrival process with marked transitions, a necessary and sufficient stability condition for the system is derived. J. R. Artalejo considers a model that represents unexpected events completely clearing the queue of all current calls. Disasters occur randomly and reset the system. Also retrial queues with disasters have been considered in [8,9].

Methods for RQ models analyzing can be generally divided into two classes: matrix methods [10,11] and asymptotic methods [12]. This article presents an asymptotic analysis method. Methods of the asymptotic analysis allow to study the behavior of models under limiting conditions, when certain system parameters tend to infinity or reach critical values. The main idea of this method is to simplify complex mathematical expressions and derive their limiting formulas, which makes it possible to obtain approximate solutions and analyze their behavior in different regimes.

The structure of the paper is as follows. Section 2 describes the considered model of a multi-server retrial queueing system with disasters, and presents Kolmogorov equations for the process under study. Section 3 is devoted to the asymptotic analysis method under the condition of a high arrival rate, which is applied to solve the equations. In Sect. 4, we present several numerical examples that demonstrate the accuracy of the asymptotic results. Section 5 introduces a modified approximation and estimation of "zero" state probability. Section 6 contains some conclusions.

## 2 Model Description

Let us consider a multiserver retrial queueing system with disasters (Fig. 1). The arrival process of calls which require a service ("positive calls") is supposed Poisson with parameter $\lambda$. There are $K$ servers in the system. The service time is distributed exponentially with rate $\mu$. If all servers are busy, then a call goes to an orbit where it performs a random delay distributed exponentially with parameter $\sigma$. There is a multiply access to servers for calls in the orbit.

Also, there is Poisson arrival process of "negative" calls with rate $\gamma$. When a negative call enters in the system, it "resets" all servers in the system and the orbit, so all positive calls leave the system and it becomes free ("zero" state). Such models are called as retrial queueing systems with disasters (or catastrophes).

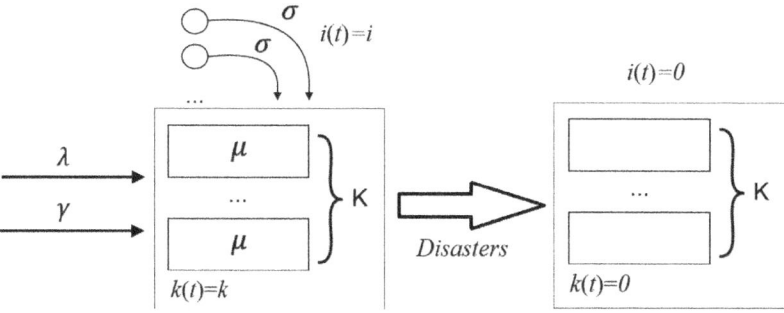

**Fig. 1.** Multiserver RQ with disasters.

Let us denote a random process of the number of calls in the orbit by $i(t)$. Process $k(t)$ determines states of the service unit as follows:

$$k(t) = \begin{cases} 0, \text{if all servers are free,} \\ 1, \text{if one server is busy,} \\ \ldots \\ k, \text{if } k \text{ servers are busy,} \\ \ldots \\ K, \text{if all servers are busy.} \end{cases}$$

Process $\{k(t), i(t)\}$ is Markovian. We denote $P\{k(t) = k, i(i) = i\} = P(k, i)$ and compose a system of Kolmogorov equations in the steady-state as follows

- for $i = 0$:

$$\begin{cases} -(\lambda + \gamma)P(0,0) + \mu P(1,0) + \gamma = 0, \\ \ldots \\ -(\lambda + k\mu + \gamma)P(k,0) + \lambda P(k-1,0) + \sigma P(k-1,1) \\ +(k+1)\mu P(k+1,0) = 0, \text{ for } k = \overline{1, K-1} \\ \ldots \\ -(\lambda + K\mu + \gamma)P(K,0) + \lambda P(K-1,0) + \sigma P(K-1,1) = 0, \end{cases} \quad (1)$$

- for $i > 0$:

$$\begin{cases} -(\lambda + \gamma)P(0,i) + \mu P(1,i) + \gamma = 0, \\ \ldots \\ -(\lambda + k\mu + i\sigma + \gamma)P(k,i) + (k+1)\mu P(k+1,i) \\ +\lambda P(k-1,i) + (i+1)\sigma P(k-1,i+1) = 0, \text{ for } k = \overline{1, K-1} \\ \ldots \\ -(\lambda + K\mu + \gamma)P(K,i) + \lambda P(K-1,i) + \lambda P(K,i-1) \\ +(i+1)\sigma P(K-1,i+1) = 0. \end{cases} \quad (2)$$

Let us introduce the following partial characteristic functions

$$H(k,u) = \sum_{i=0}^{\infty} e^{jui} P(k,i).$$

So we rewrite System (1)–(2) as follows

$$\begin{cases} -j\sigma \dfrac{\partial H(0,u)}{\partial u} = -(\lambda+\gamma)H(0,u) + \mu H(1,u) + \gamma, \\ j\sigma e^{-ju}\dfrac{\partial H(k-1,u)}{\partial u} - j\sigma\dfrac{\partial H(k,u)}{\partial u} = -(\lambda+k\mu+\gamma)H(k,u) \\ \quad + \lambda H(k-1,u) + (k+1)\mu H(k+1,u), \text{ for } k = \overline{1,K-1} \\ j\sigma e^{-ju}\dfrac{\partial H(K-1,u)}{\partial u} = -(\lambda+K\mu+\gamma)H(K,u) + \lambda e^{ju}H(K,u) \\ \quad + \lambda H(K-1,u). \end{cases} \quad (3)$$

Summing all equations of system (3), we obtain:

$$j\sigma(e^{-ju}-1)\sum_{k=0}^{K-1}\frac{\partial H(k,u)}{\partial u} = \gamma + \lambda(e^{ju}-1)H(K,u) - \gamma\sum_{k=0}^{K} H(k,u). \quad (4)$$

Due to the impossibility of directly solving of System (3)–(4), we propose the method of asymptotic analysis under the condition of a high rate of positive arrivals.

## 3  Asymptotic Analysis

First of all, we introduce an infinitesimal parameter $\varepsilon \to 0$. Let us denote

$$\lambda = \frac{\hat{\lambda}}{\varepsilon}, \ u = \varepsilon w, H(K,u) = F(K,w,\varepsilon), H(k,u) = \varepsilon F(k,w,\varepsilon).$$

So System (3)–(4) have the following form

$$\begin{cases} -j\sigma\dfrac{\partial F(0,w,\varepsilon)}{\partial w} = -\left(\dfrac{\hat{\lambda}}{\varepsilon}+\gamma\right)\varepsilon F(0,w,\varepsilon) + \mu\varepsilon F(1,w,\varepsilon) + \gamma, \\ j\sigma e^{-j\varepsilon w}\dfrac{\partial F(k-1,w,\varepsilon)}{\partial w} - j\sigma\dfrac{\partial F(k,w,\varepsilon)}{\partial w} = \hat{\lambda}F(k-1,w,\varepsilon) = \\ \quad -\left(\dfrac{\hat{\lambda}}{\varepsilon}+k\mu+\gamma\right)\varepsilon F(k,w,\varepsilon) + (k+1)\mu\varepsilon F(k+1,w,\varepsilon), \\ j\sigma e^{-j\varepsilon w}\dfrac{\partial F(K-1,w,\varepsilon)}{\partial w} = -\left(\dfrac{\hat{\lambda}}{\varepsilon}+K\mu+\gamma\right)F(K,w,\varepsilon) + \\ \quad \hat{\lambda}F(K-1,w,\varepsilon) + \dfrac{\hat{\lambda}}{\varepsilon}e^{j\varepsilon w}F(K,w,\varepsilon). \end{cases} \quad (5)$$

$$jo(e^{-j\varepsilon w}-1)\sum_{k=0}^{K-1}\frac{\partial F(k,w,\varepsilon)}{\partial w}=\gamma+\frac{\hat{\lambda}}{\varepsilon}(e^{j\varepsilon w}-1)F(K,w,\varepsilon)$$
$$-\gamma\varepsilon\sum_{k=0}^{K-1}F(k,w,\varepsilon)-\gamma F(K,w,\varepsilon).$$
(6)

Substituting Taylor series for exponents in (5)–(6), we obtain

$$\begin{cases} -j\sigma\dfrac{\partial F(0,w,\varepsilon)}{\partial w}=-(\hat{\lambda}+\gamma\varepsilon)F(0,w,\varepsilon)+\mu\varepsilon F(1,w,\varepsilon)+\gamma+O(\varepsilon^2),\\ j\sigma\dfrac{\partial F(k-1,w,\varepsilon)}{\partial w}-j\sigma(j\varepsilon w)\dfrac{\partial F(k-1,w,\varepsilon)}{\partial w}-j\sigma\dfrac{\partial F(k,w,\varepsilon)}{\partial w}\\ \quad=-\hat{\lambda}F(k,w,\varepsilon)-(k\mu+\gamma)\varepsilon F(k,w,\varepsilon)+\hat{\lambda}F(k-1,w,\varepsilon)\\ \quad+(k+1)\mu\varepsilon F(k+1,w,\varepsilon)+O(\varepsilon^2),\\ j\sigma\dfrac{\partial F(K-1,w,\varepsilon)}{\partial w}-j\sigma(j\varepsilon w)\dfrac{\partial F(K-1,w,\varepsilon)}{\partial w}\\ \quad=-\left(\dfrac{\hat{\lambda}}{\varepsilon}+K\mu+\gamma\right)F(K,w,\varepsilon)+\hat{\lambda}F(K-1,w,\varepsilon)\\ \quad+\dfrac{\hat{\lambda}}{\varepsilon}F(K,w,\varepsilon)+\hat{\lambda}(jw)F(K,w,\varepsilon)+O(\varepsilon^2). \end{cases}$$
(7)

$$-j\sigma(j\varepsilon w)\sum_{k=0}^{K-1}\frac{\partial F(k,w,\varepsilon)}{\partial w}=\gamma+\hat{\lambda}(jw)F(K,w,\varepsilon)$$
$$-\gamma\varepsilon\sum_{k=0}^{K-1}F(k,w,\varepsilon)-\gamma F(K,w,\varepsilon)+O(\varepsilon^2).$$
(8)

Under the limit $\varepsilon\to 0$, Eq. (7)–(8) have the following form:

$$\begin{cases} -j\sigma\dfrac{\partial F(0,w)}{\partial w}=-\hat{\lambda}F(0,w)+\gamma,\\ j\sigma\dfrac{\partial F(k-1,w)}{\partial w}-j\sigma\dfrac{\partial F(k,w)}{\partial w}=-\hat{\lambda}F(k,w)+\hat{\lambda}F(k-1,w),\\ j\sigma\dfrac{\partial F(K-1,w)}{\partial w}=-(K\mu+\gamma+\hat{\lambda}(jw))F(K,w)+\hat{\lambda}F(K-1,w). \end{cases}$$

and
$$0=\gamma+\hat{\lambda}(jw)F(K,w)-\gamma F(K,w). \quad (9)$$

From Eq. (9), we express $F(K,w)$:

$$F(K,w)=\frac{\gamma}{\gamma-jw\hat{\lambda}}.$$

Turning up to the notation, the characteristic function of the probability distribution of the number of calls in the orbit under the condition of a high arrival rate can be approximated as

$$h(u)=F\left(K,\frac{u}{\varepsilon}\right)+O(\varepsilon)\approx F\left(K,\frac{u}{\varepsilon}\right)=\frac{\gamma}{\gamma-ju\lambda}.$$

In this way, we have prove that the asymptotic probability distribution of the number of calls in the orbit is exponential with a parameter $\gamma/\lambda$.

## 4 Numerical Analysis

To demonstrate the accuracy of the asymptotic method, we numerically compare asymptotic distribution $PA(i)$ and exact distribution $P(i)$ obtained by a numerical algorithm for various values of the system parameters.

Examples of the comparison between the exact distribution and the asymptotic distribution are presented below (Fig. 2). The system parameters are as follows:
$$\mu = 1, \sigma = 1, \gamma = 1, K = 3.$$

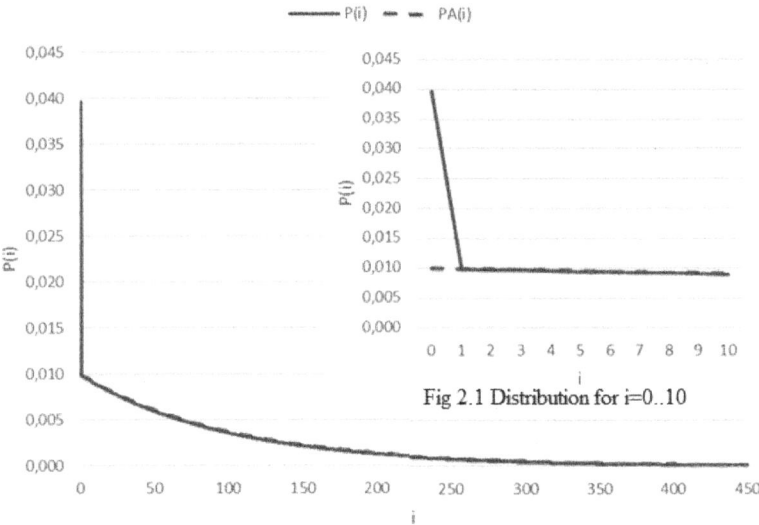

**Fig. 2.** Exact and asymptotic distribution for $\lambda = 100$.

From the Fig. 2, it can be seen that the asymptotic and exact distributions are close to each other, except at the point $P(0)$. To confirm this, let's examine the graph at its initial values in the Fig. 2.1.

We will use the Kolmogorov distance as the accuracy measure:

$$\Delta = \left| \sum_{n=0}^{i} (PA(n) - P(n)) \right|.$$

The values is presented on Table 1.

**Table 1.** Kolmogorov distances for various values of the parameter $\lambda$

| | $\lambda = 10$ | $\lambda = 50$ | $\lambda = 100$ | $\lambda = 150$ |
|---|---|---|---|---|
| $\triangle$ | 0.260 | 0.059 | 0.030 | 0.020 |

From example, we conclude that the method accuracy increases with the arrival rate growing. However, we observe that the largest difference between the distributions is at the point $i = 0$ (in "zero" state of the system). So, we further propose modified approximation to improve the results of asymptotics.

## 5 Modified Approximation

To eliminate the difference at the point $i = 0$, we propose a modified approximation, which includes a separately estimation of the zero state probability $P(0)$ and using the asymptotic distribution for others points.

### 5.1 $P(0)$ Estimation

From System (1), we can not to obtain an exact formula for calculation of $P(0) = \sum_{k=0}^{K} P(0, k)$, since the number of unknowns variables is greater than the number of equations. So let us try to estimate its value.

From Fig. 2, we see that the value of $P(1)$ is much less than the value of $P(0)$. So, we can conclude that terms $P(k, 1)$ in Eq. (1) are much less than terms $P(k, 0)$. Then we exclude them in Eq.(1) to obtain an estimation of $P(k, 0)$ denoting by $\hat{P}(k, 0)$. $\hat{P}(k, 0)$ satisfy the following equations:

$$\begin{cases} -(\lambda + \gamma)\hat{P}(0,0) + \mu\hat{P}(1,0) + \gamma = 0, \\ -(\lambda + k\mu + \gamma)\hat{P}(k,0) + \lambda\hat{P}(k-1,0) + (k+1)\mu\hat{P}(k+1,0) = 0, \ k = \overline{1, K-1} \\ -(\lambda + K\mu + \gamma)\hat{P}(K,0) + \lambda\hat{P}(K-1,0) = 0, \end{cases}$$

The system above can be easily solved. Let us numerically compare the exact values $P(0, k)$ and the estimated values $\hat{P}(k, 0)$ for different values of $\sigma$ in example (Fig. 3 a and b), and different values of $\lambda$ (Fig. 3 c and d), where $P(0, k)$ is illustrated as blue curves, $\hat{P}(k, 0)$ is red curves.

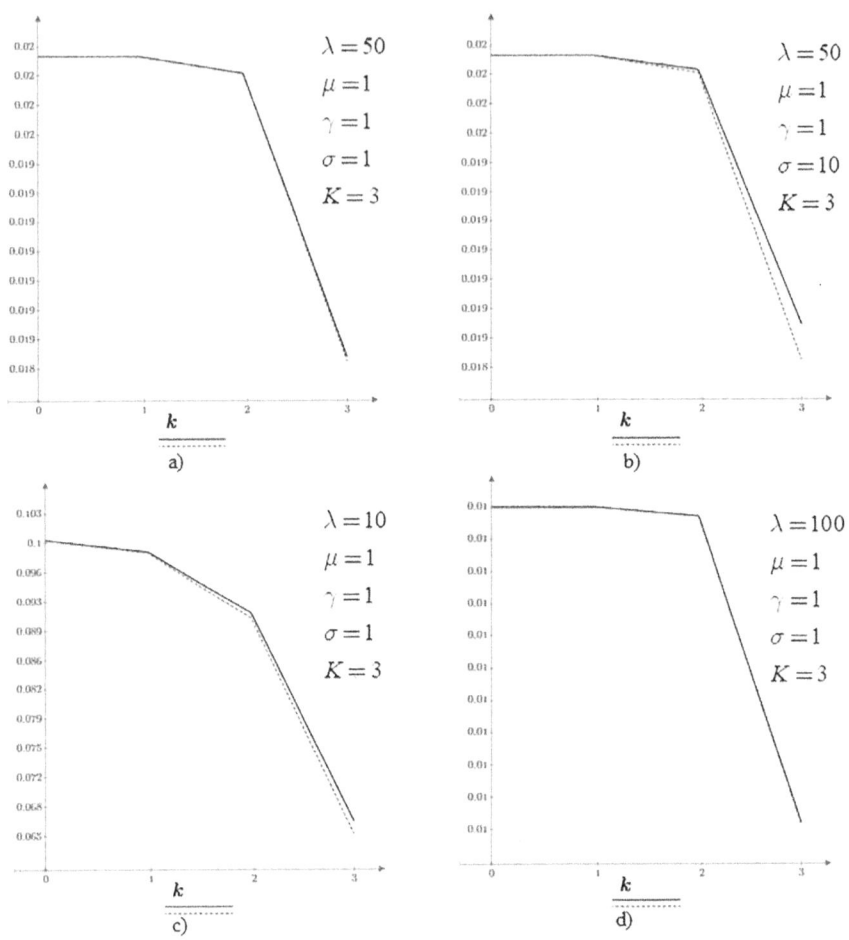

**Fig. 3.** Comparison of $P(0,k)$ and $\hat{P}(k,0)$. (Color figure online)

From example a and d, we can conclude that the estimation has quite high accuracy. Also the accuracy increases with $\lambda$ growing and $\sigma$ decreasing.

### 5.2 Modified Asymptotics

Let us propose the modified asymptotics as

$$PM(i) = \{\hat{P}(0), \text{ if } i = 0; \hat{PA}(i), \text{ otherwise}\},$$

where $\hat{PA}(i)$ is normalized values of the asymptotic exponential distribution taking into account the normalization condition.

To demonstrate the accuracy of the proposed modification, we numerically compare probability distributions $PM(i)$, $PA(i)$ and $P(i)$ for various values of the system parameters. An example of the comparison are presented on Fig. 4. The system parameters are as follows:

$$\mu = 1, \sigma = 1, \gamma = 1, K = 3.$$

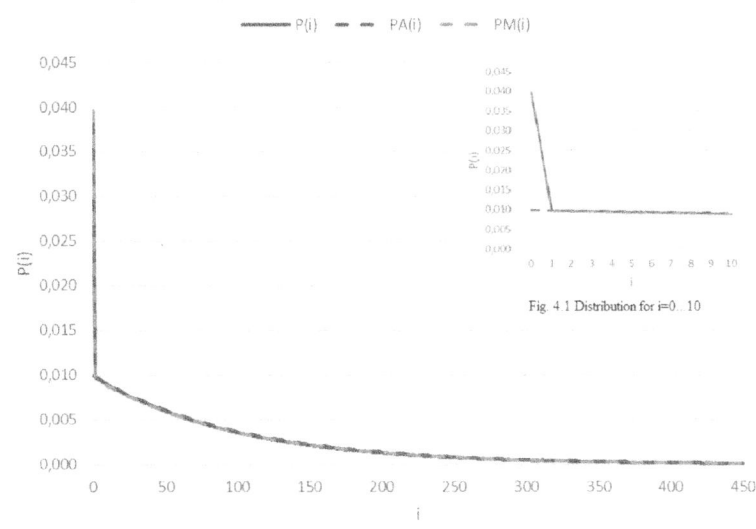

**Fig. 4.** Exact, asymptotic and modified asymptotic distribution for $\lambda = 100$.

The values of Kolmogorov distances for modified asymptotics for different values of $\lambda$ is on Table 2.

**Table 2.** Kolmogorov distances for various values of the parameter $\lambda$

|   | $\lambda = 10$ | $\lambda = 50$ | $\lambda = 100$ | $\lambda = 150$ |
|---|---|---|---|---|
| $\triangle$ | 0.057 | 0.017 | 0.006 | 0.004 |

It follows from the example that the accuracy of the modified approximation becomes higher. Note that we obtain close numerical results for other values of the system parameters.

## 6 Conclusion

The paper presents an asymptotic analysis of a multiserver retrial queueing system with disasters under a high arrival rate. It is proved that the asymptotic probability distribution of the number of calls in the orbit in the considered model is exponential with a parameter $\gamma/\lambda$. The numerical analysis has shown a good accuracy of the asymptotic analysis results for $\lambda \geq 100$. A modified approximation based on separately estimating of zero state probability is also proposed, which results are several times more accurate.

## References

1. Artalejo, J.R., Gomez-Corral, A.: Retrial Queueing Systems, 1st edn. Springer, Heidelberg (2008)
2. Falin, G.I., Templeton, J.: Retrial Queues, 1st edn. Springer, New York (1999)
3. Buyya, R., Broberg, J., Goscinski, A.: Cloud Computing: Principles and Paradigms Solutions, 1st edn. Wiley, New York (2011)
4. Gelenbe, E.: Product-form queueing networks with negative and positive customers. J. Appl. Probab. **28**(3), 656–663 (1991)
5. Do, T.V.: Bibliography on G-networks, negative customers and applications. Ann. Oper. Res. **48**, 205–212 (2011)
6. Shin, Y.W.: Multi-server retrial queue with negative customers and disasters. Queueing Syst. **55**(4), 223–337 (2007)
7. Artalejo, J.R., Gómez-Corral, A.: Computation of the limiting distribution in queueing systems with repeated attempts and disasters. RAIROO Perations Res. **33**(3), 371–382 (1999)
8. Sherif, I.A., Pakkirisamy, R.: Performance analysis of preemptive priority retrial queueing system with disaster under working breakdown services. Symmetry **11**(3), 419–434 (2019)
9. Li, K., Wang, J.: Equilibrium balking strategies in the single-server retrial queue with constant retrial rate and catastrophes. Qual. Technol. Quant. Manage. **18**(2), 156–178 (2020)
10. Neuts, M.F.: Matrix-Geometric Solutions in Stochastic Models: An Algorithmic Approach. Dover Publications, New York (1981)
11. Dudin, A.N., Klimenok, V.I.: Matrix queueing theory. Inf. Process. **2**(2), 173–175 (2002)
12. Nazarov, A.A., Moiseeva, S.P.: Method of Asymptotic Analysis in the Theory of Queuing. NTL Publishing House, Tomsk, Tomsk (2006). (In Russian)

# Analysis of a Batch Arrival Queue with Power Saving Mode

Yuta Sakai and Tuan Phung-Duc(✉)

Institute of Systems and Information Engineering, University of Tsukuba,
1-1-1 Tennodai, Tsukuba, Ibaraki 305-8573, Japan
s2320432@u.tsukuba.ac.jp, tuan@sk.tsukuba.ac.jp

**Abstract.** Queueing models with working vacations have been widely studied for their importance in optimizing both system efficiency and energy usage. This study extends the conventional M/M/1 queueing model with two distinct types of working vacations, each with a unique service rate, to a scenario where customers arrive in batches. Using the probability generating function method, we derive key performance indicators, including steady-state probabilities and the expected number of customers in the system. These indicators serve as the basis for evaluating the model's power-saving performance, providing insights into energy efficiency under various batch size distributions. Additionally, numerical experiments are conducted to examine the average number of customers for batch arrivals following arbitrary distributions. The results confirm that the analytical findings align closely with simulation outcomes, thereby validating the model's accuracy and effectiveness. This model provides a robust framework for evaluating and optimizing queue performance in systems requiring energy-saving measures, with practical applications in server management and energy-sensitive environments.

**Keywords:** Working vacation · Batch arrival · Probability generating function · Arbitrary distribution · Power-saving mode

## 1 Introduction

In recent years, the rise of IoT and AI has led to massive daily data transmissions, raising concerns over the power consumption of IT devices processing this data. To address these challenges, many devices now incorporate power-saving modes. Queueing models with working vacations have been extensively studied as tools to evaluate the performance of such systems, especially those with power-saving features. Recently, models with multiple types of working vacations have enabled the analysis of systems with multi-stage power-saving functionalities.

Foundational work by Servi and Finn [12] on M/M/1 queues with working vacations laid the groundwork for models like the $M^x/M/1$ queue studied by Xu et al. [14]. Baba [1] extended this to multiple vacations in $M^x/M/1$, while Ibe and Isijola [6] examined differentiated vacations. Phung-Duc [11] later introduced a model with two types of vacations, advancing power-saving queueing studies. Complementary research by Doshi [4] surveyed vacation policies, and Gupta and Kumar [5] analyzed retrial queues with state-dependent arrivals and vacations.

Batch arrival queues have also incorporated various vacation policies. Ye [15] studied batch arrivals with multiple vacations, while Manickam and Kalidass [10] explored bulk entry queues with differentiated vacations. Further insights on queues with feedback mechanisms or server setup times come from Krishnamoorthy and Madheswari [8], Li and Tian [9], and Wu and Takagi [13]. Ke and Chu [7] examined bulk arrival queues with N-policy interruptions, and recent studies by Wang et al. [2], Wang and Wang [3], and Liu et al. [14] have enhanced understanding of power-saving queue models.

In this paper, we extend Phung-Duc's model [11], incorporating two types of working vacations into an $M^x/M/1$ queue with batch arrivals. Using the probability generating function method, we derive performance metrics, including steady-state probabilities and expected number of customers. We validate our analytical model through numerical experiments, comparing results with simulations for average customer numbers under arbitrary batch size distributions. This study provides a robust framework for analyzing power-saving queueing systems, highlighting the balance between performance and energy efficiency in multi-stage vacation settings.

## 2 Mathematical Model

This study introduces a model that incorporates two types of working vacations with different service rates into an $M^x/M/1$ queue, where customers arrive in batches. The system includes an infinite queue to hold arriving customers, allowing the server to process each batch sequentially. Customer batches arrive at the server according to a Poisson process with rate $\lambda$. The probability that the batch size $B = i$ is $b_i$, where $i$ represents the number of customers in a batch. Service times follow an exponential distribution with an average of $1/\mu$.

When the server finishes serving a customer and there are no customers in the system, it enters vacation 1. During vacation 1, the service time follows an exponential distribution with an average of $1/\mu_1$, while the vacation duration also follows an exponential distribution with an average time of $1/\gamma_1$. If there are still no customers in the system at the end of vacation 1, the server starts vacation 2. In vacation 2, the service time follows an exponential distribution with an average of $1/\mu_2$, and the vacation duration follows an exponential distribution with an average time of $1/\gamma_2$. Vacation 2 will continue to repeat as long as the system remains empty. Once customers are available at the end of a vacation, the server exits the vacation mode and resumes regular service at rate $\mu$.

The state transition diagram of the model in this study is shown in Fig. 1.

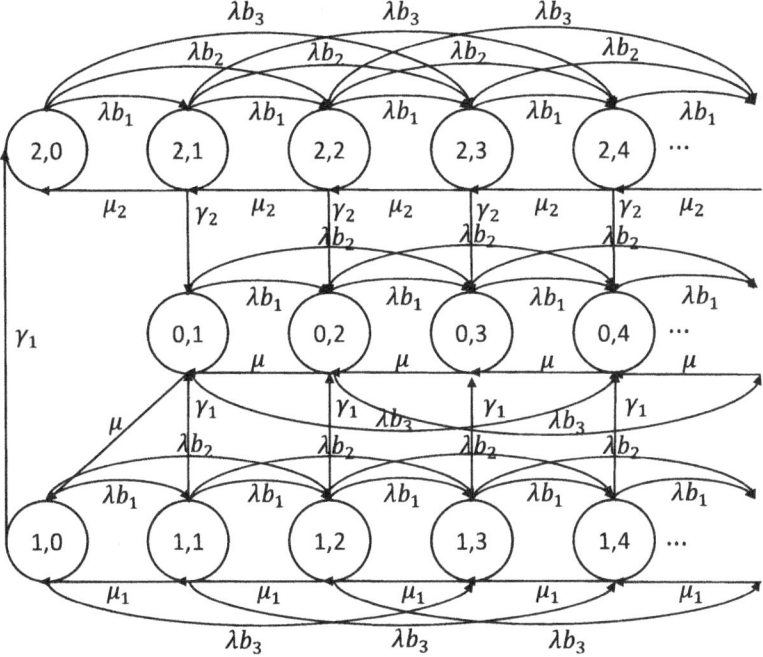

**Fig. 1.** State transition diagram ($b_1 + b_2 + b_3 = 1$).

## 3 Analysis

Let $N(t)$ be the number of customers in the server at time $t$ and $S(t)$ be the state of the server,

$$S(t) = \begin{cases} 0 & \text{if the server in normal,} \\ 1 & \text{if the server in vacation 1,} \\ 2 & \text{if the server in vacation 2.} \end{cases}$$

Then, $(S(t), N(t))$ forms a two-dimensional Markov chain on the state space $\{0, 1, 2\} \times \mathbb{Z}^+$, where $\mathbb{Z}^+ = \{0, 1, 2, \ldots\}$.

The steady-state probability of the system being in state $(i, j)$ is denoted as $\pi_{i,j}$. From the state transition diagram in Fig. 1, the following balance equations can be derived:

$$(\lambda + \mu)\pi_{0,1} = \gamma_1 \pi_{1,1} + \gamma_2 \pi_{2,1} + \mu \pi_{0,2} \qquad (1)$$

$$(\lambda + \mu)\pi_{0,n} = \gamma_1 \pi_{1,n} + \gamma_2 \pi_{2,n} + \mu \pi_{0,n+1} + \lambda \sum_{k=1}^{n-1} b_k \pi_{0,n-k} \quad (n \geq 2) \qquad (2)$$

$$(\lambda + \gamma_1)\pi_{1,0} = \mu_1 \pi_{1,1} + \mu \pi_{0,1} \qquad (3)$$

$$(\lambda + \gamma_1 + \mu_1)\pi_{1,1} = \mu_1 \pi_{1,2} + \lambda b_1 \pi_{1,0} \tag{4}$$

$$(\lambda + \gamma_1 + \mu_1)\pi_{1,n} = \mu_1 \pi_{1,n+1} + \lambda b_n \pi_{1,0} + \lambda \sum_{k=1}^{n-1} b_k \pi_{1,n-k} \quad (n \geq 2) \tag{5}$$

$$\lambda \pi_{2,0} = \mu_2 \pi_{2,1} + \gamma_1 \pi_{1,0} \tag{6}$$

$$(\lambda + \gamma_2 + \mu_2)\pi_{2,1} = \mu_2 \pi_{2,2} + \lambda b_1 \pi_{2,0} \tag{7}$$

$$(\lambda + \gamma_2 + \mu_2)\pi_{2,n} = \mu_2 \pi_{2,n+1} + \lambda b_n \pi_{2,0} + \lambda \sum_{k=1}^{n-1} b_k \pi_{2,n-k} \quad (n \geq 2) \tag{8}$$

The probability generating function of the system in state $(i,j)$ is defined as $\Pi_i(z) = \sum_{n=1}^{\infty} \pi_{i,n} z^n$ for $i = 0, 1, 2$. Additionally, the probability generating function of the batch size $B$ is defined as $B(z) = \sum_{n=1}^{\infty} b_n z^n$.

Multiplying both sides of Eqs. (1) and (2) by $z^n$, summing over $n$, and transforming into generating functions, we obtain:

$$\Pi_0(z) = \frac{z[\gamma_1 \Pi_1(z) + \gamma_2 \Pi_2(z) - \mu \pi_{0,1}]}{(\lambda + \mu)z - \mu - \lambda z B(z)}$$

Similarly, from Eqs. (4) and (5), we obtain:

$$\Pi_1(z) = \frac{z[\lambda \pi_{1,0} B(z) - \mu_1 \pi_{1,1}]}{(\lambda + \gamma_1 + \mu_1)z - \mu_1 - \lambda z B(z)}$$

Since there exists a unique $0 < z_1 < 1$ such that the denominator becomes zero, we find $z_1$. When substituting $z = z_1$, the numerator also becomes zero, yielding the following equation:

$$\pi_{1,1} = \frac{\lambda}{\mu_1} \pi_{1,0} B(z_1)$$

Similarly, from Eqs. (6) and (7), we obtain:

$$\Pi_2(z) = \frac{z[\lambda \pi_{2,0} B(z) - \mu_2 \pi_{2,1}]}{(\lambda + \gamma_2 + \mu_2)z - \mu_2 - \lambda z B(z)}$$

Similarly, we obtain the following equation:

$$\pi_{2,1} = \frac{\lambda}{\mu_2} \pi_{2,0} B(z_2)$$

where $0 < z_2 < 1$ is a unique value such that the denominator of $\Pi_2(z)$ is zero.

The steady-state probability $\lim_{z \to 1} \Pi_0(z)$ is derived as follows:

$$\Pi_0(1) = \lim_{z \to 1} \frac{z[\gamma_1 \Pi_1'(z) + \gamma_2 \Pi_2'(z)] + [\gamma_1 \Pi_1(z) + \gamma_2 \Pi_2(z) - \mu \pi_{0,1}]}{(\lambda + \mu) - \lambda z B'(z) - \lambda B(z)}$$

$$= \frac{\gamma_1 \Pi_1'(1) + \gamma_2 \Pi_2'(1) + [\gamma_1 \Pi_1(1) + \gamma_2 \Pi_2(1) - \mu \pi_{0,1}]}{\mu - \lambda B'(1)}$$

We derive the values of $\Pi'_1(1)$, $\Pi'_2(1)$, $\Pi_1(1)$, and $\Pi_2(1)$.

$$\Pi'_1(1) = \frac{\lambda B'(1)(\gamma_1 \pi_{1,0} + \lambda \pi_{1,0} - \mu_1 \pi_{1,1}) - \mu_1(\lambda \pi_{1,0} - \mu_1 \pi_{1,1})}{\gamma_1^2}$$

$$\Pi'_2(1) = \frac{\lambda B'(1)(\gamma_2 \pi_{2,0} + \lambda \pi_{2,0} - \mu_2 \pi_{2,1}) - \mu_2(\lambda \pi_{2,0} - \mu_2 \pi_{2,1})}{\gamma_2^2}$$

$$\Pi_1(1) = \frac{\lambda \pi_{1,0}(1 - B(z_1))}{\gamma_1}, \quad \Pi_2(1) = \frac{\lambda \pi_{2,0}(1 - B(z_2))}{\gamma_2}$$

From the total probability formula, we obtain:

$$\pi_{1,0} + \Pi_1(1) + \pi_{2,0} + \Pi_2(1) + \Pi_0(1) = 1$$

From the state transition diagram, the following equations are obtained:

$$\gamma_1 \pi_{1,0} = \gamma_2 \Pi_2(1), \quad \mu \pi_{0,1} = \gamma_1 \pi_{1,0} + \gamma_1 \Pi_1(1)$$

The unknowns $\pi_{1,0}$, $\pi_{2,0}$, and $\pi_{0,1}$ can be solved from the above three equations as follows:

$$A = \mu \gamma_1 (\gamma_1 + \gamma_2) + \lambda \gamma_2 (1 - B(z_1))(\mu - \mu_1) - \gamma_1^2 \mu_2$$

$$\pi_{1,0} = \frac{\lambda \gamma_1 \gamma_2 (\mu - \lambda B'(1))(1 - B(z_2))}{\lambda A(1 - B(z_2)) + \mu \gamma_1^2 \gamma_2}$$

$$\pi_{2,0} = \frac{\gamma_1^2 \gamma_2 (\mu - \lambda B'(1))}{\lambda A(1 - B(z_2)) + \mu \gamma_1^2 \gamma_2}$$

$$\pi_{0,1} = \frac{\lambda \gamma_1 \gamma_2 (\mu - \lambda B'(1))(1 - B(z_2))(\gamma_1 + \lambda(1 - B(z_1)))}{\lambda \mu A(1 - B(z_2)) + \mu^2 \gamma_1^2 \gamma_2}$$

Let $\Pi(z)$ be the generating function representing the number of customers in the system. $\Pi(z)$ is derived from the following equation:

$$\Pi(z) = \Pi_0(z) + \Pi_1(z) + \Pi_2(z) + \pi_{1,0} + \pi_{2,0}$$

The average number of customers in the system $E[L]$ is obtained from $\Pi'(1)$:

$$E[L] = \Pi'(1) = \Pi'_0(1) + \Pi'_1(1) + \Pi'_2(1)$$

The average waiting time in the system, $E[W]$, is obtained from the average number of customers in the system $E[L]$ using Little's law:

$$E[W] = \frac{E[L]}{\lambda}$$

## 4 Power-Saving Optimization

In this section, we define the total cost $TC$ of the system and aim to determine the optimal service rate $\mu$ that minimizes $TC$. The total cost function $TC$ takes into account various cost elements incurred per unit time and is defined as follows:

- $C_1$: Holding cost per customer in the system.
- $C_2$: Waiting cost per customer awaiting service.
- $C_3$: Operational cost when the server is in the normal state.
- $C_4$: Power-saving cost when the server is in vacation 1.
- $C_5$: Power-saving cost when the server is in vacation 2.
- $C_6$: Service cost associated with the service rate $\mu$ in normal state.

The total cost function is expressed as:

$$TC = C_1 E(L) + C_2 E(W) + C_3 P_N + C_4 P_{v1} + C_5 P_{v2} + C_6 \mu$$

where:

- $E(L)$: Expected number of customers in the system.
- $E(W)$: Expected waiting time of customers.
- $P_N$: Probability that the server is in the normal busy state.
- $P_{v1}$: Probability that the server is in vacation 1.
- $P_{v2}$: Probability that the server is in vacation 2.

To determine the optimal service rate $\mu$ that minimizes the total cost $TC$, we employ a direct search method. This approach systematically explores the defined range of $\mu$ and identifies the service rate value that achieves the minimum total cost. Through this direct search optimization, it is possible to balance service efficiency and power-saving costs under different operational conditions.

## 5 Numerical Examples

In this chapter, we present the behavior of the steady-state probabilities in our model with respect to the traffic density $\rho$ and the behavior of the average number of customers in the system $E[L]$ with respect to the arrival rate $\lambda$ when varying parameters. Additionally, we conduct numerical experiments to determine the optimal service rate that minimizes the total system cost $TC$ defined in Sect. 4.

### 5.1 Comparison of Analytical and Simulation Results for Steady-State Probability

In this subsection, we examine the behavior of steady-state probabilities when varying $\gamma_2$. Furthermore, we verify the accuracy of the model by comparing the analytical results with the simulation results for $\pi_{1,0}$ and $\pi_{2,0}$.

The experiments were conducted using the following parameter settings, except for the varying parameters: $\mu = 3.0$, $\mu_1 = 0.1$, $\mu_2 = 0.05$, $\gamma_1 = 1.0$, $B'(1) = 2.0$, and the batch size distribution follows a geometric distribution. Additionally, in this numerical experiment, we analyze the behavior of $\pi_{1,0}$ when changing the traffic density $\rho$.

Figure 2 illustrates the behavior of $\pi_{1,0}$ when $\gamma_2$ is varied as $0.1, 0.25, 0.5, 1.0$, indicating that a larger $\gamma_2$ results in a higher $\pi_{1,0}$.

Similarly, Fig. 3 illustrates the behavior of $\pi_{2,0}$ when $\gamma_2$ is varied as $0.1, 0.25, 0.5, 1.0$, indicating that a larger $\gamma_2$ also results in a higher $\pi_{2,0}$. Additionally, the consistency between the analytical and simulation results confirms the accuracy of the proposed model.

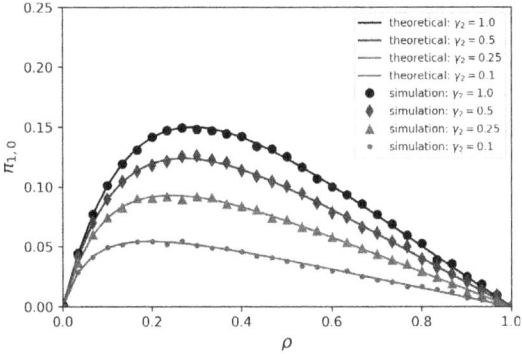

**Fig. 2.** Steady-State Probability $\pi_{1,0}$ with varying $\gamma_2$ values.

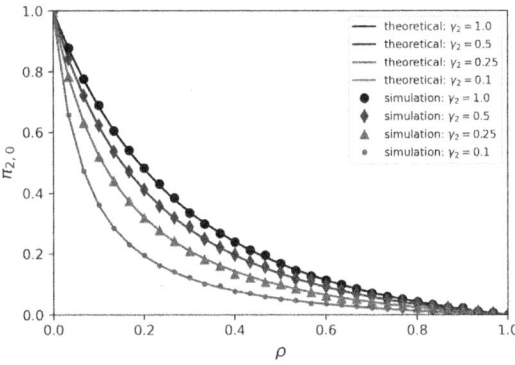

**Fig. 3.** Steady-State Probability $\pi_{2,0}$ with varying $\gamma_2$ values.

## 5.2 Comparison of Analytical and Simulation Results for the Average Number of Customers in the System $E[L]$

In this subsection, we examine the behavior of the average number of customers in the system $E[L]$ when varying different parameters. Furthermore, we verify the accuracy of the model by comparing the analytical results with the simulation results for $E[L]$.

Figure 4 illustrates the behavior of $E[L]$ when varying $\gamma_1$, $\gamma_2$, $B'(1)$, and the batch size distribution.

The experiments were conducted using the following baseline parameter settings, except for the varying parameters: $\mu = 3.0$, $\mu_1 = 0.1$, $\mu_2 = 0.05$, $\gamma_1 = 1.0$, $\gamma_2 = 1.0$, $B'(1) = 2.0$, and the batch size distribution follows a geometric distribution. Additionally, in this numerical experiment, we analyze the behavior of $E[L]$ when changing the arrival rate $\lambda$.

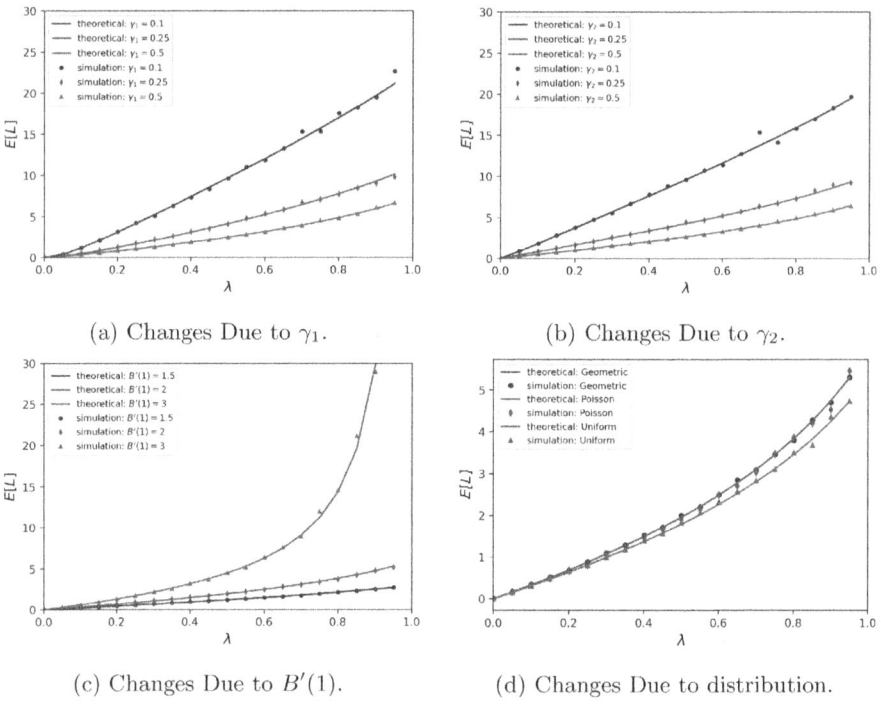

(a) Changes Due to $\gamma_1$.

(b) Changes Due to $\gamma_2$.

(c) Changes Due to $B'(1)$.

(d) Changes Due to distribution.

**Fig. 4.** Comparison Results of the Average Number of Customers in the System $E[L]$

Figure 4(a) shows the behavior of $E[L]$ when $\gamma_1$ is varied as $0.1, 0.25, 0.5$, revealing that a smaller $\gamma_1$ leads to a larger average number of customers in the system. Similarly, Fig. 4(b) illustrates the behavior of $E[L]$ when $\gamma_2$ is varied as $0.1, 0.25, 0.5$, indicating that a smaller $\gamma_2$ also results in a higher $E[L]$.

Figure 4(c) presents the behavior of $E[L]$ when the mean batch size $B'(1)$ is set to $1.5, 2.0, 3.0$. The results confirm that a larger $B'(1)$ leads to an increase in $E[L]$. Particularly, when $B'(1) = 3.0$, the traffic density $\rho = \frac{\lambda B'(1)}{\mu}$ approaches 1, leading to a noticeable increase in $E[L]$.

Furthermore, Fig. 4(d) shows $E[L]$ when varying the batch size distribution. The experiments were conducted using geometric, Poisson, and uniform distributions. The results indicate that when the batch size follows a uniform distribution, $E[L]$ is slightly smaller compared to the other distributions. This is likely due to the fact that the variance of the uniform distribution is smaller than that of the other distributions. Additionally, the consistency between the analytical and simulation results confirms the accuracy of the proposed model.

These results demonstrate that the proposed model accurately represents system behavior under different traffic conditions and power-saving settings. Moreover, they provide useful insights for evaluating the impact of power-saving mode settings on system performance.

### 5.3 Optimal Service Rate Evaluation with $\gamma_1$ and $\gamma_2$ Variations

In this experiment, we determined the optimal service rate $\mu$ by varying $\gamma_1$ and $\gamma_2$ while applying a direct search method to minimize the total cost $TC$. The parameters were set as follows: the arrival rate $\lambda = 1.0$, with service rates during vacations $\mu_1 = 0.1$ and $\mu_2 = 0.05$, and $B'(1) = 1.5$ under a geometric batch size distribution. The cost coefficients were configured as $C_1 = 20$, $C_2 = 15$, $C_3 = 10$, $C_4 = 8$, $C_5 = 5$, and $C_6 = 3$, with the search range for $\mu$ defined from 2 to 50.

First, we examined the impact of varying $\gamma_1$ on the optimal service rate $\mu$. Figure 5 shows the relationship between $TC$ and $\gamma_1$, illustrating how adjusting $\gamma_1$ can help achieve cost savings by influencing the balance between service rate $\mu$ and power-saving modes.Table 1 presents the numerical values of $E[L]$ and $TC$ when varying $\mu$ and $\gamma_1$. When $\gamma_1 = 0.25, 0.5, 1.0$, the optimal service rate was found to be $\mu = 6.59735, 6.63786, 6.68682$, yielding a minimized total cost of $TC = 227.647, 130.44, 93.4197$. This finding underscores the significant influence of $\gamma_1$ on the system's cost efficiency, as optimal tuning of $\mu$ can directly reduce $TC$.

Additionally, we explored the effect of varying $\gamma_2$ on $TC$ with similar parameter configurations. Figure 6 shows the relationship between $TC$ and $\gamma_2$, illustrating how adjusting $\gamma_2$ can help achieve cost savings by influencing the balance between service rate $\mu$ and power-saving modes. Table 2 presents the numerical values of $E[L]$ and $TC$ when varying $\mu$ and $\gamma_2$. When $\gamma_2 = 1.5, 2.0, 5.0$, the optimal service rate was found to be $\mu = 6.68173, 6.67856, 6.67165$, yielding a minimized total cost of $TC = 85.604, 82.3581, 77.9236$. This finding underscores the significant influence of $\gamma_2$ on the system's cost efficiency, as optimal tuning of $\mu$ can directly reduce $TC$.These experiments collectively demonstrate the importance of fine-tuning both $\gamma_1$ and $\gamma_2$ for achieving minimized costs in various operational modes.

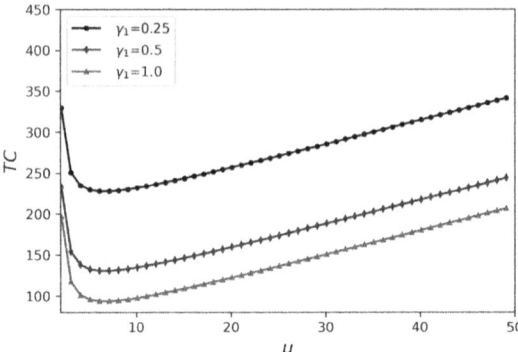

**Fig. 5.** Total cost $TC$ with varying $\gamma_1$ values.

**Table 1.** The result of $\mu$ against the cost output for different points of $\gamma_1$

| $\mu$ | $\gamma_1 = 0.25$ | | $\gamma_1 = 0.5$ | | $\gamma_1 = 1$ | |
|---|---|---|---|---|---|---|
| | $E[L]$ | $TC$ | $E[L]$ | $TC$ | $E[L]$ | $TC$ |
| 4.5 | 6.00315 | 232.061 | 3.23922 | 135.002 | 2.20082 | 98.1638 |
| 5.0 | 5.899 | 229.841 | 3.13421 | 132.735 | 2.09492 | 95.8411 |
| 5.5 | 5.82109 | 228.553 | 3.05559 | 131.409 | 2.01558 | 94.4692 |
| 6.0 | 5.76062 | 227.886 | 2.99454 | 130.71 | 1.95393 | 93.7323 |
| 6.5 | 5.71233 | 227.653 | 2.94576 | 130.451 | 1.90464 | 93.4405 |
| 7.0 | 5.67289 | 227.736 | 2.90589 | 130.511 | 1.86435 | 93.473 |
| 7.5 | 5.64007 | 228.055 | 2.87271 | 130.811 | 1.83079 | 93.7488 |

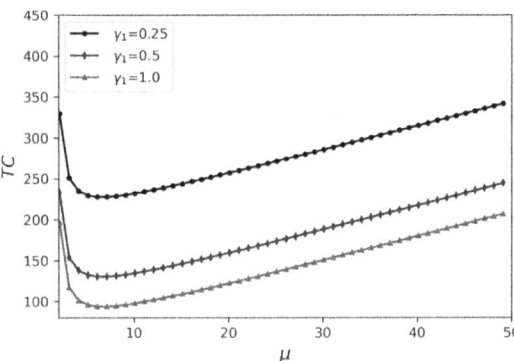

**Fig. 6.** Total cost $TC$ with varying $\gamma_2$ values.

**Table 2.** The result of $\mu$ against the cost output for different points of $\gamma_2$

| $\mu$ | $\gamma_2 = 1.5$ | | $\gamma_2 = 2.0$ | | $\gamma_2 = 5.0$ | |
|---|---|---|---|---|---|---|
| | $E[L]$ | $TC$ | $E[L]$ | $TC$ | $E[L]$ | $TC$ |
| 4.5 | 1.97433 | 90.329 | 1.87973 | 87.0713 | 1.74913 | 82.6108 |
| 5.0 | 1.86847 | 88.0122 | 1.7739 | 84.7581 | 1.64338 | 80.3056 |
| 5.5 | 1.78916 | 86.6451 | 1.69462 | 83.394 | 1.56416 | 78.948 |
| 6.0 | 1.72753 | 85.9121 | 1.63301 | 82.6635 | 1.5026 | 78.2229 |
| 6.5 | 1.67827 | 85.6237 | 1.58376 | 82.3771 | 1.4534 | 77.9412 |
| 7.0 | 1.63799 | 85.659 | 1.5435 | 82.4142 | 1.41317 | 77.9821 |
| 7.5 | 1.60445 | 85.9373 | 1.50997 | 82.694 | 1.37967 | 78.2654 |

# 6 Conclusion

In this study, we analyzed a queuing model that integrates two types of working vacations with distinct service rates into an $M^x/M/1$ system with batch arrivals. The model incorporates energy-saving considerations, aiming to minimize the total cost $TC$ by balancing service efficiency with power-saving strategies. We derived the steady-state probabilities and examined the impact of traffic density on the total cost of the system.

Through a series of numerical experiments, we determined the optimal service rate $\mu$ that minimizes $TC$ by considering variations in both $\gamma_1$ and $\gamma_2$. The results showed that adjusting $\gamma_1$ and $\gamma_2$ significantly impacts the optimal $\mu$ value, demonstrating that targeted tuning of service rates is crucial for achieving cost efficiency. Additionally, the alignment between analytical and simulation results for $TC$ and the average number of customers $E[N]$ confirmed the reliability of the model under different traffic conditions and batch arrival distributions, especially when following a geometric distribution.

The study highlighted that precise parameter adjustments in $\mu$, $\gamma_1$, and $\gamma_2$ can effectively manage both customer waiting times and power-saving costs, emphasizing the importance of optimal parameter selection to balance service performance with energy efficiency.

For future research, extending this model to an $M^x/M/C$ system with multiple servers and incorporating more complex power-saving structures could provide further insights into cost optimization strategies for multi-server queuing systems, offering enhanced scalability and practical applicability in real-world service environments.

# References

1. Baba, Y.: The $M^x/M/1$ queue with multiple working vacations. Am. J. Oper. Res. **2**(2), 217–224 (2012)
2. Bouchentouf, A., Medjahri, N.: Performance evaluation of feedback queue with differentiated vacations. Comput. Ind. Eng. **126**, 392–403 (2018)

3. Choudhury, G., Deka, S.: An M/G/1 retrial queue with Bernoulli vacation schedule and optional reservice. Appl. Math. Model. **36**(6), 2541–2553 (2012)
4. Doshi, B.T.: Queueing systems with vacations - a survey. Queueing Syst. **1**, 29–66 (1986)
5. Gupta, U., Kumar, V.: Retrial queue with differentiated vacations and state-dependent arrivals. Stoch. Model. **35**(2), 152–173 (2019)
6. Ibe, O.C., Isijola, O.A.: Markovian bulk arrival queues with differentiated vacations. J. Appl. Math. Stoch. Anal. Article ID 158247, 6 (2014)
7. Ke, J.C., Chu, C.P.: Bulk arrival queueing systems with N-policy and vacation interruption. Appl. Math. Model. **29**(6), 563–580 (2005)
8. Krishnamoorthy, A., Madheswari, R.P.: An M/M/1 feedback queue with working vacations. Math. Comput. Model. **47**(9), 1211–1218 (2008)
9. Li, H., Tian, N.: Analysis of an M/G/1 queue with batch arrivals and setup times. J. Ind. Manag. Optim. **4**(3), 413–430 (2008)
10. Manickam, V., Kalidass, K.: Optimality of bulk entry queue with differentiated hiatuses. Oper. Res. Decis. **32**(2), 138–143 (2022)
11. Phung-Duc, T.: Single-server systems with power-saving modes. In: Gribaudo, M., Manini, D., Remke, A. (eds.) ASMTA 2015. LNCS, vol. 9081, pp. 158–172. Springer, Cham (2015). https://doi.org/10.1007/978-3-319-18579-8_12
12. Servi, L.D., Finn, S.G.: M/M/1 queue with working vacations. Oper. Res. Lett. **28**(5), 223–229 (2002)
13. Wu, J., Takagi, H.: Discrete-time queues with batch arrivals and general service times. Stoch. Model. **22**(2), 185–208 (2006)
14. Xu, X.-L., Zhang, Z.-J., Tian, N.-S.: Analysis for the $M^x/M/1$ working vacation queue. Int. J. Inf. Manag. Sci. **20**(3), 379–394 (2009)
15. Ye, Q.: Batch arrival queue with multiple vacation policies. J. Ind. Manag. Optim. **14**(3), 132–150 (2021)

# Lumping and Numerical Analysis for Multi-Server Job Model

Sergey Astafiev[(✉)]

IAMR KarRC RAS, 11, Pushkinskaya str., 185035 Petrozavodsk, Russia
seryymail@mail.ru

**Abstract.** The present paper studies a model of a multi-server system operating at multiple processing speeds with random transitions between speeds as customers arrive and depart. This model is also known as the Cluster Model or the Multi-server Job Model. The possibility of reducing the phase space size of the model is substantiated. An algorithm is proposed for constructing an infinitesimal generator matrix to describe the behavior of this model with a reduced phase space. An R-language package implementing this approach, along with an example of its usage, is presented.

**Keywords:** Supercomputer Model · Matrix-Analytic Method · Quasi-Birth-Death Process · Numerical Solution

## 1 Introduction

Most modern computing systems have multiple processors, and their incoming jobs can utilize several cores simultaneously. Supercomputers are a key example of such systems, widely used for tasks requiring substantial computational resources and characterized by significant energy consumption [2]. Various energy-saving methods have been proposed, one of which is frequency scaling. Frequency scaling allows the computing cluster (i.e., the system) to switch between different operating speeds, which affects both performance and power consumption. The present paper discusses the simultaneous frequency switching between predefined speed modes at arrival and departure events, each occurring with certain probabilities. Frequency scaling is one method for balancing performance with power consumption [10].

Similar models have previously emerged in various contexts under different names, such as the multi-resource queue [17], the cluster model [18], and the multi-server job model [9]. One of the earliest studies in this area is [14]. A significant portion of early research focused on establishing the conditions for system stability, as in [22] and [1]. However, the stationary distribution was typically computed through simulation [21] or, for a small number of servers, by matrix-analytic methods [4,21].

---

The author thanks Alexander Rumyantsev for constructive criticism of the draft version of this article.

In [1], an explicit form for the rates of transitions between states was derived, enabling the numerical calculation of the stationary distribution. To the best of our knowledge, this is the most recent approach; however, it is limited by the curse of dimensionality.

The present paper focuses on a method, somewhat similar to state space lumping [23], for reducing the generally large phase space of the Quasi-Birth-Death (QBD) continuous-time Markov chain that models the behavior of a supercomputer [21]. This reduction enables the computation of the stationary distribution and related characteristics, such as the distribution of the number of customers and the mean power consumption. Our method allows for the algorithmic construction of an infinitesimal generator matrix with a reduced phase space in a single step.

The structure of the paper is as follows. Section 2 describes model and its parameters. Section 3 presents the well-known theory for computing the stationary distribution of similar systems. In Sect. 4, we prove the theoretical feasibility of reducing the phase space size. Section 5 provides a description of the algorithms for matrix construction. In Sect. 6, we describe the simulation model used for validation. Section 7 discusses R implementation of the proposed algorithm and includes a brief numerical example. The article concludes with a discussion of potential future work and a summary of key findings.

## 2 Model

We briefly describe the Multi-server Job Model (MJM) [21] below. The system consists of $c$ identical servers and a general First-Come, First-Served (FCFS) queue. Arrivals follow a Poisson process with rate $\lambda$. Each $k$-th customer is characterized by two parameters:

- A random number of required servers $N_k$ (class), drawn from a discrete distribution $\mathscr{P} = \{(j, p_j) : 1 \leqslant j \leqslant c, p_j = \mathbb{P}(N_k = j)\}$. Customer $k$ is served by exactly $N_k$ servers, which are seized and released simultaneously.
- A random job size $S_k$ which is exponentially distributed with rate $\mu_{N_k}$.

The system operates in one of $n$ regimes (or modes) which affect performance and power demand. In regime $i$, all servers run at speed (frequency) $f_i$, where $1 \leqslant i \leqslant n$. Transitions between regimes occur upon customer arrivals or departures, with transition probabilities $p_{i,j}^{(a)}$ and $p_{i,j}^{(d)}$, respectively, for a switch from regime $i$ to $j$. Consequently, the system speed $f(t)$ at time $t$ is a piecewise constant function and the service time $\tau_k$ of customer $k$ is determined as a solution of the equation

$$\int_{t_k}^{t_k+\tau_k} f(t)dt = S_k, \tag{1}$$

where $t_k$ is the time at which customer $k$ begins service. Since the service times and interarrival times in this system are exponentially distributed, the system can be considered as a Quasi-Birth-Death (QBD) process, which is described in the next section.

## 3 Quasi-Birth-Death Process

Models similar to the one presented in Sect. 2 can be analyzed using the well-known Quasi-Birth-Death (QBD) processes [14], since all transitions occur at exponentially distributed times and customers cannot arrive in groups. Consider the random process $X(t) = \{x(t), \phi(t)\}_{t \geq 0}$ in continuous time $t \in \mathbb{R}_0$ where:

- $x(t)$ is the *level* (enumerable), i.e., the number of customers in the system;
- $\phi(t)$ is the *phase* (finite), i.e., the internal state of the system. We call the set of all possible states of $\phi(t)$ as the phase space. The phase space depends on the current level $x(t)$ if $x(t) < c$. More details can be found in Sect. 4 below.

For further information on QBD processes, see e.g., [5, pp. 117-132]

The finiteness of the phase space for any level is ensured by the following fact. Let all customers in the system be enumerated using non-negative integers $\mathbb{N}^* = \{1, 2, 3, ...\}$ in the order of their arrival to the system. Then, the distribution of classes is typical for customers with numbers greater than $c$. For more details, see [14,22].

The infinitesimal generator matrix of the QBD process has a block tridiagonal structure:

$$Q = \begin{pmatrix} A_0^{(0)} & A_0^{(1)} & \mathbb{O} & \cdots & \mathbb{O} & \mathbb{O} & \mathbb{O} & \mathbb{O} & \cdots \\ A_1^{(-1)} & A_1^{(0)} & A_1^{(1)} & \ddots & \mathbb{O} & \mathbb{O} & \mathbb{O} & \mathbb{O} & \cdots \\ \mathbb{O} & A_2^{(-1)} & A_2^{(0)} & \ddots & \mathbb{O} & \mathbb{O} & \mathbb{O} & \mathbb{O} & \cdots \\ \vdots & \ddots & \ddots & \ddots & \ddots & \cdots & \cdots & \cdots \\ \mathbb{O} & \cdots & \mathbb{O} & A_c^{(-1)} & A_c^{(0)} & A^{(1)} & \mathbb{O} & \mathbb{O} & \cdots \\ \mathbb{O} & \cdots & \mathbb{O} & \mathbb{O} & A^{(-1)} & A^{(0)} & A^{(1)} & \mathbb{O} & \cdots \\ \vdots & \vdots & \vdots & \vdots & \ddots & \ddots & \ddots & \ddots & \cdots \end{pmatrix}, \quad (2)$$

where $\mathbb{O}$ represents zero matrices, and the matrices $A_*^{(1)}$, $A_*^{(-1)}$, and $A_*^{(0)}$ are the rate matrices for the transitions between levels and phases.

The notation of the matrices $A_j^{(k)}$ is described as follows. The index $k$ represents increment of the level after transition. Specifically, if $k = 1$, the matrix describes the transition to one level up; if $k = -1$ it corresponds to a transition to one level down; and if $k = 0$, it corresponds to a transition within the same level. The index $j$ represents the level from which the transition occurs. These indices are no longer needed for levels greater than $c$ as the transition sub-matrices between levels become identical. There is no matrix $A_0^{(-1)}$ because level 0 is the lowest. The rows of the sub-matrices correspond to the phase from which the transition occurs, and the columns correspond to the phase into which the transition occurs.

### 3.1 Stationary Distribution of QBD Process

The stationary distribution $\boldsymbol{\pi}$ of the QBD process is expressed as the solution to an infinite system of linear equations $\boldsymbol{\pi} Q = \boldsymbol{0}$ with the normalization condition

$\sum \pi_i = 1$. It can be calculated using the well-known matrix-analytic method [11,16]. A brief description of this method is provided below.

The stationary distributions of levels with numbers less than $c+1$ are expressed as the solution to the following system of equations:

$$(\pi_0, \pi_1, ..., \pi_c) \begin{pmatrix} A_0^{(0)} & A_0^{(1)} & \mathbb{O} & \cdots & & \mathbb{O} \\ A_1^{(-1)} & A_1^{(0)} & A_1^{(1)} & \ddots & & \mathbb{O} \\ \mathbb{O} & A_2^{(-1)} & A_2^{(0)} & \ddots & & \mathbb{O} \\ \vdots & \ddots & \ddots & \ddots & & \ddots \\ \mathbb{O} & \cdots & \mathbb{O} & A_c^{(-1)} & A_c^{(0)} + RA^{(-1)} \end{pmatrix} = \mathbf{0}, \quad (3)$$

$$\pi_0 \mathbf{1}^T + \pi_1 \mathbf{1}^T + \pi_2 \mathbf{1}^T + ... + \pi_{c-1} \mathbf{1}^T + \pi_c (I - R)^{-1} \mathbf{1}^T = 1$$

where $\mathbf{1}$ are a vectors of ones and $\mathbf{0}$ is a vector of zeroes. The stationary distribution at the model levels with a number greater than $c$ can be expressed in matrix-geometric form as:

$$\pi_{c+i} = \pi_c R^i, i > 0, \quad (4)$$

where the matrix $R$ is minimal (in the spectral sense) non-negative (componentwise) root of matrix equation:

$$R^2 A^{(-1)} + RA^{(0)} + A^{(1)} = \mathbb{O}. \quad (5)$$

More details about method can be found in [11,16].

An analytical solution to this equation is usually only possible in the case of a small number of phases and was obtained for such models (for particular cases) in [4,21]. Since the size of the rate matrices is not known in advance because it depends on system parameters, the present paper uses the logarithmic reduction method [3, pp. 188-189], which allows us to find the matrix $R$ numerically.

The same physical system can be described using different Markov chains, depending on the goals of the researcher. One of the main challenges in analyzing QBD processes is the curse of dimensionality. Even if the matrices $A^{(1)}$, $A^{(-1)}$ and $A^{(0)}$ are sparse, this does not guarantee that the matrix $R$ will also be sparse. The size of the matrices $A^{(1)}$, $A^{(-1)}$ and $A^{(0)}$ and, consequently, the matrix $R$ depends on the method used to describe the process phases. There are methods for reducing the phase space of CTMCs, such as lumpability [13,23]. Our method for phase description and its reduction is discussed in the next section.

## 4 Description of the Phase

There are several ways to describe the phase $\phi(t)$ of the process $X(t)$. In the following, we use the notation in the form of tuples to describe the phase. Since the set of possible phase states depends on the level $x(t)$ and its cardinality increases with the level (from 0 to $c$), in this section, we focus on the set of phase states at the level $x(t) = c$.

Consider the pair $(\mathfrak{m}, \mathfrak{u})$ first, where $\mathfrak{m}$ is a current regime and $\mathfrak{u}$ is a class for the first $c$ customers in the system. This method is used to explicitly express the stationary distribution in [21] for small systems ($c \leqslant 2$). The size of the phase space is

$$|\{(\mathfrak{m}, \mathfrak{u}) : \mathfrak{m} \in 1..n, \mathfrak{u} \in \{1..c\}^{\times c}\}| = nc^c, \tag{6}$$

where $X^{\times c}$ a is $c$-ary Cartesian power of the set $X$, for levels greater than or equal to $c$. This is one of the simplest methods, and it makes it quite easy to obtain the transition matrices explicitly (e.g., see [21]).

The present paper focuses on the pair $(\mathfrak{m}, \mathfrak{s})$ where $\mathfrak{m}$ is a current regime, $\mathfrak{s}$ is a vector containing the number of customers being served in each class. This approach was first used for a single-speed version of the MJM in [22] for obtain the stability condition. Its easy to see that

$$|\{(\mathfrak{m}, \mathfrak{s}) : \mathfrak{m} \in 1..n, \mathfrak{s} \in \{\{0..c\}^{\times c} : 0 < \sum i\mathfrak{s}_i \leqslant c\}\}| \leq nc^c, \tag{7}$$

but an explicit expression for the size of this set is unknown (see [19, A026905] and [19, A000041]). As of the time of writing, to the best of our knowledge, this is the description of the phase with the smallest set of states for which it was possible to obtain transition matrices.

A comparison of the cardinalities for the first ten values of $c$ can be found in Sect. A for different phase descriptions. Further in this section theoretical foundations are described that allow one to move from $(\mathfrak{m}, \mathfrak{u})$ description to $(\mathfrak{m}, \mathfrak{s})$.

### 4.1 Reducing Phase Description

Consider the first type of description, $(\mathfrak{m}, \mathfrak{u})$. The article [14] demonstrates that this description can be reduced: the distribution of customer classes is the same for every customer in the queue, except for the first one. We denote this state description as $(\mathfrak{m}, \hat{\mathfrak{u}}, \mathfrak{q})$ where $\mathfrak{m}$ is a current regime, $\hat{\mathfrak{u}}$ is a list of currently served customers and $\mathfrak{q}$ is the first customer in queue. Note that the vector $\hat{\mathfrak{u}}$ has a variable length, unlike the vector $\mathfrak{u}$.

Due to the memoryless property [20, pp. 191] of the exponential distribution, the order of the vector $\hat{\mathfrak{u}}$ is not important. This allows us to move to a set $\{(\mathfrak{m}, \bar{\mathfrak{u}}, \mathfrak{q})\} \subseteq \{(\mathfrak{m}, \hat{\mathfrak{u}}, \mathfrak{q})\}$ where the vectors $\bar{u}$ are unique in terms of the permutation of elements for all fixed $\mathfrak{m}$ and $\mathfrak{q}$. However, for greater convenience, we can use the following property of the exponential distribution: let $\xi_1 \sim Exp(\lambda_1)$ and $\xi_2 \sim Exp(\lambda_2)$, then $\min(\xi_1, \xi_2) \sim Exp(\lambda_1 + \lambda_2)$ (see e.g. [20, p. 193]). This property allows us to group individual customers of the same class together, resulting in the form $(\mathfrak{m}, \mathfrak{s}, \mathfrak{q})$, where $\mathfrak{m}$, $\mathfrak{s}$ and $\mathfrak{q}$ were described earlier, and $|\{(\mathfrak{m}, \bar{\mathfrak{u}}, \mathfrak{q})\}| = |\{(\mathfrak{m}, \mathfrak{s}, \mathfrak{q})\}|$. Note that $\mathfrak{q}$ has distribution $\mathscr{P}$, when all servers are busy. We use notation $\mathfrak{q} = 0$ in this case, and if the queue is empty. This phase description was used in [8] for simplified saturated system. Further reduction is possible by removing the $\mathfrak{q}$ component, which is based on the following theorem:

**Theorem 1.** *Let the initial state $X(0)$ of the Markov chain $X(\tau) = (x(\tau), \phi(\tau))$, where $\phi(\tau) \in \{(\mathfrak{m}, \mathfrak{s}, \mathfrak{q})\}$ be such that $\phi(0) = (m(0), s(0), 0)$ (i.e. $q(0) = 0$). Let the number of idle servers at time $t$ be $v(t) = c - \sum_{i=1}^{c} i s(t)$. If queue is not empty at time $t$ then $\mathbb{P}(q(t) = i|s(t)) = \mathbb{P}(q(t) = i|v(t)) = \frac{p_i}{w}$ for all $v(t) < i \leq c$ where $w = \sum_{i=v(t)+1}^{c} p_i$.*

*Proof.* To prove Theorem 1 we consider the changes in the component $q(t)$ and the number of idle servers step by step, when a customer leaves or attempts to start service. We can analyze the behavior of the system only at these moments, since only then do the components $q(t)$ and $s(t)$ change. Note that the condition $q(0) = 0$ is equivalent to one of the following three conditions:

1. If the queue is empty, the next arriving customer will be drawn from the distribution $\mathscr{P}$;
2. If all servers are busy and the queue is not empty, the first customer in the queue will have the distribution $\mathscr{P}$;
3. If all servers are busy and the queue is empty, and the next event is the arrival of a customer, this will lead to condition 2. If the next event is a departure, this will lead to condition 1.

For convenience, the proof is divided into two parts:

I) Let the class of customers attempting to enter service have the initial distribution $\mathscr{P}$, i.e. $q^{(0)} = 0$ where $q^{(0)}$ is the state of the component $q(t)$ before the trial. Let $v^{(0)}$ be a number of idle servers before the trial, i.e. $v^{(0)} = c - \sum_{i=1}^{c} i s_i^{(0)}$ where $s^{(0)} \in \mathfrak{s}$ represents the currently served customers. Then

$$q^{(1)} = \begin{cases} 0, & \text{w.p. } \sum_{i=1}^{v^{(0)}} p_i, \\ k, & \text{w.p. } p_k, k > v^{(0)}, \end{cases} \quad v^{(1)} = \begin{cases} v^{(0)} - k, & \text{w.p. } p_k, k \leq v^{(0)}, \\ v^{(0)}, & \text{w.p. } \sum_{i=v^{(0)}+1}^{c} p_i, \end{cases}$$

where $q^{(1)}, v^{(1)}$ are the class of the first customer in queue and the number of idle servers respectively after the trial. Then, by the Bayes theorem [20, pp. 79-86]:

$$\mathbb{P}(q^{(1)} = k|v^{(1)} = v^{(0)}) = \frac{\mathbb{P}(v^{(1)} = v^{(0)}|q^{(1)} = k)\mathbb{P}(q^{(1)} = k)}{\mathbb{P}(v^{(1)} = v^{(0)})} = \frac{p_k}{\sum_{i=v^{(1)}+1}^{c} p_i},$$

where $k > v^{(1)}$. $\mathbb{P}(v^{(1)} = v^{(0)}|q^{(1)} = k) = 1$ because customers enter service sequentially, and $q^{(1)} \neq 0$ means that the next customer joined or remained in the queue. Denote

$$\mathscr{D}(v) = \left\{ (k, d_k) : v < k \leq c, d_k = \frac{p_k}{\sum_{j=v+1}^{c} p_j} \right\} \tag{8}$$

Note that $\mathscr{D}(0) = \mathscr{P}$, which corresponds to all servers being busy.

II) Let $\overline{v}^{(-1)}$ and $\overline{v}^{(0)}$ denote the number of idle servers before and after a customer leaves the system, respectively. Let the queue have been non-empty before the customer left, and $\bar{q}^{(-1)} \sim \mathscr{D}(\bar{v}^{(-1)})$. Then, after the customer leaves, we have: $\bar{q}^{(0)} \sim \mathscr{D}(\bar{v}^{(-1)})$ and

$$\bar{q}^{(1)} = \begin{cases} 0, & \text{w.p. } \sum_{i=\bar{v}^{(-1)}+1}^{\bar{v}^{(0)}} d_i, \\ k, & \text{w.p. } d_k, k > \bar{v}^{(0)}, \end{cases}, \bar{v}^{(1)} = \begin{cases} \bar{v}^{(0)}, & \text{w.p. } \sum_{i=\bar{v}^{(0)}+1}^{c} d_i, \\ \bar{v}^{(0)} - k, & \text{w.p. } d_k, \ k \leq \bar{v}^{(0)}, \end{cases},$$

where $\bar{q}^{(1)}$ and $\bar{v}^{(1)}$ are the class of the first customer in the queue and the number of idle servers, respectively, after the trial. Then, by Bayes theorem:

$$\mathbb{P}(\bar{q}^{(1)} = k | \bar{v}^{(1)} = \bar{v}^{(0)}) = \frac{\mathbb{P}(\bar{v}^{(1)} = \bar{v}^{(0)} | \bar{q}^{(1)} = k) \mathbb{P}(\bar{q}^{(1)} = k)}{\mathbb{P}(\bar{v}^{(1)} = \bar{v}^{(0)})} =$$

$$\frac{d_k}{\sum_{i=\bar{v}^{(0)}+1}^{c} d_i} = \frac{p_k}{\sum_{j=\bar{v}^{(-1)}+1}^{c} p_j} \div \sum_{i=\bar{v}^{(0)}+1}^{c} \frac{p_i}{\sum_{j=\bar{v}^{(-1)}+1}^{c} p_j} = \frac{p_k}{\sum_{i=\bar{v}^{(1)}+1}^{c} p_i}.$$

The probability $\mathbb{P}(\bar{v}^{(1)} = \bar{v}^{(0)} | \bar{q}^{(1)} = k) = 1$ because $k > 0$ means that the next customer remains in the queue and the number of idle servers has not changed. That is, if not enough servers are idle after a customer leaves, then $\mathscr{D}(\bar{v}^{(1)})$ is the distribution of the first customer in the queue (with the new value of $v$); otherwise, its distribution reduces to the first case considered. This completes the proof of Theorem 1.

The proof is shown schematically in Fig 1. Note that the distribution of the time customers spent in service taken into account in $v(t)$, and thus the Theorem 1 should also apply in the non-exponential case.

When calculating the stationary distribution of the system, the initial conditions of Theorem 1 can be ignored if the stability condition [5, p. 122] is satisfied. Since there is only one stationary distribution, independent of the initial distribution, we can assume that the initial distribution satisfies the conditions of the theorem. When calculating the transition distribution, the initial condition cannot be ignored, but it can be expanded: at the initial moment of time, the queue must be empty, all servers must be busy, or $q(0) \sim \mathscr{D}(v(0))$.

If the distribution of customer classes $\mathscr{P}$ contains zero probabilities, then the phase space can be further reduced by excluding states whose probability is a priori zero. That is, if $p_k = 0$ then all states in $\{(\mathfrak{m}, \mathfrak{s})\}$ for which $\mathfrak{s}_k > 0$ can be removed (see Sect. A).

## 5 Algorithms for Matrices Construction

This section provides a brief description of the algorithms for constructing transition rate matrices between levels. The C++ code is available as a package in the R programming language. For further details, see Sect. 7.

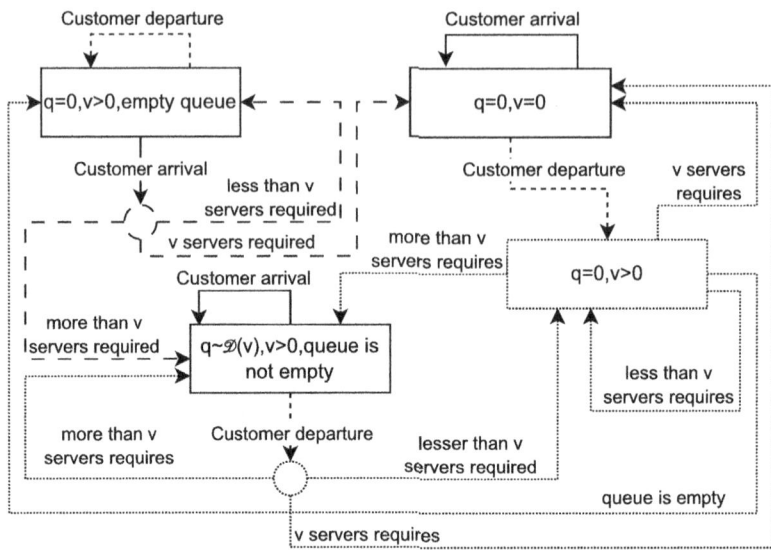

**Fig. 1.** Change in the state of the queue and the number of idle servers, provided that the initial conditions of Theorem 1 are satisfied.

Since there are no transitions within levels in this model, the transition matrices for each level are defined as follows [21]:

$$A_*^{(0)} = -diag(A_*^{(-1)}\mathbf{1}^T + A_*^{(1)}\mathbf{1}^T). \tag{9}$$

Therefore, it is sufficient to construct matrices for transitions one level down and one level up for all levels with unique sets of states.

Before constructing the transition matrices between levels, it is necessary to determine the sets of internal states (phases) for both the source and destination levels of each transition. The simplest, though highly inefficient, approach would be to generate the set $\{\{0..c\}^{\times c} \setminus \{0,..,0\}\}$, then filter out elements that exceed the number of available servers or the total number of customers in the system. However, in our case, a technique similar to the branch-and-bound method [15, pp. 246-249] can be applied. The key difference is that, instead of finding a single optimal solution, we generate a set of feasible solutions, with early pruning of subsets that cannot contain feasible solutions.

The GENERATE_ALL_STATES function from Algorithm 1 utilizes a branch-and-bound approach to enumerate all possible phase states for a specified level. This is a straightforward recursive algorithm that generates all feasible states while pruning subsets that do not contain valid states. It requires the number of regimes $n$, the number of servers $c$, a specific *level* for which the set of possible states must be generated, and probabilities $p_j \in \mathscr{P}$. The distribution $\mathscr{P}$ is used to exclude states that are infeasible in the current system configuration due to zero probabilities for customer arrivals with certain classes. Additionally, ordering the set $S$ is necessary for the future application of binary search [15, pp. 34-36] when constructing the transition matrices.

---

**Algorithm 1.** Branch-and-bound enumeration of all possible phases for level

  **procedure** GEN_S_COMBINATIONS($\mathfrak{s}$, *level*, *c*, *lcl*, &*set*, *V*)
    **var** $busy = \sum_{i=1}^{c} i\mathfrak{s}_i$     ▷ How many servers are busy
    **var** $customers = \sum_{i=1}^{c} \mathfrak{s}_i$     ▷ How many customers are in service
    **if** $busy = c$ **OR** $customers = level$ **then**
      Add state $\mathfrak{s}$ to *set*.
    **else**
      **for** $k = lcl..\text{LENGTH}(V)$ **do**     ▷ Starting with last class to remove repetitions
        **var** $\mathfrak{t} = \mathfrak{s}$
        $\mathfrak{t}[V_k]+=1$
        **var** $busy = \sum_{i=1}^{c} i\mathfrak{t}_i$
        **if** $busy > c$ **then**     ▷ Add unfilled state to *set*
          Add state $\mathfrak{s}$ to *set*.
          **break** ▷ Skipping larger customers if the current one is already too large.
        **end if**
        GEN_S_COMBINATIONS($\mathfrak{t}$, *level*, *c*, *k*, *set*, *V*)
      **end for**
    **end if**
  **end procedure**
  **procedure** GENERATE_ALL_STATES($n$, $c$, *level*, $\mathscr{P}$)
    Create vector $f$ of length $c$
    Set $f_i = 0, i = 1..c$
    Create empty set of vectors of length $c$ with name $S$
    Create an empty array $V$
    **for** $j = 1..c$ **do**
      **if** $p_j \neq 0, p_j \in \mathscr{P}$ **then**
        Add $j$ to the end of $V$
      **end if**
    **end for**
    GEN_S_COMBINATIONS(f, *level*, *c*, 1, *S*, *V*)
    Transform the set $S$ into an ordered set by sorting its elements.
    Return the ordered set of tuples $\{1..n\} \times S$
  **end procedure**

---

Since the matrix construction algorithms are both lengthy and relatively straightforward, we will outline only their key principles. Essentially, both matrix generation algorithms function as simple enumeration methods that first obtain an ordered set of states for the current level and the target level to which the transition occurs. For each state in the current level:

- all possible changes for a single event of customer arrival or departure are calculated,
- resulting in a set of potential states that the current state can transition to; simultaneously, the transition rate and the total probability that all required events for a specific transition will occur are determined,
- the target states are then located in the set of states for the next level, and the product of the transition rate and probability is recorded in the matrix cell $a_{i,j}$, where $i$ is the index of the originating state, and $j$ is the index of the target state.

In the following discussion, we denote the ordered set of states for the source level as $F$, and for the target level as $T$. Both ordered sets are generated using Algorithm 1.

The algorithm for constructing transition rate matrices to one level up is organized as follows. The function accepts as parameters the ordered set $F$ of phase states for the initial level and the ordered set $T$ for the target level, the arrival rate $\lambda$, the matrix $P^{(a)}$, the distribution $\mathscr{P}$, the number of speed modes $n$ and the initial level number $l$. This algorithm is relatively straightforward, as only two events are possible upon a customer's arrival: switching the service mode or adding a new customer to service (if the queue is empty).

The algorithm for constructing transition matrices to one level down is slightly more complex than the algorithm for constructing matrices to one level up. The function takes as parameters the ordered set $F$ of phase states for the initial level, the ordered set $T$ for the target level, the matrix $P^{(d)}$, the vector of speeds for each mode $V = f_1..f_n$, the distributions $\mathscr{D}(k), 0 \leqslant k < c$ as defined by (8), the vector $M = \mu_1..\mu_c$, the initial level number $lev$, the number of servers $c$ and the number of modes $n$. This algorithm is notably more complicated because, when a customer departs, in addition to switching modes, up to $c$ events may occur where customers move from the queue to service. Since it is necessary to enumerate all possible classes of incoming customers, the most simple approach is to use recursion.

The algorithms presented in this section allow for the construction of transition matrices between levels and, ultimately, the matrix $Q$. The following section introduces the simulation model that complements the numerical algorithm and is used for validation, as well as for calculating the model's characteristics when the state space is too large.

## 6 Regenerative Simulation Estimation

The validation of the obtained numerical method was carried out by comparing the results of the numerical solution with those from simulation modeling of the same system. We employed an approach based on Generalized Semi-Markov Processes (GSMP) [6,7]. A brief description of the simulation used to validate the method is provided below.

The Generalized Semi-Markov Processes for our system can be represented as a stochastic process $\Theta(t) = \{X(t), T(t)\}$ where $X(t)$ is defined as in Sect. 3

and $T(t) \in \mathbb{R}_{\geq 0}^u$ is a vector of system timers (with $u$ representing the number of timers). Events that change the value of the discrete component $X(t)$ occur only when at least one of the timers reaches zero. After such an event, the corresponding timer is either reset or set to $+\infty$ if the event it controls cannot occur for any reason. Additionally, events can also affect the timers. The next state $\Theta(t+h)$ is calculated as follows:

$$h = \min_{i=1..u}\left(\frac{T(t)_i}{r(X(t))_i}\right), \quad T(t+h) = T(t) - hr(X(t)), \tag{10}$$

where $r(X(t))$ is a vector of timers speed. We use the representation of $\phi(t)$ in the form $\{(\mathfrak{m}, \hat{\mathfrak{u}}, \mathfrak{q})\}$ because this approach allows to validate our phase space reduction method.

The regenerative estimation method [12] a allows for the estimation of stationary characteristics of the process $\Theta(t)$ if it possesses regeneration points. Classic regeneration points occur when a new customer arrives in an empty system. A brief description of the method follows. Let $W(\Theta(t))$ be the required characteristic of the process $\Theta(t)$. We now consider the statistics:

$$Y_j = \sum_{k=\beta_j}^{\beta_{j+1}-1} W(\Theta(t_k))(t_{k+1} - t_k), \quad R_j = \sum_{k=\beta_j}^{\beta_{j+1}-1}(t_{k+1} - t_k), \tag{11}$$

where $\beta_j$ are the indices of $t_k$ when a new customer arrives into an empty system, i.e. $\beta_{j+1} = \min\{k > \beta_j : x(t_k) = 0\}$. Then, the point estimate $\overline{y}$ of the required stationary characteristic $y = \lim_{t=\infty} \mathbb{E}W(\Theta(t))$ can be estimated as:

$$y = \frac{\mathbb{E}Y_1}{\mathbb{E}R_1} \approx \overline{y} = \frac{Y_1 + \cdots + Y_l}{R_1 + \cdots + R_l} = \frac{S_Y}{S_R}, \tag{12}$$

where $l$ is the number of regeneration cycles. The $\alpha$-level confidence interval can be constructed as:

$$\overline{y} \pm z_{1-\frac{\alpha}{2}} \frac{\sqrt{\overline{\text{var}}(l)}}{\sqrt{l}S_R} \tag{13}$$

where $z_\gamma$ is the $\gamma$-quantile of the standard normal distribution, and

$$\overline{\text{var}} = \frac{lS_{Y^2} - (S_Y)^2 - 2\overline{y}(lS_{YR} - S_Y S_R) + \overline{y}^2(lS_{R^2} - (S_R)^2)}{l(l-1)}, \tag{14}$$

where

$$S_{Y^2} = Y_1^2 + \cdots + Y_l^2, \quad S_{YR} = Y_1 R_1 + \cdots + Y_l R_l, \quad S_{R^2} = R_1^2 + \cdots + R_l^2. \tag{15}$$

In the next section, the proposed method is applied to obtain the stationary distribution of the number of clients in the system, followed by its validation.

## 7 Implementation and Validation

The methods described in the article were implemented in the `MJMrss` package[1] for the R language[2]. Listing 1.1 demonstrates the use of this package. The matrix $R$ was calculated by logarithmic reduction algorithm [3, pp. 188-189].

**Listing 1.1.** Example of using MJMrss package

```
remotes :: install_github("ProgGrey/MJMrss")
library(MJMrss)
lambda = 10 # Input rate
N = 16 # Number of servers
f = c(1, 2.2) # Speeds in modes 1 and 2
# Matrix of mode transition probabilities on customer
# arrivals:
P_a = matrix(c(0.1, 0.9,
               0.0, 1.0), nrow = 2, byrow = TRUE)
# Matrix of mode transition probabilities on customer
# departure:
P_d = matrix(c(1.0, 0.0,
               0.2, 0.8), nrow = 2, byrow = TRUE)
# Classes description:
classes = matrix(c(1,    2,    3,    4, # j
                   0.5,  0.25, 0.15, 0.1, # p_j
                   1,    1.9,  2.5,  3),  # mu_j
                 nrow = 3, byrow = TRUE)
# Create transition matrices using model parameters:
m = build_model(lambda, N, classes, f, P_a, P_d)
# Compute distribution for levels 0 to 20:
dist = m$distribution(20)
```

To validate the calculation results, a discrete-event simulation model of the system was used. Confidence intervals for the probabilities of being at each level were calculated using the regenerative estimation method [12]. In total, $10^3$ independent trajectories were run, each with a length of $10^4$ events of the customers arrivals.

The distribution of the number of customers in the system for the model parameters from Listing 1.1 is shown in Fig. 2 (for levels 0..20 only). The normalized root-mean-square deviation (NRMSD) is approximately 0.433420 % for this case and is determined by the following formula:

$$\text{NRMSD} = \frac{21}{\sum_{j=0}^{20} \overline{y}_j} \sqrt{\frac{\sum_{j=0}^{20} (\overline{y}_j - y_j)^2}{21}}, \tag{16}$$

---

[1] https://github.com/ProgGrey/MJMrss.
[2] https://www.r-project.org/.

where $y_j$ is the probability that the Markov chain is in level $j$, obtained by the numerical solution, and $\overline{y}_j$ is the simulation estimation of $y_j$ on average.

Figure 3 shows the relative deviations of the 99% confidence intervals and mean values obtained from the simulation. The relative deviation $Y_j^{(*)}$ is determined as follows:

$$Y_j^{(*)} = \frac{\overline{y}_j^{(*)}}{y_j} - 1, \qquad (17)$$

where $y_j$ is the probability that the Markov chain is in level $j$ obtained by the numerical solution, $\overline{y}_j^{(*)}$ is a regenerative estimate for mean, upper or lower bound of confidence interval for $y_j$.

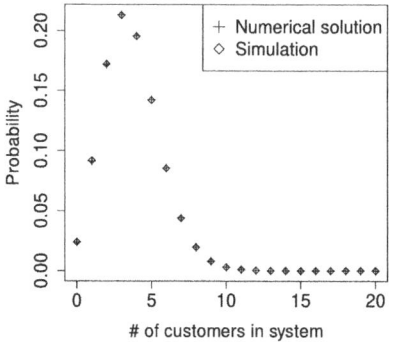
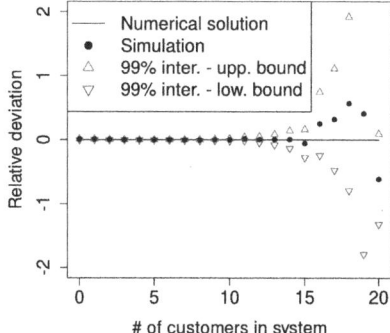

**Fig. 2.** Distribution of the number of customers in the system.

**Fig. 3.** Relative deviation for the 99% confidence intervals.

The Figures and NRMSD show that the values obtained numerically are close to those obtained by simulation. The widened confidence intervals on the right side of Fig. 3 is associated with the proximity of the numerical values to zero.

## 8 Conclusion

The method presented in this article allows for obtaining the stationary distribution of the supercomputer model through numerical solutions when the number of servers is not too large (up to several dozen, depending on available RAM and computation time limits). This limitation is primarily due to the rapid growth in the number of possible phase states of a two-dimensional Markov chain. However, this is one order of magnitude higher than in earlier work, which relied solely on simulations and/or experimental data to calculate the stationary distribution. Thus, the MJMrss R package presented in this paper can be used as an additional tool for validating models of larger systems.

Further work is possible in the direction of advanced reduction of the phase space size (provided that this can be achieved), or by using non-standard methods for calculating the stationary distribution using the infinitesimal generator matrix.

## A  Appendix: Cardinality for Some Phase Descriptions

The following two tables contain the number of phase states for different values of $c$. Table 1a shows the values for different phase descriptions. Table 1b shows the number of phase states when one of the classes does not exist, and the $\{(\mathfrak{m},\mathfrak{s})\}$ description is used. It is assumed that $n = 1$. If $n > 1$, then the cardinality of the corresponding set will be equal to the value in the table multiplied by $n$.

Table 1. Cardinality of phase descriptions, $c = 1..10$

(a) Without zero probabilities

| $c$ | $|\{(\mathfrak{m},\mathfrak{u})\}|$ | $|\{(\mathfrak{m},\mathfrak{s},\mathfrak{q})\}|$ | $|\{(\mathfrak{m},\mathfrak{s})\}|$ |
|---|---|---|---|
| 1 | 1 | 1 | 1 |
| 2 | 4 | 3 | 3 |
| 3 | 27 | 8 | 6 |
| 4 | 256 | 19 | 11 |
| 5 | 3125 | 41 | 18 |
| 6 | 46656 | 80 | 29 |
| 7 | 823543 | 150 | 44 |
| 8 | 16777216 | 262 | 66 |
| 9 | 387420489 | 446 | 96 |
| 10 | 10000000000 | 728 | 138 |

(b) $|\{(\mathfrak{m},\mathfrak{s})\}|$ with $p_k = 0$

| $c$ \ $k$ | 1 | 2 | 3 | 4 | 5 | 6 | 7 | 8 | 9 | 10 |
|---|---|---|---|---|---|---|---|---|---|---|
| 2 | 1 | 1 | | | | | | | | |
| 3 | 2 | 4 | 4 | | | | | | | |
| 4 | 4 | 7 | 9 | 9 | | | | | | |
| 5 | 6 | 11 | 14 | 16 | 16 | | | | | |
| 6 | 10 | 17 | 22 | 25 | 27 | 27 | | | | |
| 7 | 14 | 25 | 32 | 37 | 40 | 42 | 42 | | | |
| 8 | 21 | 36 | 47 | 54 | 59 | 62 | 64 | 64 | | |
| 9 | 29 | 51 | 66 | 77 | 84 | 89 | 92 | 94 | 94 | |
| 10 | 41 | 71 | 93 | 108 | 119 | 126 | 131 | 134 | 136 | 136 |

## References

1. Afanaseva, L., Bashtova, E., Grishunina, S.: Stability analysis of a multi-server model with simultaneous service and a regenerative input flow. Methodol. Comput. Appl. Probab. (2019)
2. Basmadjian, R.: Flexibility-based energy and demand management in data centers: a case study for cloud computing. Energies **12**(17) (2019). https://doi.org/10.3390/en12173301
3. Bini, D., Latouche, G., Meini, B.: Numerical Methods for Structured Markov Chains. Numerical Mathematics and Scientific Computation. Oxford University Press, Oxford (2005). OCLC: ocm56807167
4. Garimella, R.M., Rumyantsev, A.: On an exact solution of the rate matrix of G/M/1-type Markov process with small number of phases. J. Parallel Distrib. Comput. **119**, 172–178 (2018). https://doi.org/10.1016/j.jpdc.2018.04.013
5. Gautam, N.: Analysis of Queues: Methods and Applications. Operations research series. CRC Press, Boca Raton (2012)
6. Glynn, P.W.: A gsmp formalism for discrete event systems. Proc. IEEE **77**(1), 14–23 (1989). https://doi.org/10.1109/5.21067
7. Glynn, P.W., Haas, P.J.: Laws of large numbers and functional central limit theorems for generalized Semi-Markov processes. Stoch. Models **22**(2), 201–231 (2006). https://doi.org/10.1080/15326340600648997, publisher: Taylor & Francis

8. Grosof, I., Hong, Y., Harchol-Balter, M., Scheller-Wolf, A.: The reset and marc techniques, with application to multiserver-job analysis. Perform. Eval. **162**, 102378 (2023). https://doi.org/10.1016/j.peva.2023.102378
9. Harchol-Balter, M.: The multiserver job queueing model. Queueing Syst. (2022). https://doi.org/10.1007/s11134-022-09762-x
10. Harrison, P.G., Patel, N.M., Knottenbelt, W.J.: Energy–performance trade-offs via the ep queue. ACM Trans. Model. Perform. Eval. Comput. Syst. **1**(2), 6:1–6:31 (2016). https://doi.org/10.1145/2818726
11. He, Q.M.: Fundamentals of Matrix-Analytic Methods. Springer, New York, NY (2014). https://doi.org/10.1007/978-1-4614-7330-5
12. Henderson, S.G., Glynn, P.W.: Regenerative steady-state simulation of discrete-event systems. ACM Trans. Model. Comput. Simul. **11**(4), 313–345 (2001). https://doi.org/10.1145/508366.508367
13. Katehakis, M., Smit, L.: A successive lumping procedure for a class of Markov chains. Probab. Eng. Inf. Sci. **26**, 483–508 (2012). https://doi.org/10.1017/S0269964812000150
14. Kim, S.: M/M/s Queueing System Where Customers Demand Multiple Server Use. PhD Thesis, Southern Methodist University (1979)
15. Kurt Mehlhorn, P.S.: Algorithms and Data Structures. The Basic Toolbox. Springer (2010)
16. Latouche, G., Ramaswami, V.: Introduction to Matrix Analytic Methods in Stochastic Modeling. Society for Industrial and Applied Mathematics, Philadelphia (January 1999). https://doi.org/10.1137/1.9780898719734
17. Melikov, A.Z.: Computation and optimization methods for multiresource queues. Cybern. Syst. Anal. **32**(6), 821–836 (1996). https://doi.org/10.1007/BF02366862
18. Morozov, E., Rumyantsev, A.: Stability analysis of a $MAP/M/s$ cluster model by matrix-analytic method. In: Fiems, D., Paolieri, M., Platis, A.N. (eds.) EPEW 2016. LNCS, vol. 9951, pp. 63–76. Springer, Cham (2016). https://doi.org/10.1007/978-3-319-46433-6_5
19. OEIS Foundation Inc.: The On-Line Encyclopedia of Integer Sequences (2024). published electronically at http://oeis.org
20. Ross, S.M.: Introduction to Probability and Statistics for Engineers and Scientists, 6 edn. Academic Press, Cambridge (2020)
21. Rumyantsev, A., Basmadjian, R., Astafiev, S., Golovin, A.: Three-level modeling of a speed-scaling supercomputer. Ann. Oper. Res. (2022). https://doi.org/10.1007/s10479-022-04830-0
22. Rumyantsev, A., Morozov, E.: Stability criterion of a multiserver model with simultaneous service. Ann. Oper. Res. **252**(1), 29–39 (2017). https://doi.org/10.1007/s10479-015-1917-2
23. Tian, J.P., Kannan, D.: Lumpability and commutativity of Markov processes. Stoch. Anal. Appl. **24**(3), 685–702 (2006). https://doi.org/10.1080/07362990600632045

# Unloading Time Martingale Relations in a Cyclic Queueing System in Random Environment

Andrei V. Zorine

Lobachevsky University, Nizhni Novgorod, Russian Federation
andrei.zorine@itmm.unn.ru

**Abstract.** A queuing system with a finite number of input flows controlled by a fixed-time cyclic algorithm is considered. The input flows are formed in a random external environment with finite number of states. The environment is synchronized with the server. In each environment state the input flows are Poisson flows of groups with intensities and size distributions depending on the state. A mathematical model is constructed as a multivariate denumerable Markov chain, necessary and sufficient conditions for the existence of a stationary distributions are found. Martingale sequences are introduced. Using these one can study the mean emptying time for a particular queue.

**Keywords:** Conflicting flows · random environment · cyclic service algorithm · queue emptying time · Markov chain · stationarity condition · martingales

## 1 Introduction

It's of practical interest to research the time needed to a queue to reach zero level in queuing systems with conflicting flows. In the present work we provide some relations for the expected time to zero level in a queuing sysytem in a class of cyclic switching algorithm when the input flows are modulated by a random external environment. Keeping in mind traffic control as a possible area of application, the random external environment may represent weather conditions influencing both arrival rates and batch size probabilities. Since weather conditions vary considerably slower than traffic light signals switching, we take a finite-state Markov chain as a mathematical model for the external environment ans assume that its state may change only at signal change epochs.

Control of input flows with a similar varying probability structure was considered in [1,2] where some conditions for the existence of a stationary probability distribution were found under an additional assumption of constant overall intensity for each flow at different states of the external environment. In the framework of queuing theory it is of a theoretical and practical value to solve queuing systems with input flows modulated by a random environment (see, e.g. [7]). A problem of mean emptying time for a queue has been studied in [3]. Martingale technique goes back to [4,5].

## 2  The Problem Statement and the Mathematical Model

There are $m < \infty$ conflicting input flows to the system. The probability structure of the flows depends on the state of a random external environment with a finite set of possible states $\{e^{(1)}, e^{(2)}, \ldots, e^{(d)}\}$, $d < \infty$. In the state of the environment $e^{(k)}$ the customers in the flow $\Pi_j$ arrive in batches with known parameters: the batch arrival intensities $\lambda_j^{(k)} > 0$ and the probabilities $g_j^{(k)}(b)$ of a batch of the size $b = 1, 2, \ldots$; here $j = 1, 2, \ldots, m$. Let us assume that for each pair $(j, k)$ the finest lattice which supports the probability distribution $\{g_j^{(k)}(b); b = 1, 2, \ldots\}$ has the unit step. Environment may change its state only at epochs of the server state changes and denote by $a_{k,l}$ the probability of a jump from the state $e^{(k)}$ to the state $e^{(l)}$. Let the random environment states be linked in an ergodic Markov chain. Customers from the flow $\Pi_j$ get buffered in a queue $O_j$ with infinite capacity. The server has $2m$ states $\Gamma^{(1)}$, $\Gamma^{(2)}$, $\ldots$, $\Gamma^{(2m)}$. The duration of the state $\Gamma^{(r)}$ is nonrandom and equals $T_r$, $r = 1, 2, \ldots, 2m$. In the states $\Gamma^{(2j)}$, $j = 1, 2, \ldots, m$, no customers are serviced. In the state $\Gamma^{(2j-1)}$ only vustomers from the queue $O_j$ are services. The largest number of customers that may be serviced in this state is set to a nonrandom threshold $\ell_j \in \{1, 2, \ldots\}$. Serviced customers leave the queuing system.

We assume that all random variables and random elements defined below are defined on a common probability space $(\Omega, \mathfrak{F}, \mathbb{P})$. Denote by $\tau_i$, $i = 0, 1, \ldots$ the instants when the server changes its state, set $\tau_0 = 0$. Let $\Gamma_i \in \{\Gamma^{(1)}, \Gamma^{(2)}, \ldots, \Gamma^{(2m)}\}$ be the server state during the interval $(\tau_{i-1}, \tau_i]$, let $X_{j,i}$ denote the number of customers in the queue $O_j$ at the time instant $\tau_i$, let $\eta_{j,i}$ denote the number of customers arrived in the input flow $\Pi_j$ during the time interval $(\tau_i, \tau_{i+1}]$, let $\xi_{j,i}$ be the number of customers in the $j$-th saturation flow [] during the time interval $(\tau_i, \tau_{i+1}]$, let $\bar{\xi}_{j,i}$ be the number of services customers from the queue $O_j$ during the time interval $(\tau_i, \tau_{i+1}]$. Set $r \oplus 1 = r + 1$ for $r = 1, 2, \ldots, 2m - 1$, $(2m) \oplus 1 = 1$, $r \oplus x = (r \oplus (x - 1)) \oplus 1$ for all non-negative integers $x$. Finally, let $\chi_i$ denote the random environment state during the time interval $(\tau_i, \tau_{i+1}]$. Following functional relations can be obtained from the physical problem description:

$$X_{j,i+1} = X_{j,i} + \eta_{j,i} - \bar{\xi}_{j,i} = \max\{0, X_{j,i} + \eta_{j,i} - \xi_{j,i}\}, \tag{1}$$

$$\bar{\xi}_{j,i} = \min\{\xi_{j,i}, X_{j,i} + \eta_{j,i}\}, \tag{2}$$

$$\Gamma_{i+1} = u(\Gamma_i) \quad \text{where } u(\Gamma^{(r)}) = \Gamma^{(r \oplus 1)}, \tag{3}$$

$$\tau_{i+1} = \tau_i + v(\Gamma_i) \quad \text{where } v(\Gamma^{(r)}) = T_{r \oplus 1}. \tag{4}$$

It follows also from the problem description that for a fixed value $(\Gamma^{(r)}, e^{(k)})$ of the vector $(\Gamma_i, \chi_i)$ the set $(\chi_i, \eta_{1,i}, \eta_{2,i}, \ldots, \eta_{m,i})$ is conditionally independent of $(\xi_{1,i}, \xi_{2,i}, \ldots, \xi_{m,i})$ and that both sets are independent of the events up to time $\tau_i$. Furthermore, для for arbitrary non-negative integers $b_1, b_2, \ldots, b_m, \bar{b}_1, \bar{b}_2, \ldots, \bar{b}_m$ the conditional probability of an event

$$\{\omega \colon \chi_{i+1} = e^{(l)}, \eta_{1,i} = b_1, \eta_{2,i} = b_2, \ldots, \eta_{m,i} = b_m\} \cap$$

$$\cup\{\omega\colon \xi_{1,i}=\bar{b}_1,\xi_{2,i}=\bar{b}_2,\ldots,\xi_{m,i}=\bar{b}_m\}$$

equals

$$a_{k,l}\varphi_1(b_1;T_{r\oplus 1},k)\varphi_2(b_2;T_{r\oplus 1},k)\times \varphi_m(b_m;T_{r\oplus 1},\ k)$$

either if $r$ is odd and $\bar{b}_1 = \bar{b}_2 = \ldots = \bar{b}_m = 0$, or if $r = 2j$ and $\bar{b}_s = \ell_{j\oplus 1}\delta_{s,j\oplus 1}$, $s = 1, 2, \ldots, m$; in other cases the conditional probability equals 0. Here $\delta_{s,j}$ is the Kronecker's delta taking on the value 1 for $s \ne j$ and the value 1 for $s = j$, and the functions $\varphi_j(\cdot;\cdot,\cdot)$, $j = 1, 2, \ldots, m$, are defined from Taylor expansions of probability generating functions

$$\sum_{b=0}^{\infty} z^b \varphi_j(b;t,k) = \exp\Big\{\lambda_j^{(k)} t\Big(\sum_{b=1}^{\infty} g_j^{(k)}(b) z^b - 1\Big)\Big\}, \qquad t>0, |z|<1,$$

corresponding to a non-ordinary Poisson flow.

**Theorem 1.** *Given a probability distribution of* $(\Gamma_0, \chi_0, X_{1,i}, X_{2,0}, \ldots, X_{m,0})$, *a multivariate stochastic sequence*

$$\{(\Gamma_i,\chi_i,X_{1,i},X_{2,i},\ldots,X_{m,i}); i=0,1,\ldots\} \tag{5}$$

*is a time-homogeneous Markov chain with a single class of essential communicating periodic states with period* $2m$.

Let $\alpha_k$ be the stationary probability of the environment state $e^{(k)}$, $k = 1, 2, \ldots, d$. Let $\mu_j^{(k)} = \sum_{b=1}^{\infty} b g_j^{(k)}(b) < \infty$ denote the expected batch size for the flow $\Pi_j$ in the environment state $e^{(k)}$, and put $\bar{\lambda}_j^{(k)} = \lambda_j^{(k)} \mu_j^{(k)}$.

**Theorem 2.** *For the existence of a unique stationary probability distribution of the Markov chain* (5) *it is necessary and sufficient that following inequality holds:*

$$\max_{1\leqslant j\leqslant m}\Big\{(T_1+\ldots+T_{2m})\sum_{k=1}^{d}\alpha_k\bar{\lambda}_j^{(k)} - \ell_j\Big\} < 0. \tag{6}$$

A proof of Theorem 2 relies on Lemma 1 giving recurrent equations for the probability generating functions

$$\Psi_{j,i}(z;r,k) = \mathbb{E}\big(z^{X_{j,i}} I(\Gamma_i = \Gamma^{(r)}, \chi_i = e^{(k)})\big), \qquad |z|\leqslant 1.$$

Let

$$f_j^{(k)}(z) = \sum_{b=1}^{\infty} z^b g_j^{(k)}(b), \quad q_j(z;t,e^{(k)}) = \exp\big\{\lambda_j^{(k)} t (f_j^{(k)}(z) - 1)\big\}, \quad |z|\leqslant 1.$$

**Lemma 1.** *The following equations w.r.t. $i = 0, 1, \ldots$ hold:*

$$\Psi_{j,i+1}(z; r \oplus 1, l) = \sum_{k=1}^{d} a_{k,l} q_j(z; T_{r\oplus 1}, e^{(k)}) \Psi_{j,i}(z; r, k), \quad r \oplus 1 \neq 2j - 1;$$

$$\Psi_{j,i+1}(z; r \oplus 1, l) = \sum_{k=1}^{d} a_{k,l} \left( z^{-\ell_j} q_j(z; T_{r\oplus}, e^{(k)}) \Psi_{j,i}(z; r, k) + \right.$$
$$+ \sum_{x=0}^{\ell_j} \mathbb{P}(\Gamma_i = \Gamma^{(r)}, X_{j,i} = x, \chi_i = e^{(k)}) \sum_{b=0}^{\ell_j - 1 - x} \varphi_j(b; T_{r\oplus 1}, k)(1 - z^{x+b-\ell_j}) \bigg).$$

In the remaining part of this paper we will assume the stationarity conditions fulfilled.

## 3 The Expected Queue Emtying Time

For the rest of the talk without loss of generality set $j = 1$. Introduce a random time

$$\nu = \begin{cases} \infty & \text{if } X_{1,i} > 0 \text{ for all } i = 1, 2, \ldots; \\ \min\{i \geqslant 1 \colon X_{1,i} = 0\} & \text{otherwise.} \end{cases}$$

Then the unloading period for the queue $O_1$ is defined as the random time interval $[0, \tau_\nu)$. If $\Gamma_0 = \Gamma^{(r)}$ then

$$\tau_\nu = T_{r\oplus 1} + T_{r\oplus 2} + \ldots + T_{r\oplus \nu},$$

and $r \oplus \nu = 1$ since zero level can be reached only at epochs of end of service for the queue $O_1$, $\Gamma_\nu = \Gamma^{(1)}$. Moreover,

$$\tau_\nu = T_{r\oplus 1} + \ldots + T_{2m} + T_1 + T \cdot \left[\frac{\nu - 1}{2m}\right]$$

where $[\cdot]$ denotes the integer part of a number in the square brackets. So, the study of the expected emptying time can be reduced to the study of the mathematical expectation of the random quantity $[(\nu - 1)/2m]$.

Let us introduce matrices

$$A(z; t) = \begin{pmatrix} a_{1,1} q_1^{(1)}(z; t) & a_{1,2} q_1^{(1)}(z; t) & \ldots & a_{1,d} q_1^{(1)}(z; t) \\ a_{2,1} q_1^{(2)}(z; t) & a_{2,2} q_1^{(2)}(z; t) & \ldots & a_{2,d} q_1^{(2)}(z; t) \\ \vdots & \vdots & \ddots & \vdots \\ a_{d,1} q_1^{(d)}(z; t) & a_{d,2} q_1^{(d)}(z; t) & \ldots & a_{d,d} q_1^{(d)}(z; t) \end{pmatrix}, \quad t > 0.$$

When $0 < z < 1$ the matrix has non-negative entries and it is irreducible (mind the ergodicity assumption for the external environment). Hence by Frobenius's theorem [6] it has a simple eigenvalue $\zeta_r(z)$ which dominates the absolute values of all other eigenvalues, and such that $\zeta_r(1) = 1$. Consider a continuation of this eigenvalue to the disk $|z| \leqslant 1$ in the complex plane.

**Lemma 2.** Let $\zeta_r(z)$ be a regular branch in the domain $|z| < 1$ of the function satisfying a characteristic equation

$$\det(A(z, T_r) - \zeta I_d) = 0, \qquad (I_d = (\delta_{k,l})_{k,l=\overline{1,d}}), \tag{7}$$

such that $\zeta_r(1) = 1$. Then $|\zeta_r(z)| < 1$ for $|z| \leqslant 1$, $z \neq 1$ and

$$\zeta_r'(1) = T_r \sum_{k=1}^{d} \alpha_k \bar{\lambda}_1^{(k)}.$$

*Proof.* According to Gershgorin's circles theorem [6], all solutions to equation (7) lay inside one of the circles

$$|\zeta - a_{l,l} q_1^{(l)}(z; T_r)| \leqslant \sum_{\substack{k=1 \\ k \neq l}}^{d} |a_{k,l} q_1^{(l)}(z; T_r)| = (1 - a_{l,l}) |q_1^{(l)}(z; T_r)|$$

From the geometry point of view, the points $\zeta$ satisfying this equation lay inside the circle with radius $(1 - a_{l,l})|q_1^{(l)}(z; T_r)|$, and the circle center belongs to a segment of the length less than one that joins the origin and the point corresponding to the complex number $q_1^{(l)}(z; T_r)$. The most distant point from the origin is the point $q_1^{(l)}(z; T_r)$, so the whole circle is inside the circle $|z| < 1$.

Let $\bar{1} = (1, 1, \ldots, 1)^T$ be a column-vector of units. Let $\tilde{\psi}_r(z)$ be the left eigenvector of the matrix $A(z; T_r)$ corresponding to the eigenvalue $\zeta_r(z)$ and normalized by the equation $\tilde{\psi}_r(z)\bar{1} = 1$. Let us assume that it is chosen so that for $z = 1$ it turns into the vector $\bar{\alpha}$ of the stationary probabilities for the random environment. Then, taking the derivatives w.r.t. $z$ the equation

$$\zeta_r(z) = \zeta_r(z)\tilde{\psi}_r(z)\bar{1} = \tilde{\psi}_r(z) A(z; T_r)\bar{1}$$

and passing to the limit $z \to 1$, $|z| < 1$, we get

$$\zeta_r'(1) = \tilde{\psi}_r'(1) A(1; T_r)\bar{1} + \tilde{\psi}_r(1) A'(1; T_r)\bar{1} =$$

$$= \tilde{\psi}_r'(1)\bar{1} + \bar{\alpha} A'(1; T_r)\bar{1} = \sum_{k=1}^{d} \alpha_k \frac{d}{dz} q_1^{(k)}(z; T_r)\Big|_{z=1},$$

where $\tilde{\psi}_r'(1)\bar{1} = 0$ due to the derivative of the normalization condition for the vector $\tilde{\psi}_r(z)$.

Remark that that inequality of the inequalities (6) which corresponds to the first queue ($j = 1$) becomes $\sum_{r=1}^{2m} \zeta_r'(1) < \ell_1$.

**Lemma 3.** Equation $z^{\ell_1} = \zeta_1(z)\zeta_2(z) \times \ldots \times \zeta_{2m}(z)$ has $\ell_1 - 1$ zeros $\beta_1, \beta_2, \ldots, \beta_{\ell_1 - 1}$ in the disk $|z| < 1$ and one zero $\beta_{\ell_1} = 1$ on its boundary.

*Proof.* Let us use the generalized Rouche's theorem from [8]. Regard the equation from the lemma as $\theta_1(z) + \theta_2(z) = 0$ where $\theta_1(z) = z^{\ell_1}$, $\theta_2(z) = -\zeta_1(z)\zeta_2(z) \times \ldots \times \zeta_{2m}(z)$. It follows from Lemma 2 that $|\theta_1(z)| > |\theta_2(z)|$ for $|z| = 1$, $z \neq 1$; then, $\theta_1(1) = -\theta_2(1)$. Finally,

$$\frac{\theta_1'(1) + \theta_2'(1)}{\theta_1(1)} = \ell_1 - \zeta_1'(1) - \zeta_2'(1) - \ldots - \zeta_{2m}'(1) > 0.$$

All conditions from the generalized Rouche's theorem are fulfilled. Hence the claim of the current lemma is true.

Let $\psi^{(r)}(z)$ be the right eigenvector of the matrix $A(z, T_r)$ corresponding to the eigenvalue $\zeta_r(z)$ and satisfying the equation $\psi^{(r)}(1) = \bar{1}$. Let us remark that in the paper [5] another normalization was proposed but that is less useful in analysis of the martingale equations.

**Theorem 3.** *If $|z| \leq 1$, a stochastic sequence*

$$M_i(z) = z^{X_{1,i} + \sum_{s=0}^{i-1}\min\{\xi_{1,s}, X_{1,s}+\eta_{1,s}\}} \frac{\prod_{s=0}^{i} \sum_{r=1}^{2m} I(\Gamma_s = \Gamma^{(r\oplus 1)})\psi_{\chi_s}^{(r\oplus 1)}(z)}{\prod_{s=0}^{i-1} \sum_{r=1}^{2m} I(\Gamma_s = \Gamma^{(r)})\zeta_{r\oplus 1}(z)\psi_{\chi_s}^{(r\oplus 1)}(z)}, \quad (8)$$

$i = 0, 1, \ldots,$ *is a martingale w.r.t. a filtration of $\sigma$-algebras*

$$\mathfrak{F}_i = \sigma(X_{1,0}, \Gamma_0, \eta_{1,0}, \eta_{1,1}, \ldots, \eta_{1,i-1}, \chi_0, \chi_1, \ldots, \chi_i), \quad i = 0, 1, \ldots.$$

*Proof.* Consider for each $|z| \leq 1$ stochastic sequences

$$\widetilde{M}_{i,r,k}(z) = z^{X_{1,i} + \sum_{s=0}^{i-1}\min\{\xi_{1,s}, X_{1,s}+\eta_{1,s}\}} I(\Gamma_i = \Gamma^{(r)}, \chi_i = e^{(k)}), i = 0, 1, \ldots$$

for $r = 1, 2, \ldots, 2m$, $k = 1, 2, \ldots d$. The random variable $\widetilde{M}_{i,r,k}(z)$ is $\mathfrak{F}_i$-measurable. Take a row-vector

$$\widetilde{M}_{i,r}(z) = (\widetilde{M}_{i,r,1}(z), \widetilde{M}_{i,r,2}(z), \ldots, \widetilde{M}_{i,r,d}(z)).$$

It follows from the properties of conditional expectation and Eq. (1). that

$$\mathbb{E}(\widetilde{M}_{i+1, r\oplus 1, l}(z) \mid \mathfrak{F}_i) =$$

$$= \mathbb{E}\left(z^{X_{1,i+1} + \sum_{s=0}^{i}\min\{\xi_{1,s}, X_{1,s}+\eta_{1,s}\}} I(\Gamma_{i+1} = \Gamma^{(r\oplus 1)}, \chi_{i+1} = e^{(l)}) \mid \mathfrak{F}_i\right)$$

$$= z^{X_{1,i} + \sum_{s=0}^{i-1}\min\{\xi_{1,s}, X_{1,s}+\eta_{1,s}\}} \mathbb{E}\left(z^{\eta_{1,i}} I(\Gamma_{i+1} = \Gamma^{(r\oplus 1)}, \chi_{i+1} = e^{(l)}) \mid \mathfrak{F}_i\right)$$

$$= z^{X_{1,i} + \sum_{s=0}^{i-1}\min\{\xi_{1,s}, X_{1,s}+\eta_{1,s}\}} \sum_{k=1}^{d} I(\Gamma_i = \Gamma^{(r)}, \chi_i = e^{(k)}) a_{k,l} q_1^{(k)}(z; T_{r\oplus 1}).$$

Recalling the notations we get

$$\mathbb{E}(\widetilde{M}_{i+1,r\oplus 1}(z) \mid \mathfrak{F}_i) = \widetilde{M}_{i,r}(z) A(z; T_{r\oplus 1}).$$

Multiplication from right by the eigenvector $\psi^{(r\oplus 1)}(z)$ we get

$$\mathbb{E}(\widetilde{M}_{i+1,r\oplus 1}(z)\psi^{(r\oplus 1)}(z) \mid \mathfrak{F}_i) = \zeta_{r\oplus 1}(z)\widetilde{M}_{i,r}(z)\psi^{(r\oplus 1)}(z).$$

Here

$$\widetilde{M}_{i+1,r\oplus 1}(z)\psi^{(r\oplus 1)}(z) = z^{X_{1,i+1}+\sum_{s=0}^{i}\min\{\xi_{1,s},X_{1,s}+\eta_{1,s}\}}$$
$$\times \sum_{k=1}^{d} I(\Gamma_{i+1} = \Gamma^{(r\oplus 1)}, \chi_{i+1} = e^{(k)})\psi_k^{(r\oplus 1)}(z)$$
$$= z^{X_{1,i+1}+\sum_{s=0}^{i}\min\{\xi_{1,s},X_{1,s}+\eta_{1,s}\}} I(\Gamma_{i+1} = \Gamma^{(r\oplus 1)})\psi_{\chi_{i+1}}^{(r\oplus 1)}(z).$$

Similarily,

$$\zeta_{r\oplus 1}(z)\widetilde{M}_{i,r}(z)\psi^{(r\oplus 1)}(z) =$$
$$= \zeta_{r\oplus 1}(z) z^{X_{1,i}+\sum_{s=0}^{i-1}\min\{\xi_{1,s},X_{1,s}+\eta_{1,s}\}} I(\Gamma_i = \Gamma^{(r)})\psi_{\chi_i}^{(r\oplus 1)}(z).$$

Summing over $r = 1, 2, \ldots, 2m$ we get

$$\mathbb{E}\left(z^{X_{1,i+1}+\sum_{s=0}^{i}\min\{\xi_{1,s},X_{1,s}+\eta_{1,s}\}} \sum_{r=1}^{2m} I(\Gamma_{i+1} = \Gamma^{(r\oplus 1)})\psi_{\chi_{i+1}}^{(r\oplus 1)}(z) \,\Big|\, \mathfrak{F}_i\right)$$
$$= z^{X_{1,i}+\sum_{s=0}^{i-1}\min\{\xi_{1,s},X_{1,s}+\eta_{1,s}\}} \sum_{r=1}^{2m} I(\Gamma_i = \Gamma^{(r)})\zeta_{r\oplus 1}(z)\psi_{\chi_i}^{(r\oplus 1)}(z).$$

Then we multiply the last equation by a $\mathfrak{F}_i$-measurable random variable

$$\frac{\prod_{s=0}^{i}\sum_{r=1}^{2m} I(\Gamma_s = \Gamma^{(r\oplus 1)})\psi_{\chi_s}^{(r\oplus 1)}(z)}{\prod_{s=0}^{i}\sum_{r=1}^{2m} I(\Gamma_s = \Gamma^{(r)})\zeta_{r\oplus 1}(z)\psi_{\chi_s}^{(r\oplus 1)}(z)}$$

and obtain the theorem's claim.

**Corollary 1.** *Let* $x = 1, 2, \ldots$. *Then*

$$\mathbb{E}\left(z^{X_{1,\nu-1}+\eta_{1,\nu-1}}\left(\frac{z^{\ell_1}}{\zeta_1(z)\times \ldots \times \zeta_{2m}(z)}\right)^{[\frac{\nu-1}{2m}]}\right.$$

$$\times \frac{\prod_{s=0}^{\nu} \psi_{\chi_s}^{(r \oplus s)}(z)}{\prod_{s=0}^{\nu-1} \psi_{\chi_s}^{(r \oplus (s+1))}(z)} \bigg| \{\Gamma_0 = \Gamma^{(r)}, \chi_0 = e^{(k)}, X_{1,0} = x\}\bigg)$$

$$= \begin{cases} z^x \psi_k^{(r)}(z) \zeta_1(z), & r = 2m, \\ z^x \psi_k^{(r)}(z) \zeta_1(z) \zeta_{r \oplus 1}(z) \times \ldots \times \zeta_{2m}(z), & r = 1, 2, \ldots, 2m-1. \end{cases}$$

Setting $z = \beta_k$ we get

$$\mathbb{E}\bigg(\beta_k^{X_{1,\nu-1}+\eta_{1,\nu-1}} \frac{\prod_{s=0}^{\nu} \psi_{\chi_s}^{(r \oplus s)}(\beta_k)}{\prod_{s=0}^{\nu-1} \psi_{\chi_s}^{(r \oplus (s+1))}(\beta_k)} \bigg| \{\Gamma_0 = \Gamma^{(r)}, \chi_0 = e^{(k)}, X_{1,0} = x\}\bigg)$$

$$= \begin{cases} \beta_k^x \psi_k^{(r)}(\beta_k) \zeta_1(\beta_k), & r = 2m, \\ \beta_k^x \psi_k^{(r)}(\beta_k) \zeta_1(\beta_k) \zeta_{r \oplus 1}(\beta_k) \times \ldots \times \zeta_{2m}(\beta_k), & r = 1, 2, \ldots, 2m-1. \end{cases}$$

The proof of the lemma uses the Doob's optional stopping theorem and specializes to a particular event from $\mathfrak{F}_0$.

Assume the normalization $\psi^{(r)}(1) = \bar{1}$. Then by taking a derivative w.r.t. $z$ at $z = 1$ we get:

$$\mathbb{E}\bigg(X_{1,\nu-1} + \eta_{1,\nu-1} + \bigg(\ell_1 - \sum_{k=1}^{d} \alpha_k \bar{\lambda}_1^{(k)}\bigg)(T_1 + \ldots + T_{2m}))\bigg[\frac{\nu-1}{2m}\bigg]$$

$$+ \frac{d}{dz}\bigg(\frac{\prod_{s=0}^{\nu} \psi_{\chi_s}^{(r \oplus s)}(z)}{\prod_{s=0}^{\nu-1} \psi_{\chi_s}^{(r \oplus (s+1))}(z)}\bigg)\bigg|_{z=1} \bigg| \{\Gamma_0 = \Gamma^{(r)}, \chi_0 = e^{(k)}, X_{1,0} = x\}\bigg)$$

$$= x + (\psi_k^{(r)})'(1) + \begin{cases} \sum_{k=1}^{d} \alpha_k \bar{\lambda}_1^{(k)} T_1, & r = 2m, \\ \sum_{k=1}^{d} \alpha_k \bar{\lambda}_1^{(k)}(T_{r \oplus 1} + \ldots + T_{2m} + T_1), & r = 1, 2, \ldots, 2m-1. \end{cases}$$

The last equation allows to obtain the conditional expectation

$$\mathbb{E}\bigg(\bigg[\frac{\nu-1}{2m}\bigg]\bigg| \{\Gamma_0 = \Gamma^{(r)}, \chi_0 = e^{(k)}, X_{1,0} = x\}\bigg)$$

once we get similar conditional expectations for $X_{1,\nu-1} + \eta_{1,\nu-1}$ and for

$$\frac{d}{dz}\bigg(\prod_{s=0}^{\nu} \psi_{\chi_s}^{(r \oplus s)}(z) \bigg/ \prod_{s=0}^{\nu-1} \psi_{\chi_s}^{(r \oplus (s+1))}(z)\bigg)\bigg|_{z=1}.$$

Unfortunately the former is hard to obtain from Corollary 1. Let's find more martingales in the problem.

Let us introduce a matrix

$$\mathbb{A}(z) = \begin{pmatrix} 0 & A(z;T_2) & 0 & \cdots & 0 \\ 0 & 0 & A(z;T_3) & \cdots & 0 \\ \vdots & \vdots & \vdots & \ddots & \vdots \\ 0 & 0 & 0 & \cdots & A(z;T_{2m}) \\ A(z;T_1) & 0 & 0 & \cdots & 0 \end{pmatrix},$$

and let $\hat{\zeta}(z)$ be an eigenvalue of $\mathbb{A}(z)$, $|z| < 1$. Let $\hat{\psi}(z)$ be the corresponding right eigenvector.

**Lemma 4.** $\zeta(z)$ *is an eigenvalue of the matrix* $(z)$ *if and only if* $\zeta(z)^{2m}$ *is an eigenvalue of* $A(z;T_2) \times \ldots \times A(z;T_{2m})A(z;T_1)$.

*Proof.* Let us remark that a matrix identity $\mathbb{A}(z)\hat{\psi}(z) = \zeta(z)\hat{\psi}(z)$ with a column-vector $\hat{\psi}(z)$ of block-columns $\hat{\psi}^{(1)}(z)$, $\hat{\psi}^{(2)}(z)$, ..., $\hat{\psi}^{(2m)}(z)$ is equivalent to a chain of equations

$$A(z;T_2)\hat{\psi}^{(2)}(z) = \zeta(z)\hat{\psi}^{(1)}(z),$$
$$A(z;T_3)\hat{\psi}^{(3)}(z) = \zeta(z)\hat{\psi}^{(2)}(z),$$
$$\vdots$$
$$A(z;T_1)\hat{\psi}^{(1)}(z) = \zeta(z)\hat{\psi}^{(2m)}(z).$$

Hence a necessary condition for $\zeta(z)$ to be an eigenvalue of $\mathbb{A}(z)$ is

$$A(z;T_2) \times \ldots \times A(z;T_{2m})A(z;T_1)\hat{\psi}^{(1)}(z) = \zeta(z)^{2m}\hat{\psi}^{(1)}(z),$$

i.e. the power $\zeta(z)^{2m}$ needs to be an eigenvalue of the matrix

$$A(z;T_2) \times \ldots \times A(z;T_{2m})A(z;T_1).$$

Sufficiency follows since if $\tilde{\zeta}(z)$ and $\tilde{\psi}(z)$ are eigenvalue and right eigenvector of $A(z;T_2) \times \ldots \times A(z;T_{2m})A(z;T_1)$ then any of the roots $(\tilde{\zeta}(z))^{1/(2m)}$ and a column-vector $\hat{\psi}(z)$ with block-columns $\hat{\psi}^{(r)} = (\tilde{\zeta}(z))^{1/(2m)}(A(z;T_r))^{-1}\hat{\psi}^{r-1}(z)$, $\hat{\psi}^{(1)}(z) = \tilde{\psi}(z)$ is a corresponding right eigenvector.

Let us denote the elements of a column-vector $\hat{\psi}^{(r)}(z)$ as $\hat{\psi}_k^{(r)}(z)$, $k = 1, 2, \ldots, d$. The proof of next theorem is similar to the proof of Theorem 3.

**Theorem 4.** *Let* $\hat{\zeta}(z)$ *be an eigenvalue of the matrix* $\mathbb{A}(z)$ *and* $\hat{\psi}(z)$ *be a corresponding right eigenvector. For* $|z| \leqslant 1$ *a stochastic sequence*

$$\hat{M}_i(z) = (\hat{\zeta}(z))^{-i}\hat{\psi}_{\chi_i}^{(\Gamma_i)}(z)z^{X_{1,i}+\sum_{s=0}^{i-1}\min\{\xi_{1,s},X_{1,s}+\eta_{1,s}\}}, \quad i = 0,1,\ldots \quad (9)$$

*is a martingale w.r.t.* $\{\mathfrak{F}_i; i = 0,1,\ldots\}$.

Again, stopping the martingale (9) at the optional stopping time $\nu$ gives

**Corollary 2.** *Let $x = 1, 2, \ldots$. We have the equalities*

$$\mathbb{E}\left(z^{X_{1,\nu-1}+\eta_{1,\nu-1}}\left(\frac{z^{\ell_1}}{\hat{\zeta}(z)^{2m}}\right)^{\left[\frac{\nu-1}{2m}\right]}\hat{\psi}^{(1)}_{\chi_\nu}(z)\bigg|\,\Gamma_0 = \Gamma^{(r)}, \chi_0 = e^{(k)}, X_{1,0} = x\right)$$
$$= z^x \hat{\zeta}(z)^{2m-r}\hat{\psi}^{(r)}_k(z). \quad (10)$$

Let $\beta$ be a root of $z^{\ell_1} = \hat{\zeta}(z)^{2m}$. Setting $z = \beta$ in Eq. (10) we get

$$\sum_{g=1}^{d}\hat{\psi}^{(1)}_g(\beta)L_g(\beta) = \beta^x \hat{\zeta}(\beta)^{2m-r}\hat{\psi}^{(r)}_k(\beta) \quad (11)$$

where

$$L_g(\beta) = \sum_{w=0}^{\ell_1}\beta^w$$
$$\times \mathbb{P}(X_{1,\nu-1}+\eta_{1,\nu-1} = w, \chi_\nu = e^{(g)} \mid \Gamma_0 = \Gamma^{(r)}, \chi_0 = e^{(k)}, X_{1,0} = x)$$

can be considered as a polynomial in $\beta$. Now, assuming that $z^{\ell_1} = \hat{\zeta}(z)^{2m}$ has $\ell_1$ zeros for each of $d$ simple eigenvalues $\zeta(z)$ of $A(z;T_2)\times\ldots\times A(z;T_{2m})A(z;T_1)$ we obtain a system of linear equations for the probabilities

$$\mathbb{P}(X_{1,\nu-1}+\eta_{1,\nu-1} = w, \chi_\nu = e^{(g)} \mid \Gamma_0 = \Gamma^{(r)}, \chi_0 = e^{(k)}, X_{1,0} = x)$$

for $w = 0, 1, \ldots, \ell_1;\ g = 1, 2, \ldots, d$.

**Lemma 5.** *Let $\tilde{\zeta}(z)$ and $\tilde{\psi}(z) = (\tilde{\psi}_1(z), \tilde{\psi}_2(z), \ldots, \tilde{\psi}_d(z))^T$ be the eigenvalue and the right eigenvector of the matrix $A(z;T_2)\times\ldots\times A(z;T_{2m})A(z;T_1)$ such that $\tilde{\zeta}(1) = 1$ and $\tilde{\psi}(z) = \bar{1}$. Then*

$$\tilde{\zeta}'(1) = (T_1 + T_2 + \ldots + T_{2m})\sum_{k=1}^{d}\alpha_k \bar{\lambda}^{(k)}_1.$$

The proof goes along the lines of the proof of Lemma 2.

Finally, substituting $\tilde{\zeta}(z)$ for $\hat{\zeta}(z)^{2m}$ and $\tilde{\psi}(z)$ for $\hat{\psi}^{(1)}(z)$ and taking the derivative w.r.t. $z$ at $z = 1$, we get a simple equation for the mean value of $[(\nu-1)/(2m)]$:

$$\mathbb{E}(X_{1,\nu-1} + \eta_{1,\nu-1} + \mid \Gamma_0 = \Gamma^{(r)}, \chi_0 = e^{(k)}, X_{1,0} = x)$$
$$+ \Big(\ell_1 - (T_1 + \ldots + T_{2m})\sum_{k=1}^{d}\alpha\bar{\lambda}_1^{(k)}\Big)\mathbb{E}\Big(\Big[\frac{\nu-1}{2m}\Big] \,\Big|\, \Gamma_0 = \Gamma^{(r)}, \chi_0 = e^{(k)}, X_{1,0} = x\Big)$$
$$+ \sum_{g=1}^{d}\frac{d}{dz}\tilde{\psi}_d(z)\Big|_{z=1}\mathbb{P}(\chi_\nu = e^{(d)} \mid \Gamma_0 = \Gamma^{(r)}, \chi_0 = e^{(k)}, X_{1,0} = x)$$
$$= x + \Big(1 - \frac{r}{2m}\Big)\Big(\ell_1 - (T_1 + \ldots + T_{2m})\sum_{k=1}^{d}\alpha\bar{\lambda}_1^{(k)}\Big)\hat{\psi}_k^{(r)}(1) + \frac{d}{dz}\hat{\psi}_k^{(r)}(z)\Big|_{z=1}.$$

## 4 Conclusion

In the present paper we developed an algorithm to evaluate the expected emptying times for a single queue in a cyclic queuing system. This yet cannot be considered as ready for numerical implementation until solving the equation $z^{\ell_1} = \hat{\zeta}(z)^{2m}$ in the disk $|z| \leqslant 1$ is made feasible by numerical methods. However, once this problem is solved, all necessary checks concerning the Eqs. (11) can be done by a computer and the numerical evaluation of the conditional expected emptying time can be carried out.

## References

1. Kudelin, A.N., Fedotkin, M.A.: Conflict flow management in a random environment on information about the presence of a queue. VINITI, No. 1717-V96, p. 22 (1996)
2. Kudelin, A.N., Fedotkin, M.A.: Limit theorems for flow control systems in a random environment in the class of algorithms with anticipation. VINITI, No. 2593-V96, p. 40 (1996)
3. Zorine, A.V.: Towards a definition of a busy period under nonlocal description of input flows. Inform. Appl. **18**(3), 45–51 (2024)
4. Baccelli, F., Makowski, A.M.: Dynamic, transient and stationary behavior of the M/GI/1 queue via martingales. Ann. Probab. **17**(4), 1691–1699 (1989)
5. Baccelli, F., Makowski, A.M.: Matringale relations for the M/GI/1 queue with Markov modulated Poisson input. Stoch. Process. Appl. **38**, 99–133 (1991)
6. Gantmacher, F.R.: The Theory of Matrices. AMS Chelsea Publishing, New York (1959)
7. Dudin, A.N., Klimenok, V.I., Vishnevsky, V.M.: The theory of queuing systems with correlated flows. Springer, Cham (2020)
8. Klimenok, V.L.: On the modification of Rouche's theorem for the queueing theory problems. Queueing Syst. **30**, 431–434 (2001)

# On Estimation of Some Functional of the Distribution Function for Dependent Incomplete Observations in Queuing Theory

Rustamjon S. Muradov[1](✉) and Nurlan T. Dushatov[2]

[1] Namangan State Technical University, Namangan, Uzbekistan
rustamjonmuradov@gmail.com
[2] Almalyk branch of Tashkent State Technical University named after Islam Karimov, Almalyk, Uzbekistan

**Abstract.** In queuing theory (QT), the problem of estimating the mean residual lifetime functions in dependent models with random right censoring of observations often arises. This is important for analyzing the time parameters of a queuing system, such as waiting time, service time, and other characteristics. The article considers estimating the mean residual lifetime function in a dependent model with random right censoring of observations. The estimate is constructed using the Archimedean copula of functions. The property of consistency of the estimate is proved.

**Keywords:** Incomplete-censored data · Mean residual lifetime · Archimedean copula functions · Risk functions · Queuing theory

## 1 Introduction

### 1.1 Brief Overview of Queuing Theory and Study Background

QT is a branch of operations research and mathematics that focuses on analyzing and optimizing waiting lines or queues. Originating from the work of Danish mathematician A.K. Erlang in the early 20th century, it provides insights into the behavior of queues and helps design efficient systems that minimize waiting times, costs, and resource utilization. QT is widely used in multiple fields, for example, telecommunications (to manage call traffic, reduce congestion, and optimize bandwidth usage), healthcare (for patient flow management in hospitals and clinics), manufacturing (to optimize assembly lines and reduce bottlenecks), computer science (in load balancing, CPU scheduling, and network traffic), retail (to reduce customer wait times in stores and optimize staff allocation). Kendall (1953) introduced notation and a systematized way to describe queuing models, which became foundational to QT. Gross and Harris (1998) provided an in-depth exploration of fundamental queuing models and their applications (see, [5,6]). Cox and Smith (1961) detailed the statistical analysis of queues, including distribution functions relevant to different queuing models (see Fig. 1).

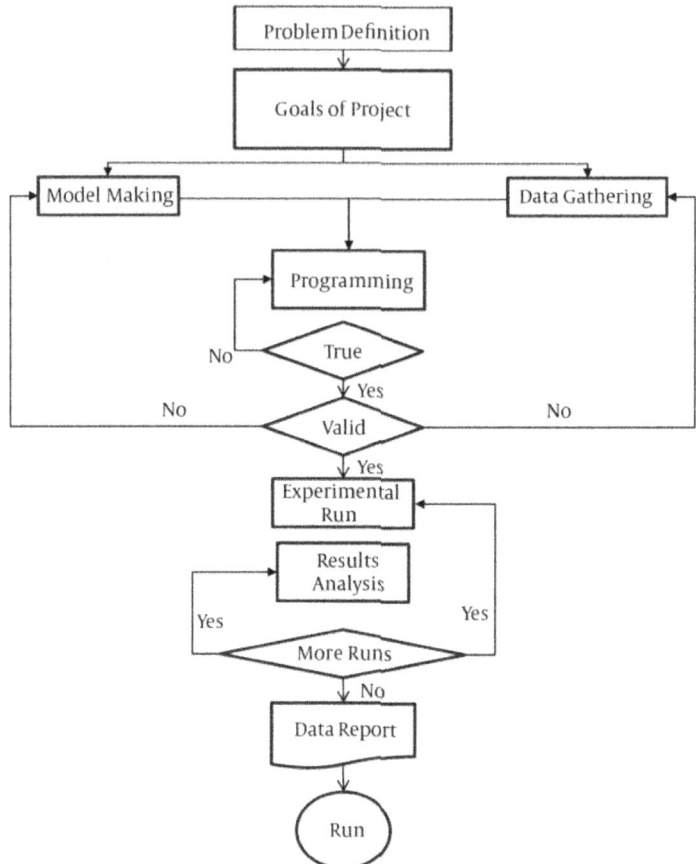

**Fig. 1.** General algorithm of queuing model analysis.

We know that, queuing systems often exhibit dependencies. For example, the service time of one customer might depend on the previous customer's service time, leading to dependency in observations (see Fig. 2). Dependencies add complexity to estimation problems because traditional methods often assume independence. Whitt (1982) explained dependencies in queues, especially in systems where arrivals or services depend on previous events [10]. In real-world applications, data from queuing systems may be incomplete due to limited observation windows or other constraints. Incomplete observations are often managed through estimation techniques that attempt to infer missing information based on observed data. Little and Rubin (2002) explored statistical methods for handling missing data, relevant to scenarios in queuing theory where observations are incomplete. Estimation theory for dependent data is well-covered by Anderson (1971), who addressed the estimation in cases with auto-correlated observations. When dealing with queues specifically, estimating functionals of dependent data

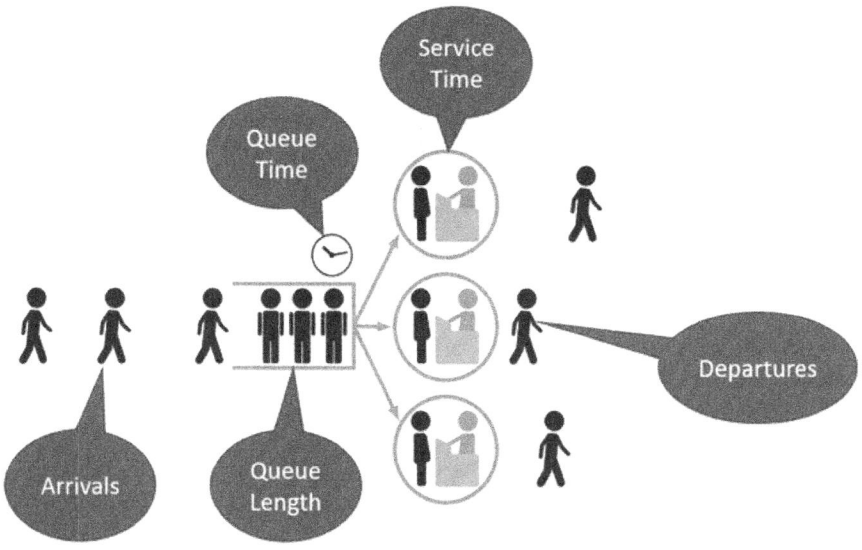

**Fig. 2.** Queuing-example.

requires modifications of classical techniques, often drawing from both time series analysis and survival analysis methods.

### 1.2 Definition of the Mean Residual Life (MRL) Function

The MRL function in QT is a concept used to measure the expected remaining time until an event occurs, given that a certain amount of time has already elapsed. In QT, this concept is especially useful for analyzing and predicting waiting times, service times, and system reliability.

For a non-negative random variable $X$ that represents the time until an event (such as service completion) occurs, the MRL function at time $x$ denoted by $m(x)$, is defined as

$$m(x) = \mathbb{E}[X - x \mid X \geq x], \tag{1}$$

where $X$ is the total time until an event occurs, $X - x$ is the remaining time given that $X \geq x$, $\mathbb{E}[X]$ denotes the expectation or mean. This represents the expected remaining time until the event occurs, given that $x$ units of time have already passed. This function is particularly useful in QT because it helps assess the expected additional waiting time for a customer who has already waited for $x$ time units.

MRL function is of interest in many fields such as reliability research, survival analysis, actuarial study, and so forth. For instance, an insurance company may be concerned about how much longer a product can be used, given that the product has been used normally for, say, $x$ years. Estimation of $E(x)$ was first considered by Yang (1978), replacing the survival function with its empirical survival function. The properties of this estimator were discussed by Csörgo and Zitikis (1996). A natural extension of this estimator to randomly right-censored data is replacing $S(x)$ with its Kaplan-Meier estimator (see, for example, Hall and Wellner (1981)). Chaubey and Sen, (1999,2008) proposed an alternative smoothed estimator based on complete and right-censored data respectively (see, [1–4,9]).

For a random variable $X$ with probability density function $f(x)$ and distribution function $F(x)$, the MRL function $m(x)$ can be expressed as

$$m(x) = \frac{\int_x^\infty (z-x)f(z)\,dz}{1 - F(x)},$$

$f(x)$ is the probability density function of $X$, $F(x) = P(X \leq x)$ is the distribution function of $X$, $S(x) = 1 - F(x) = P(X > x)$ is the survival function, which gives the probability that the event time $X$ exceeds $x$. This formula calculates the expected time remaining by integrating the excess time $(z - x)$ weighted by the probability density $f(z)$ over all times $z$ greater than $x$. The MRL function can also be expressed in terms of the reliability function $R(x) = 1 - F(x)$

$$m(x) = \frac{1}{R(x)} \int_x^\infty R(z)\,dz.$$

This version is often convenient in queuing and reliability applications.

**Examples of MRL Functions.** To illustrate the MRL function, let's examine a few common distributions used in queuing theory.

**Example 1.** (Exponential Distribution). The exponential distribution is commonly used in queuing theory because of its "memory-less" property, meaning that the waiting time until the next event is independent of how much time has already elapsed. For an exponentially distributed random variable $X$ with rate $\lambda$, the probability density function $f(x)$ and cumulative distribution function $F(x)$ are

$$f(x) = \lambda e^{-\lambda x}, \quad F(x) = 1 - e^{-\lambda x}.$$

The MRL function for an exponential distribution is simply $m(x) = \frac{1}{\lambda}$. This is constant and independent of $x$, reflecting the memoryless property of the exponential distribution.

**Example 2.** (Weibull Distribution). The Weibull distribution is frequently used in reliability analysis and queuing systems with a shape parameter $k$ and a scale parameter $\lambda$. For a Weibull random variable $X$, the MRL function $m(x)$ can be computed as

$$m(x) = \frac{\lambda \Gamma\left(1 + \frac{1}{k}\right) - x}{R(x)},$$

where $\Gamma(\cdot)$ is the Gamma function and $R(x) = e^{-(x/\lambda)^k}$. For $k > 1$, the MRL increases with $x$, suggesting that the expected additional waiting time increases the longer a customer has waited. For $k < 1$, the MRL decreases with $x$.

**Example 3.** (Uniform Distribution). For a uniformly distributed random variable $X$ on the interval $[0, a]$, the MRL function is

$$m(x) = \frac{a-x}{2}.$$

This linear function implies that the remaining expected time decreases as more time has elapsed since the start.

## 2 The Copula Model in Queuing Theory

The copula model in queuing theory is a statistical tool used to capture dependencies between random variables (see, [2,7,8]), such as inter-arrival times, service times, or waiting times in queues. Traditional queuing models often assume that these times are independent, however, in real-world applications, dependencies frequently exist. Copula models provide a way to represent and analyze these dependencies, making them valuable for accurate performance modeling of complex queuing systems. The multivariate normal distribution is not a good model for describing the joint distribution of many economic and financial variables. This leads to the problem of finding more adequate multidimensional models. The theory of copula functions is one of the possible ways to solve it. The beginning of the theory of copula functions was proposed by the works of Hoeffding (1940) and Sklar (1959)[1], but statistical modeling using copula functions in applications appeared relatively recently and dates back to the late 1990s. As studies of the last fifteen years show, the capabilities of copula functions in statistics are enormous and they can be successfully used in solving a variety of statistical problems. Early monographs on copulas are Joe (1997) with focus on novel probabilistic notions around copulas, Nelsen (1999, 2006) a well-known, readable introduction. An interesting historical perspective and introduction can be found in Durante and Sempi(2010). A more advanced probabilistic treatment of copulas is the recent Durante and Sempi (2016). A copula is a mathematical function that links multivariate joint distributions to their marginal distributions. Abe Sklar proved a very important statement, which became the basis of the copula methodology, that joint distributions are formed by joining together one-dimensional marginal distributions using joint distributions on the unit cube. According to Sklar's Theorem [7], any multivariate joint distribution $H_{X,Y}(x,y)$ with continuous marginal distributions $F_X(x)$ and $G_Y(y)$ can be expressed as

$$H_{X,Y}(x,y) = C(F_X(x), G_Y(y)), \qquad (2)$$

where $C$ is the copula function, which captures the dependency structure between $X$ and $Y$, $F_X(x)$ and $G_Y(y)$ are the marginal distributions of $X$ and $Y$, respectively.

---

[1] **Abe Sklar-** (November 25, 1925 - October 30, 2020) was an American mathematician and a professor of applied mathematics at the Illinois Institute of Technology and the inventor of copulas in probability theory.

## 2.1 Why Use Copulas in Queuing Theory?

In many queuing systems, service times, inter-arrival times, or waiting times are not independent. For example:

- Customer service times may be correlated with arrival times (e.g., busier periods might lead to shorter or longer service times);
- In network systems, packet arrival times and processing times may be influenced by network congestion, causing dependencies. Copulas provide a flexible framework to model such dependencies without assuming a specific joint distribution. They allow us to model the marginal distributions independently and then use the copula to introduce the dependencies between these variables (see Fig. 3).

## 2.2 Types of Copulas Commonly Used in QT

**Example 4.** (Gaussian Copula.) The Gaussian copula is based on a multivariate normal distribution and is often used when dependencies are linear or approximately linear. Its copula function $C$ is derived from the cumulative distribution function of the multivariate normal distribution with correlation matrix $\Sigma$. For two random variables $U$ and $V$ with standard normal marginals, the Gaussian copula is

$$C(u, v; \rho) = \Phi_\rho(\Phi^{-1}(u), \Phi^{-1}(v)),$$

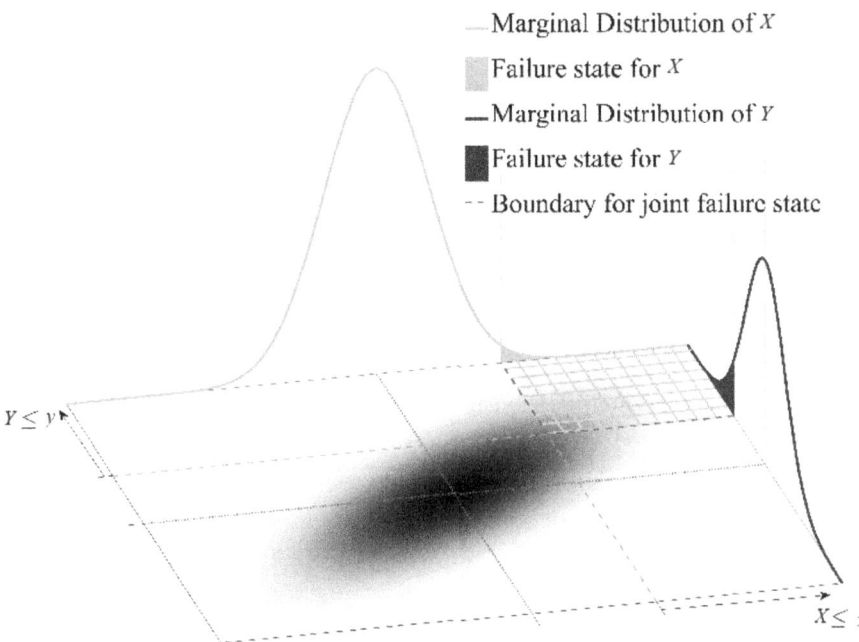

Fig. 3. Visualization of copula method.

where $\Phi_\rho$ is the bivariate normal cumulative distribution function with correlation coefficient $\rho$, $\Phi^{-1}$ is the inverse of the standard normal distribution function.

**Example 5.** (Archimedean Copulas). Archimedean copulas, such as the Clayton, Gumbel, and Frank copulas (see Table 1), are popular for modeling non-linear dependencies. They provide a way to capture asymmetry in dependencies, where the strength of dependence might differ in different tail regions of the distribution. Each of these copulas has a parameter $\theta$ that controls the degree of dependence.

Table 1. Some popular Archimedean Copulas.

| Name of copula | Function | Used for |
|---|---|---|
| Clayton | $C(u,v;\theta) = \max\left((u^{-\theta} + v^{-\theta} - 1)^{-\frac{1}{\theta}}, 0\right)$ | Often used for modeling positive lower-tail dependence |
| Gumbel | $C(u,v;\theta) = \exp\left(-[(-\ln u)^\theta + (-\ln v)^\theta]^{\frac{1}{\theta}}\right)$ | Suitable for modeling upper-tail dependence |
| Frank | $C(u,v;\theta) = -\frac{1}{\theta}\ln\left(1 + \frac{(e^{-\theta u}-1)(e^{-\theta v}-1)}{e^{-\theta}-1}\right)$ | Appropriate for both positive and negative dependencies |

## 3 Dependent Censoring Model and Estimation Method

Consider a sequence $\{(X_k, Y_k), k \geq 1\}$- independent and identically distributed (i.i.d.) pairs of non-negative random variables (r.v.) with a common joint distribution function (d.f.)

$$H(x,y) = P(X_1 \leq x, Y_1 \leq y), (x,y) \in \overline{R}^{+2} = [0,\infty] \times [0,\infty].$$

Let the marginal d.f. $F(x) = P(X_1 \leq x) = H(x, +\infty)$ and $G(y) = P(Y_1 \leq y) = H(+\infty, y)$, $x, y \in \overline{R}^+$, be continuous, $F(0) = G(0) = 0$. Consider the case where the sequence $\mathbb{X} = \{X_k, k \geq 1\}$ is right-censored by the sequence $\mathbb{Y} = \{Y_k, k \geq 1\}$ and in the $n-$ th step of the experiment the sample

$$\mathbb{V}^{(n)} = \{(Z_k, \delta_k), 1 \leq k \leq n\},$$

is observed, where $Z_k = \min(X_k, Y_k)$ and $\delta_k = I(Z_k = X_k)$, i.e. the r.v. $X_k$ that interest us are observable only in the case $\delta_k = 1$.

Let $C(u;v)$ be the copula function corresponding to the pair $(X_1, Y_1)$ and

$$C^*(u;v) = u + v - 1 + C(1-u, 1-v), \quad (u,v) \in [0,1]^2,$$

the corresponding survival copula is itself an Archimedean copula

$$C^*(u;v) = \varphi^{-1}[\varphi(u) + \varphi(v)].$$

Here $\varphi$ is the generator function of the copula $\varphi : [0,1] \to \overline{R}^+$ — continuous, strictly decreasing function such that $\varphi(1) = 0$, $\varphi(0) = \infty$ and $\varphi^{-1}$ is the inverse function for $\varphi$. Note that in the case of independent random censoring, i.e. when the sequences $\mathbb{X}$ and $\mathbb{Y}$ are independent and the generator copula $\varphi(u) = -\ln u$, $u \in (0,1)$, $\varphi^{-1}(t) = \exp(-t)$, $t \geq 0$.

Let us define the estimate $F_n$ by the formula $F_n(x) = 1 - S_n(x)$, where

$$S_n^X(x) = \varphi^{-1}\left[\varphi\left(\widehat{S}_n^Z(x)\right)\mu_n(x)\right] = 1 - F_n(x), \tag{3}$$

$$\mu_n(x) = \varphi\left(S_n^X(x)\right)/\varphi\left(\widetilde{S}_n^Z(x)\right),$$

$$\varphi\left(S_n^X(x)\right) = -\int_0^x \varphi'\left(S_n^Z(u)\right) dH_n^{(1)}(u),$$

$$\varphi\left(\widehat{S}_n^Z(x)\right) = -\int_0^x n\left[\varphi\left(S_n^Z(u)\right) - \varphi\left(S_n^Z(u) - \frac{1}{n}\right)\right] dH_n^{(1)}(u),$$

$$\varphi\left(\widetilde{S}_n^Z(x)\right) = -\int_0^x \varphi'\left(S_n^Z(u)\right) dH_n(u), \; H_n(x) = \frac{1}{n}\sum_{i=1}^n I(Z_i \leq x) = 1 - \widetilde{S}_n^Z(x),$$

$$H_n^{(1)}(x) = \frac{1}{n}\sum_{i=1}^n I(Z_i \leq x, \; \delta_i = 1).$$

Let's consider the functional

$$E(x) = E(x;F) = E(X_1 - x/X_1 > x) = \left(S^X(x)\right)^{-1} \cdot \int_x^{+\infty} S^X(u) du, \tag{4}$$

where $T_X = \inf\{x \geq 0 : S^X(x) = 0\}$, $S^X = 1 - F$ (see, [3]). When analyzing lifetime data, the functional $E(\bullet; F)$ is called the mean residual lifetime, which exists under the condition $\mu = E(0; F) = EX_1 < +\infty$. Here, in the dependent random right censoring model, we construct the following estimator for $E(x)$:

$$E_n(x) = E(x; F_n) = \begin{cases} \left(S_n^X(x)\right)^{-1} \cdot \int_x^\infty S_n^X(u) du, & x \in [0, Z^{(n)}), \\ 0, & x \geq Z^{(n)}, \end{cases} \tag{5}$$

where $Z^{(n)} = \inf\{x \geq 0 : S_n^Z(x) = 0\} = \max\{Z_1, ..., Z_n\}$. Let

$$T_Z = \inf\{x \geq 0 : S^Z(x) = 0\},$$

$S^Z(x) = P(Z_1 > x)$. Then $T_Z = \min\{T_X, T_Y\}$. Since estimates (3) and (5) are defined on the interval $[0, Z^{(n)}]$, as well as for $n \to \infty$

$$Z^{(n)} \overset{\text{a.s.}}{\to} T_Z, \tag{6}$$

then we define the function (4), setting

$$E(x) = \begin{cases} E(x; F), & x < T_Z, \\ 0, & x \geq T_Z. \end{cases} \tag{7}$$

We define a sequence of r.v. $\varepsilon_n(F) = \sup\limits_{0 \leq x < \infty} \chi(F(x)) |E_n(x) - E(x)|$, where the weight function $\chi : [0,1] \to \overline{R}^+$ satisfies the conditions:

(A) $\chi$ is a measurable function and for each $\eta > 0$:
$$\sup \{\chi(u) : u \in [0, 1-\eta]\} < \infty;$$

(B) The function $\chi^*(u) = \chi(u)(1-u)^{-1}$ does not decrease in the neighborhood of $0, 1$;

(C) $\int_0^{T_X} \left\{ (S^X(x))^{-1} \cdot \int_x^{T_X} \chi(F(y)) dy \right\} dF(x) < \infty.$

In the conditions below on the distribution of $H$ and the copula $\varphi$:

(D) The function $\varphi$ strictly decreases on $(0,1]$ and is is sufficiently smooth in the following sense: the first two derivatives of $\varphi(x)$ and $\psi(x) = -x\varphi'(x)$ are bounded for $x \in [\varepsilon, 1]$, where $\varepsilon > 0$ is an arbitrary number. Moreover, the derivative $\varphi'$ is bounded and separated from zero on $[0,1]$;

(E) $0 < \int_0^{T_Z} \left[\psi(S^Z(x))\right]^2 d\Lambda(x) < \infty;$

(F) $\int_0^{T_Z} |\psi'(S^Z(x))| d\Lambda(x) < \infty$. The following statement about the uniform consistency of $E_n(x)$ with weight holds.

**Theorem 1.** Let $\mu = EX_1 < \infty$ and conditions (A)-(F) are satisfied. Then for $n \to \infty$
$$\varepsilon_n(F) \xrightarrow{P} 0. \tag{8}$$

*Proof.* We have
$$\varepsilon_n(F) \leq \sup_{x \in [0; Z^{(n)}]} \chi(F(x)) |E_n(x) - E(x)| +$$
$$+ \sup_{x \in [Z^{(n)}; T_Z]} \chi(F(x)) |E_n(x) - E(x)| = \varepsilon_{1n}(F) + \varepsilon_{2n}(F). \tag{9}$$

To prove (8), it is necessary to prove that for $n \to \infty$, $\varepsilon_{mn}(F) \xrightarrow{P} 0$, $m = 1, 2$. Since $E_n(x) = 0$ for all $x \geq Z^{(n)}$, then by (7) for $n \to \infty$
$$\varepsilon_{2n}(F) = \sup_{x \in [Z^{(n)}; T_Z]} \chi(F(x)) E(x) \xrightarrow{a.s.} 0. \tag{10}$$

By the other way, for a given number $c > 1$ and almost all of the elementary events $\omega$, we can find a number $n_0 = n_0(\omega; c)$ such that for all $x \in [0; Z^{(n)})$ and $n \geq n_0$:
$$\frac{S_n^X(x)}{S^X(x)} \geq c. \tag{11}$$

According to (10) and for all $x \in [0; Z^{(n)})$
$$|E_n(x) - E(x)| \leq c\Phi_{1n}(x) + c\Phi_{2n}(x) \tag{12}$$

where
$$\Phi_{1n}(x) = \frac{E(x)}{S^X(x)} \left| S_n^X(x) - S^X(x) \right|,$$

$$\Phi_{2n}(x) = \frac{1}{S^X(x)} \int_x^{+\infty} \left| S_n^X(u) - S^X(u) \right| du.$$

Since for $n \to \infty$, $\varepsilon_{1n}(F) \le c(\varepsilon_{3n}(F) + \varepsilon_{4n}(F))$, then we show that

$$\varepsilon_{3n}(F) = \sup_{x \in [0; Z^{(n)}]} \chi(F(x)) \Phi_{1n}(x) \xrightarrow{P} 0, \qquad (13)$$

$$\varepsilon_{4n}(F) = \sup_{x \in [0; Z^{(n)}]} \chi(F(x)) \Phi_{2n}(x) \xrightarrow{P} 0. \qquad (14)$$

For given number $\eta > 0$ and sufficiently large number $n$

$$\varepsilon_{3n}(F) \le \sup_{x \in [0; F^{-1}(1-\eta))} \chi(F(x)) \Phi_{1n}(x) +$$
$$+ \sup_{x \in [F^{-1}(1-\eta); Z^{(n)}]} \chi(F(x)) \Phi_{1n}(x) = \varepsilon_{5n}(F) + \varepsilon_{6n}(F), \qquad (15)$$

where for $n \to \infty$, according to (5) and conditions (A), (B) we have

$$\varepsilon_{mn}(F) \xrightarrow{P} 0, \ m = 5, 6. \qquad (16)$$

From (15) and (16) we obtain (13). Similarly, for sufficiently large $n$

$$\varepsilon_{4n}(F) \le \sup_{x \in [0; F^{-1}(1-\eta))} \chi(F(x)) \Phi_{2n}(x) +$$
$$+ \sup_{x \in [F^{-1}(1-\eta); Z^{(n)}]} \chi(F(x)) \Phi_{2n}(x) = \varepsilon_{7n}(F) + \varepsilon_{8n}(F). \qquad (17)$$

In view of condition $\mu < \infty$, for $n \to \infty$ exist a number $c_0 > 0$ such that

$$\varepsilon_{7n}(F) \le c_0 \cdot \int_0^{+\infty} \left| S_n^X(x) - S^X(x) \right| dx \xrightarrow{P} 0. \qquad (18)$$

Let $\chi^*(F(x)) = \chi(F(x))/S^X(x)$. Then according to the conditions (A), (B) and (C) for $n \to \infty$

$$\varepsilon_{8n}(F) \le \int_{F^{-1}(1-\eta)}^{T_X} \chi^*(F(x)) \left| S_n^X(x) - S^X(x) \right| dx \xrightarrow{P} 0. \qquad (19)$$

From (17)–(19) follows (14). The theorem is proved.

## 4 Conclusion

The copula model in queuing theory offers a powerful tool for accurately modeling dependencies in arrival and service processes, leading to more realistic queuing models. By capturing complex dependencies between variables, copulas provide insights that are essential for optimizing system performance, improving reliability, and better handling variability in queuing systems. Estimating the mean remaining lifetime function in queuing theory within dependent models with incomplete data requires integrating copula-based dependency modeling with survival analysis methods. This approach allows for more accurate assessment of system reliability and performance, taking into account the complex dependencies inherent in queuing systems.

**Acknowledgments.** The authors would like to thank the reviewers and Professor Abdurahim Abdushukurov of the Tashkent branch of Moscow State University for their comments and suggestions for improving this article.

**Disclosure of Interests.** The authors have no competing interests to declare that relevant to the content of this article.

## References

1. Abdushukurov, A.: Nonparametric estimation of distribution function based on relative-risk function. Commun. Statist. Theory Methods. **27**(8), 1991–2012 (1998)
2. Abdushukurov, A.: Estimation of conditional survival function under dependent right random censored data. Austrian J. Statist. **49**(1), 1–8 (2020)
3. Abdushukurov, A.: Generalization of the relative risk power estimator for dependent randomly censored data. Mod. Math. Fund. Direct. **68**(1), 1–13 (2022). (In Russian)
4. Abdushukurov, A., Muradov, R.: Estimation of conditional jointly survival function under dependent right random censored data. Lobachevskii J. Math. **72**(4), 2111–2122 (2022)
5. Gross, D., Harris, C.: Fundamentals of Queueing Theory. Wiley Series in Probability and Statistics (1998)
6. Kendall, D.G.: Stochastic processes occurring in the theory of queues and their analysis by the method of the imbedded Markov chain. Ann. Math. Stat. **24**(1), 338–354 (1953)
7. Nelsen, R.: An Introduction to Copulas, 2nd edn. Springer, New York (2006)
8. Muradov, R.-S., Dushatov N.-T. : Estimation of some functionals of the distribution function for dependent incomplete observations in queuing theory. In: XXIII th International Conference Proceedings ITMM2024, pp. 452–455. TSU, Karshi (2024). (In Russian)
9. Abdushukurov, A., Dushatov, N., Muradov, R.: Estimation of functionals of multivariate distribution by censored observation via copula function. J. Math. Sci. **267**(1), 108–116 (2022)
10. Whitt, W.: Approximations for departure processes and queues in series. Nav. Res. Logist. **31**(4), 499–521 (1984)

# Analysis of a Queueing System Providing Service to Regular and Ad Hoc Clients

Sergei Dudin, Alexander Dudin(✉), and Olga Dudina

Department of Applied Mathematics and Computer Science,
Belarusian State University, Minsk 220030, Belarus
{dudins,dudin,dudina}@bsu.by

**Abstract.** A multi-server queueing system is under study. Two kinds of requests are processed in the system. One kind of requests is generated by regular clients in exponentially distributed intervals of time. A whole sequence of requests can be generated by one regular client. Another kind of requests is generated by ad hoc clients. Each such client generates only one request. Regular and ad hoc client arrivals are defined by two independent Markov arrival processes. Service times for both kinds of requests are exponentially distributed, with the rate independent of the kind of request. A regular client resides in the system during an exponentially distributed time and departs from the system. All requests generated by this client when there are idle servers will obtain full service even the client has departed from the system. Ad hoc clients are less valuable for the system than the regular ones. Therefore, a request by a regular client is admitted to the system if there is at least one idle server at the moment of arrival. Ad hoc clients are admitted to the system only if the number of busy servers is less than a certain threshold value. The behaviour of this system is described by the four-dimensional continuous-time Markov chain with the state-inhomogeneous transition rates. Analysis of this chain is implemented, including demonstration of the feasibility of the proposed algorithms for computation of its stationary distribution. The problem of optimal choice of the number of servers and the threshold defining the policy of ad hoc clients admission is numerically solved.

**Keywords:** regular and ad hoc customers · $MAP$ · admission control

## 1 Introduction

The goal of the operation of many real-world systems described in terms of a queueing system is to earn maximal revenue by providing service to arriving clients. Therefore, the system manager is interested in the admission of all arriving clients if the quality of their service will be satisfactory.

Sometimes, besides the regular clients, which usually are serviced in this system (e.g., have a membership card, season ticket, subscription, etc.), there exists some additional stream of occasional clients that would like to opportunistically receive service. We call these clients as ad hoc clients because the system does not anticipate their arrival and does not have strong obligations with respect to them. The system can admit such a client for service to earn an additional amount of money if the system is currently underutilised, i.e., some servers are idle and, therefore, do not bring profit. If the system is more or less heavily loaded (while not all servers are necessarily busy), the ad hoc client has to be rejected in anticipation that regular clients can arrive and require service soon.

In this paper, we consider a queueing system with two types of clients. The arrivals of both types are defined by the two independent Markov arrival processes ($MAP$s). The arrival of the regular clients occurs in such a way that the arrival of a client causes further arrival of a whole sequence of the requests that have to be processed in the system. Ad hoc clients arrive in the individual fashion; the arrival of one client does not trigger the arrival of more requests.

The regular client can be considered the frequenter client of the system that more or less regularly visits this system and, during the visit, sequentially requires some kind of repeated operation (services). The mechanism of the regular clients arrival is similar to the session (or train) clients arrival previously investigated, e.g., in [1–5]. The difference is that the session arrival to the system in [1,2] assumed that the first request from a session is generated immediately upon the session arrival. Here, we assume that the first request from a client will arrive after the exponentially distributed time. The ad hoc client occasionally visits this system; it instantaneously generates a request, and its service requires only a single operation. It is worth mentioning that analysis of session arrivals is implemented in [1,2] in the assumption that only a finite number of sessions can be processed in the system simultaneously, and the focus of the research is on defining the optimal number of sessions. Here, we assume that the number of regular clients simultaneously processed in the system is not limited.

The choice of the $MAP$s, see, e.g., [6–12], as the models of arrivals is easily explained by their popularity in the literature and our personal experience of good fitting of the traces of several important real-world flows by the $MAP$. Fitting the same traces by the stationary Poisson arrival process implies an extremely optimistic prediction of the system performance measures. This is because the $MAP$ allows to catch the fluctuation of the instantaneous arrival rate in contrast to the stationary Poisson arrival process that accounts only for the mean arrival rate. High fluctuation implies the existence of periods of time when the clients arrive very rarely (and starvation of the servers occurs) and periods of very frequent arrivals when the congestion occurs.

The rest of the paper is organised as follows. The detailed description of the queueing system under study is presented in Sect. 2. The behaviour of the system is described by the four-dimensional continuous-time Markov chain ($CTMC$) in Sect. 3. The generator of this $CTMC$ is presented and explained. The existence of its stationary distribution is stated, and the way for computation of its steady-state distribution is outlined. Formulas for calculation of the main performance characteristics of the system are presented in Sect. 4. Section 5 is devoted to a description of the results of the numerical experiment. Section 6 briefly summarises the results of the paper.

## 2 Mathematical Model

Let us consider a queuing system consisting of $N$ independent identical servers, the structure of which is shown in Fig. 1.

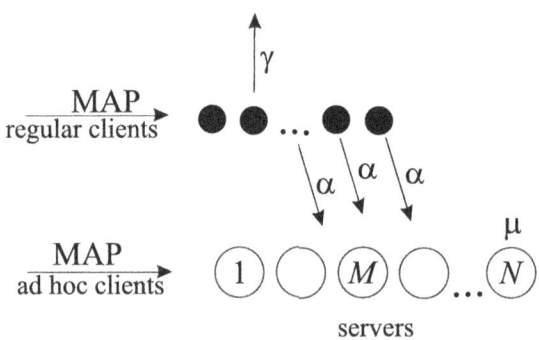

**Fig. 1.** The structure of the system

A $MAP$ flow of regular clients enters the system. This input flow is specified by the underlying process $d_t$, $t \geq 0$, which is an irreducible $CTMC$ with a finite state space $\{1, 2, \ldots, W\}$, and the matrices $D_0$ and $D_1$. The average arrival intensity of regular clients is denoted as $\lambda_d$ and is calculated as $\lambda_d = \boldsymbol{\theta} D_1 \mathbf{e}$ where $\boldsymbol{\theta} = (\theta_1, \ldots, \theta_W)$ is the invariant probability vector of the $CTMC$ $d_t$. It is defined as the unique solution of the system $\boldsymbol{\theta}(D_0 + D_1) = \mathbf{0}$, $\boldsymbol{\theta}\mathbf{e} = 1$. Here and throughout the paper, $\mathbf{e}$ is a column vector of suitable size consisting of ones, and $\mathbf{0}$ is a row vector of suitable size consisting of zeros. A more detailed description of the $MAP$, probabilistic meaning of the matrices $D_0$ and $D_1$ and formulas for determining its characteristics, in particular such as the correlation and variation coefficients, can be found, e.g., in [6–12].

The number of regular clients that can present in the system simultaneously is not limited. Each regular client, independently of others, after arrival sequentially generates requests for service in exponentially distributed time with the intensity $\alpha$, $\alpha > 0$. If a regular client has generated a request and there is a free server, then this request occupies the server and starts service. The service is implemented during an exponentially distributed time with the parameter $\mu$. If all servers are busy at a request generation epoch, then the request is lost. In this case, the client that generated the lost request leaves the system with the probability $q$, $1 \geq q > 0$, and with the complementary probability remains in the system and will generate another request in exponentially distributed time with the rate $\alpha$. We assume that the client may not wait for the end of servicing the generated request (as it is assumed in so-called systems with the finite number of sources) and generate new requests. The time a regular client spends in the system has an exponential distribution with the parameter $\gamma$, $\gamma \geq 0$. After this time expires, the client leaves the system permanently, and all requests generated by this client that are being serviced remain in the system until the end of service. Note that the parameters $\gamma$ and $q$ cannot be equal to zero at the same time, since otherwise, regular clients will not be able to leave the system, which will lead to their infinite accumulation.

In addition to requests from regular clients, the system receives a $MAP$-flow of ad hoc clients (single ad hoc requests). This input flow is defined by the underlying process $h_t$, $t \geq 0$, which is an irreducible $CTMC$ $h_t$, $t \geq 0$, with a finite state space $\{1, 2, \ldots, V\}$, and the matrices $H_0$ and $H_1$. The average arrival intensity of ad hoc requests is denoted as $\lambda_h$. Ad hoc requests are accepted for servicing only if, at the time of their arrival, the number of busy servers is less than the threshold value $M$. Otherwise, the ad hoc request leaves the system forever.

The service times of an ad hoc request and a request generated by a regular client do not differ and have an exponential distribution with the parameter $\mu$, $\mu > 0$.

## 3 The Process of System States and its Stationary Distribution

Let

$i_t$, $i_t \geq 0$, be the number of regular clients in the system;

$n_t$, $n_t = \overline{0, N}$, be the number of busy servers;

$d_t$, $d_t = \overline{1, W}$, and $h_t$, $h_t = \overline{1, V}$, be the states of the underlying processes of the $MAP$s of the regular clients and ad hoc requests, respectively, at time $t$, $t \geq 0$.

Then the process $\xi_t = \{i_t, n_t, d_t, h_t\}$, $t \geq 0$, is an irreducible $CTMC$. Let us enumerate the states of the $CTMC$ $\xi_t$ in the lexicographic order and call the set of the states having the values $i$ of the first component as level $i$ of the chain.

**Theorem 1.** *The generator $Q$ of the $CTMC$ $\xi_t$, $t \geq 0$, has a block-tridiagonal structure*

$$Q = \begin{pmatrix} Q_{0,0} & Q_{0,1} & O & O & \cdots \\ Q_{1,0} & Q_{1,1} & Q_{1,2} & O & \cdots \\ O & Q_{2,1} & Q_{2,2} & Q_{2,3} & \cdots \\ \vdots & \vdots & \vdots & \vdots & \ddots \end{pmatrix}$$

*where the matrices $Q_{i,j}$, $i, j \geq 0$, $|i - j| \leq 1$, contain the transition rates from the states belonging to the level $i$ to the states belonging to the level $j$ and are defined as follows:*

$$Q_{i,i} = I_{N+1} \otimes (D_0 \oplus H_0) - i(\gamma + \alpha)I_{(N+1)WV} - \mu C_{N+1} \otimes I_{WV} +$$
$$+\mu C_{N+1} \Phi^{-}_{N+1} \otimes I_{WV} + i\alpha \Phi^{+}_{N+1} \otimes I_{WV} + (1-q)i\alpha \tilde{I}_{N+1} \otimes I_{WV} + \tilde{\Phi}_{N+1} \otimes I_W \otimes H_0, \, i \geq 0,$$
$$Q_{i,i-1} = (i\gamma I_{N+1} + qi\alpha \tilde{I}_{N+1}) \otimes I_{WV}, \, i > 0,$$
$$Q_{i,i+1} = I_{N+1} \otimes D_1 \otimes I_V, \, i \geq 0.$$

Here,

$I_m$ is an identity matrix of size $m$;

$\oplus$ and $\otimes$ are the Kronecker symbols of sum and product, respectively, see, e.g., [13];

$C_{N+1}$ is a diagonal matrix of size $N+1$ with the diagonal elements $0, 1, 2, \ldots, N$;

$\Phi^-_{N+1}$ and $\Phi^+_{N+1}$ are the square matrices of size $N+1$ of the form:

$$\Phi^-_{N+1} = \begin{pmatrix} 0 & 0 & 0 & \ldots & 0 & 0 \\ 1 & 0 & 0 & \ldots & 0 & 0 \\ 0 & 1 & 0 & \ldots & 0 & 0 \\ \vdots & \vdots & \vdots & \ddots & \vdots & \vdots \\ 0 & 0 & 0 & \ldots & 1 & 0 \end{pmatrix}, \Phi^+_{N+1} = \begin{pmatrix} 0 & 1 & 0 & \ldots & 0 & 0 \\ 0 & 0 & 1 & \ldots & 0 & 0 \\ \vdots & \vdots & \vdots & \ddots & \vdots & \vdots \\ 0 & 0 & 0 & \ldots & 0 & 1 \\ 0 & 0 & 0 & \ldots & 0 & 0 \end{pmatrix};$$

$\tilde{I}_{N+1}$ is a square matrix of size $N+1$ with all zero elements except the element $(\tilde{I}_{N+1})_{N,N}$, equal to one (the entries are enumerated from zero);

$\tilde{\Phi}_{N+1}$ is a square matrix of size $N+1$ with all zero elements except for the elements $(\tilde{\Phi}_{N+1})_{m,m+1}$, $m = \overline{0, M-1}$, and $(\tilde{\Phi}_{N+1})_{m,m}$, $m = \overline{M, N}$, equal to one.

*Proof.* The theorem is proved by analysing the intensities of all possible transitions of the $CTMC$ $\xi_t$ over an infinitely small time interval. The block-tridiagonal form of the generator $Q$ is easily explained by the fact that the regular clients enter and leave the system one by one.

First, let us consider the blocks $Q_{i,i}$, $i \geq 0$.

$Q_{i,i}$ is a diagonal block of the generator, so all its diagonal elements are negative, and the moduli of these elements determine the intensities of the exit of the $CTMC$ $\xi_t$ from the corresponding states. The exit of the $CTMC$ $\xi_t$ from the current state is possible in the following cases:

- The underlying process $d_t$ of the regular client arrivals exits the current state. The corresponding transition intensities of the process $\{n_t, d_t, h_t\}$ are determined up to the sign by the diagonal elements of the matrix $I_{N+1} \otimes (D_0 \otimes I_V)$.
- The underlying process $h_t$ of the ad hoc request arrivals exits the current state. The corresponding transition intensities are determined up to the sign by the diagonal elements of the matrix $I_{N+1} \otimes (I_W \otimes H_0)$.
  Note that if a new ad hoc request arrives to the system when the number of busy servers in the system is greater than $M - 1$, the request abandons the system, thus the $CTMC$ $\xi_t$ does not exit from the current state. So, the diagonal elements of the matrix $\tilde{\Phi}_{N+1} \otimes I_W \otimes H_1$ are added to the diagonal elements of the matrix $Q_{i,i}$.
- Service is completed in one of the busy servers. The matrix $\mu C_{N+1} \otimes I_{WV}$ defines the corresponding intensities.
- A regular client leaves the system due to some reason, e.g., impatience or the end of the subscription term. The corresponding intensities are defined by the matrix $i\gamma I_{(N+1)WV}$.

- A regular client generates a new request. The matrix $i\alpha I_{(N+1)WV}$ contains the corresponding intensities. Note that if a regular client generates a new request when all servers are busy, this request is lost but the client remains in the system, then the $CTMC$ $\xi_t$ does not exit from the current state, and the matrix $Q_{i,i}$ has the summand $(1-q)i\alpha \tilde{I}_{N+1} \otimes I_{WV}$.

The non-diagonal elements of the matrices $Q_{i,i}$, $i \geq 0$, determine the intensities of the transitions of the $CTMC$ $\xi_t$ without changing the value $i$ of the first component. These transitions are determined by the following elements:

- Non-diagonal elements of the matrix $I_{N+1} \otimes D_0 \otimes I_V$ when the underlying process $d_t$ makes a transition without generating a regular client.
- Non-diagonal elements of the matrix $I_{N+1} \otimes I_W \otimes H_0$ when the underlying process $h_t$ makes a transition without generating an ad hoc request.
- Elements of the matrix $\mu C_{N+1}\Phi^-_{N+1} \otimes I_{WV}$ when service ends in one of the busy servers.
- Elements of the matrix $i\alpha \Phi^+_{N+1} \otimes I_{WV}$ when a regular client generates a new request that is accepted to the system.
- Non-diagonal elements of the matrix $\tilde{\Phi}_{N+1} \otimes I_W \otimes H_1$ when an ad hoc request arrives to the system (and is accepted or lost).

As a result, we obtain the blocks $Q_{i,i}$, $i \geq 0$, presented above.

The form of the blocks $Q_{i,i-1}$, $i \geq 0$, is explained as follows. These blocks contain the intensities of the transitions of the $CTMC$ $\xi_t$ which lead to a decrease in the number of regular clients in the system by one. Such transitions are possible only in two cases: the time which a regular client spends in the system finishes or a regular client leaves the system after the request generated by him/her is lost due to the busyness of all servers. The intensities of these events are determined by the elements of the matrices $i\gamma I_{(N+1)WV}$ and $qi\alpha \tilde{I}_{N+1} \otimes I_{WV}$, respectively.

The blocks $Q_{i,i+1}$, $i \geq 0$, contain the intensities of transitions of the $CTMC$ $\xi_t$, which lead to an increase in the number of regular clients in the system by one. Such an increase occurs when the new regular client arrives in the system. Therefore, the mentioned intensities are determined by the elements of the matrix $I_{N+1} \otimes D_1 \otimes I_V$. The theorem is proven.

It is clear from the form of the blocks of the generator that this generator does not possess a quasi-Toeplitz property, i.e., the blocks $Q_{i,j}$ do not depend on only the difference $j-i$ but depend on $i$ as well. Therefore, the $CTMC$ $\xi_t$ does not belong to the class of space homogeneous Quasi-Birth-and-Death processes, and well-known results by M. Neuts from [14] related to the ergodicity condition and existence of the matrix geometric form of the stationary distribution of the states are not applicable here. However, it is possible to show that the $CTMC$ $\xi_t$ belongs to the class of asymptotically Quasi-Toeplitz Markov chains, see, e.g., [10,15]. Using the results from [10,15], it is possible to show that, due to the positivity of the rate $\gamma$ or the probability $q$, the $CTMC$ $\xi_t$ is ergodic for any values of the system parameters.

Thus, there exist the stationary probabilities of the system states

$$\pi(i, n, d, h) = \lim_{t \to \infty} P\{i_t = i, n_t = n, d_t = d, h_t = h\},$$

$i \geq 0$, $n = \overline{0, N}$, $d = \overline{1, W}$, $h = \overline{1, V}$.

We denote by $\boldsymbol{\pi}_i$ the row vectors of the stationary probabilities of the states that belong to the level $i$, numbered in the lexicographic order of the components $n$, $d$, $h$. It is well known that these vectors satisfy the system of linear algebraic equations:

$$(\boldsymbol{\pi}_0, \boldsymbol{\pi}_1, \ldots, \boldsymbol{\pi}_i, \ldots) Q = \mathbf{0}, \quad (\boldsymbol{\pi}_0, \boldsymbol{\pi}_1, \ldots, \boldsymbol{\pi}_i, \ldots) \mathbf{e} = 1$$

where $Q$ is the infinitesimal generator of the $CTMC$ $\xi_t$, $t \geq 0$. To solve this infinite system with the generator that does not possess the quasi-Toeplitz structure, we recommend using an efficient and numerically stable algorithm proposed in [16].

## 4 Performance Measures

Having found the vectors defining the stationary distribution of the system states, it is possible to compute the values of the key performance measures of the system under any value of the system parameters and the threshold of the ad hoc clients admission policy.

The average number of regular clients in the system is calculated using the formula

$$N_{reg-client} = \sum_{i=1}^{\infty} i \boldsymbol{\pi}_i \mathbf{e}.$$

The average number of busy servers is calculated using the formula

$$N_{serv} = \sum_{i=0}^{\infty} \sum_{n=1}^{N} n \boldsymbol{\pi}(i, n) \mathbf{e}.$$

The average intensity of incoming requests is defined as

$$\lambda_{in} = \lambda_h + \alpha N_{reg-client}.$$

The average rate of the serviced requests is defined as

$$\lambda_{serv} = \mu N_{serv}.$$

The loss probability of an arbitrary ad hoc request is calculated as

$$P_{adhoc-loss} = \frac{1}{\lambda_h} \sum_{i=0}^{\infty} \sum_{n=M}^{N} \boldsymbol{\pi}(i, n)(I_W \otimes H_1) \mathbf{e}.$$

The loss probability of an arbitrary regular client's request is calculated as

$$P_{reg-loss} = \frac{\sum_{i=1}^{\infty} i\pi(i,N)\mathbf{e}}{N_{reg-client}}.$$

The loss probability of an arbitrary request is calculated as

$$P_{loss} = 1 - \frac{\lambda_{serv}}{\lambda_{in}} = \frac{\lambda_h P_{adhoc-loss} + \sum_{i=1}^{\infty} i\alpha\pi(i,N)\mathbf{e}}{\lambda_{in}}.$$

The loss probability of a regular client is calculated as

$$P_{client-loss} = \frac{1}{\lambda_d} q\alpha \sum_{i=1}^{\infty} i\pi(i,N) = \frac{\lambda_d - \gamma N_{reg-client}}{\lambda_d}.$$

The existence of these two alternative formulas for computation of a regular client loss probability is helpful for verification of the analytical and numerical results.

## 5 Numerical Experiment

The goals of this numerical experiment are to confirm the feasibility of the presented way for computation of the main performance indicators of the system, to highlight the dependence of these indicators on the number $N$ of the servers and the control threshold $M$ and to show an opportunity to use the obtained results for optimisation of the system operation.

We suppose that the arrival flow of regular clients is defined by the $MAP$ arrival flow that is described by the following matrices:

$$D_0 = \begin{pmatrix} -0.3542862 & 0.0114286 \\ 0.0114286 & -0.125714 \end{pmatrix}, D_1 = \begin{pmatrix} 0.331429 & 0.0114286 \\ 0.0022854 & 0.112 \end{pmatrix}.$$

This arrival flow has the average arrival intensity $\lambda_d = 0.2$, and the coefficients of correlation and variation of inter-arrival times $c_{cor} = 0.14325$, $c_{var} = 1.522$.

The $MAP$ arrival flow of ad hoc requests is described by the matrices

$$H_0 = \begin{pmatrix} -2.982104 & 0.298211 \\ 0.0477136 & -0.894632 \end{pmatrix}, H_1 = \begin{pmatrix} 2.32604 & 0.357853 \\ 0.0119284 & 0.83499 \end{pmatrix}$$

with the following metrics: $\lambda_h = 1$, $c_{cor} = 0.05255$, $c_{var} = 1.16813$.

The rest of the system's parameters are defined as follows:

The average intensity of request generation by a regular client is $\alpha = 0.1$.

The parameter of the exponential distribution of the sojourn time of a regular client in the system is $\gamma = 0.002$.

The parameter of the exponential distribution of the request's service time is $\mu = 0.5$.

The probability of the regular client abandonment due to a request loss is $q = 0.2$.

Let us vary the number of the used servers $N$ over the interval $[1, 60]$ with step 1, and the parameter $M$ over the interval $[1, N]$ also with step 1.

Figures 2, 3 and 4 illustrate the dependence of the average number of busy servers $N_{serv}$, the average number of regular clients $N_{reg-client}$, and the average request arrival rate $\lambda_{in}$ on the parameters $N$ and $M$.

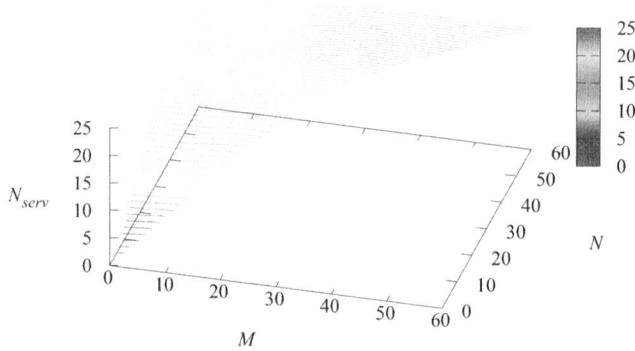

**Fig. 2.** Dependence of the average number $N_{serv}$ of busy servers on the parameters $N$ and $M$

Figures 5, 6, 7 and 8 illustrate the dependence of the loss probability $P_{reg-loss}$ of a request generated by a regular client, the loss probability $P_{adhoc-loss}$ of an ad hoc request, the loss probability $P_{loss}$ of an arbitrary request, and the client loss probability $P_{client-loss}$ on the parameters $N$ and $M$.

From Fig. 2, one can conclude that the average number of busy servers $N_{serv}$ grows with the increase in the number of servers $N$. It is evident since the increase in the number of servers leads to a decrease in the loss probability of arriving requests; see Fig. 7. The dependence of the average number of busy servers $N_{serv}$ on the parameter $M$ is non-monotonic in this numerical experiment. At first, the average number of busy servers increases with the increase in $M$. This is explained by the fact that with an increase in $M$, the probability of ad hoc request loss essentially decreases (see Fig. 6). However, with further increase in $M$, the average number $N_{serv}$ begins to decrease. This can be explained by the increase in client loss probability $P_{client-loss}$ with an increase in $M$ (see Fig. 8). That leads to a decrease in the number of clients in the system; see Fig. 3. Thus, fewer regular clients generate requests, which implies a decrease in the average number of busy servers $N_{serv}$, see Fig. 2. The dependence of the average request arrival rate on $N$ and $M$ is similar due to the same reasons.

From these figures, one can conclude that the growth in the number of servers $N$ under the fixed parameter $M$ leads to better system performance. Namely, the loss probabilities of clients and all types of requests decrease with the increase in $N$. Thus, for better system's performance, the manager should increase the number of servers. However, in real-word systems, servers cost money, and it can be

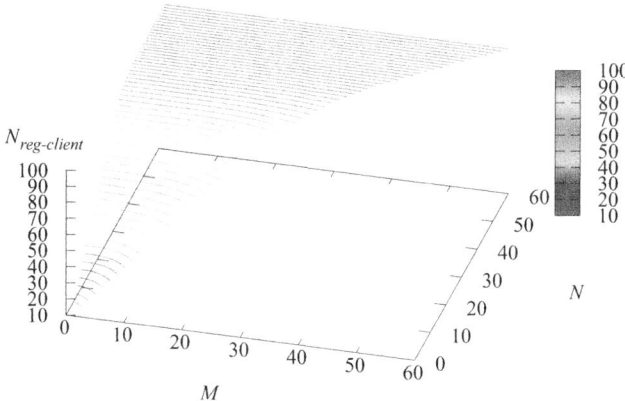

**Fig. 3.** Dependence of the average number $N_{reg-client}$ of regular clients in the system on the parameters $N$ and $M$

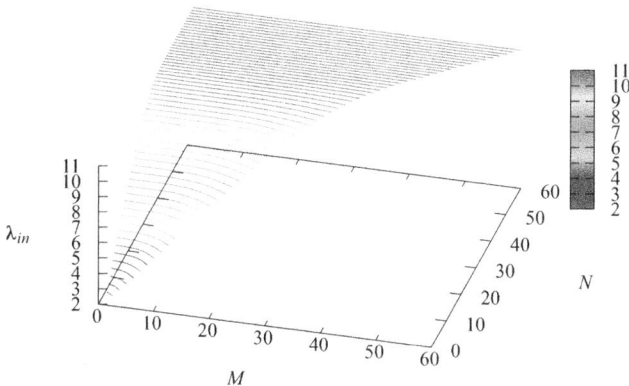

**Fig. 4.** Dependence of the average rate $\lambda_{in}$ of incoming requests on the parameters $N$ and $M$

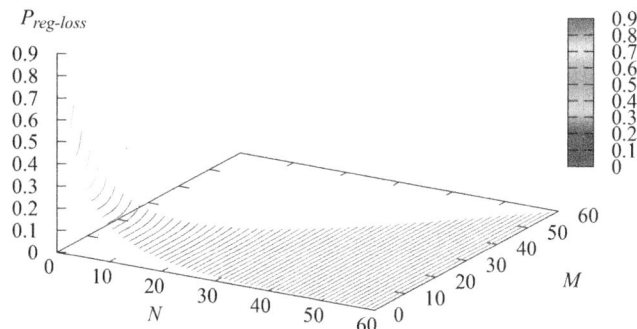

**Fig. 5.** Dependence of the probability $P_{reg-loss}$ on the parameters $N$ and $M$

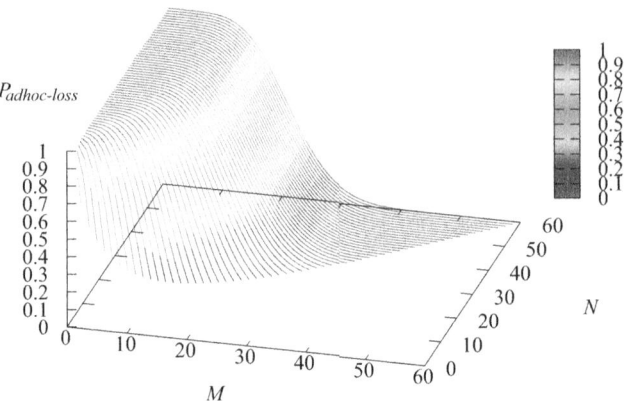

**Fig. 6.** Dependence of the probability $P_{adhoc-loss}$ on the parameters $N$ and $M$

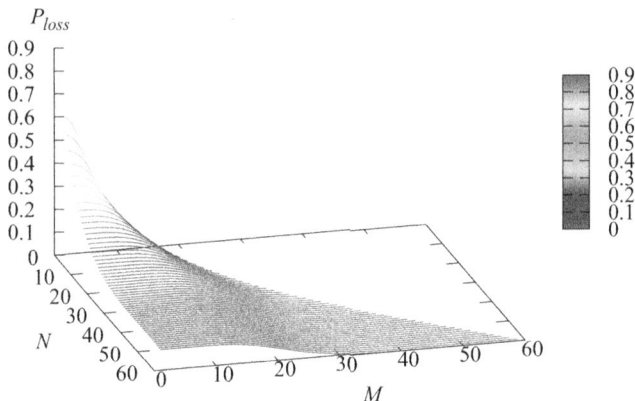

**Fig. 7.** Dependence of the probability $P_{loss}$ of an arbitrary request loss on the parameters $N$ and $M$

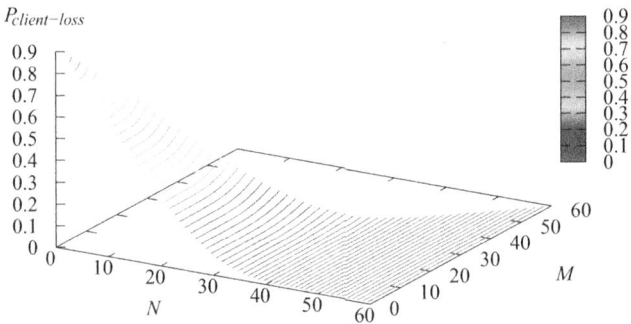

**Fig. 8.** Dependence of $P_{client-loss}$ on the parameters $N$ and $M$

not profitable to maintain a large number of servers $N$. With the decrease in the parameter $M$, the performance measures related to regular clients are improved, but the loss probability of the ad hoc request increases. The parameter $M$ defines the balance between the interests of ad hoc requests and regular clients. So, the problem of optimal choice of the parameters $N$ and $M$ is important.

Let us assume that the quality of the system operation is evaluated via the following cost criterion:

$$E = E(N, M) =$$
$$a\lambda_{serv} - c_1 \lambda_h P_{adhoc-loss} - c_2 \alpha N_{client} P_{reg-loss} - c_3 \lambda_d P_{client-loss} - c_4 N,$$

where $a$ is the profit gained by the system for one request service, $c_1$ is the charge for one ad hoc request loss, $c_2$ is the charge for one regular client's request loss, $c_3$ is the charge for one regular client loss, and $c_4$ is the cost of one server maintenance.

The cost criterion $E$ defines the average profit (revenue) of the system per unit time. The goal is to find the values of $N$ and $M$ that provide the maximum value to this criterion.

In this numerical example, we fix the following cost coefficients:

$$a = 1; \ c_1 = 1; c_2 = 5; c_3 = 50; c_4 = 0.1.$$

Figure 9 illustrates the shape of the surface that shows the dependence of the cost criterion $E$ on the parameters $N$ and $M$.

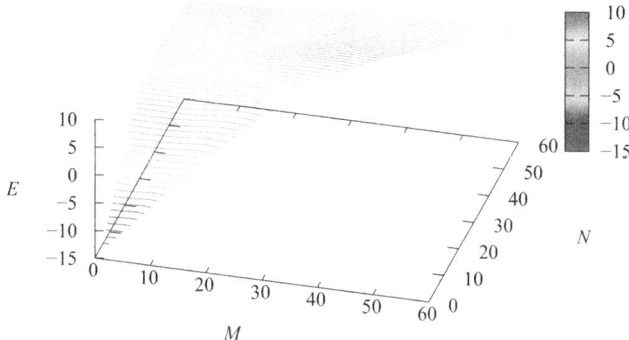

**Fig. 9.** Dependence of the cost criterion $E$ on the parameters $N$ and $M$.

One can conclude that the system profit is below zero if the number of servers $N$ is less than 20. To have the maximal positive profit with the number of servers $N = 20$ one has to choose the parameter $M$ equal to 17 that provides the 0.1433 value of the profit. With an increase in $N$ above 20, the values of the cost criterion increase until $N \leq 39$. Further increase in $N$ is not profitable since the addition of extra servers does not lead to essential improvement of the system performance, but the servers maintenance price increases.

Therefore, the optimal value of the cost criterion is $E^* = 6.79296$ and is achieved for $N^* = 39$ and $M^* = 34$. If we fix the value of $N = 60$ and set $M = 60$ (ad hoc clients are never rejected when there are idle servers), the value of the cost criterion will be $E(60, 60) = 5$. Thus, the optimal choice of the parameters $N$ and $M$ compared to the use of all available 60 servers and not applying admission control by the ad hoc clients can bring the essential economical profit.

## 6 Conclusion

We have analysed the multi-server queueing system that provides service to the regular clients, each of which sequentially generates a bunch of requests, and ad hoc clients, each of which generates only one request. Due to the higher importance of the regular clients, a certain server reservation scheme is applied: service of ad hoc clients is denied when the number of busy servers at an ad hoc client arrival epoch exceeds a preassigned threshold. All distributions characterising the behaviour of the system are assumed to be exponential except the arrival processes that are assumed to be Markov arrival processes suitable for the adequate description of arrival flows in various real-world systems, including telecommunication systems and call centres.

The behaviour of this system is described by the multidimensional $CTMC$. The explicit expression of the generator of this chain is obtained. Formulas for computation of the key performance indicators are given. The dependence of these indicators on the number of servers and the threshold defining the policy of ad hoc client admission is numerically highlighted. The possibility of using these results for managerial goals is illustrated.

The system with many types of ad hoc requests having different priorities with multi-threshold control strategy, see e.g., [17], can be considered as the extension of the results of this paper. Practically important modification of the model when the flows of the regular and ad hoc clients are mutually correlated and defined by the Marked Markov Arrival process, see, e.g., [18–21], can be analysed as well. Extension of the results to the model with more general than the exponential, phase type ($PH$) distribution of service times, see, e.g., [14, 22–24].

## References

1. Kim, C., Dudin, S., Klimenok, V.: The $MAP/PH/1/N$ queue with flows of customers as a model for traffic control in telecommunication networks. Perform. Eval. **66**(9–10), 564–579 (2009)
2. Kim, C., Dudin, S.: A multi-server queueing model with retrial connection arrivals as a model for optimisation of the traffic control. Int. J. Syst. Sci. **43**(8), 1555–1567 (2012)
3. Kist, A.A., Lloyd-Smith, B., Harris, R.J.: A simple IP flow blocking model: performance challenges for efficient next generation networks. In: Proceedings of 19th International Teletraffic Congress, pp. 355–364 (2005)

4. De Vuyst, S., Wittevrongel, S., Bruneel, H.: Statistical multiplexing of correlated variable-length packet trains: an analytic performance study. J. Oper. Res. Soc. **52**, 318–327 (2001)
5. De Vuyst, S., Wittevrongel, S., Bruneel, H.: Mean value and tail distribution of the message delay in statistical multiplexers with correlated train arrivals. Perform. Eval. **48**, 103–129 (2002)
6. Chakravarthy, S.R.: The batch Markovian arrival process: a review and future work. Adv. Probabil. Theory Stochastic Processes **1**(1), 21–49 (2001)
7. Chakravarthy, S.R.: Introduction to Matrix-Analytic Methods in Queues 1: Analytical and Simulation Approach - Basics. ISTE Ltd, London and John Wiley and Sons, New York (2022)
8. Chakravarthy, S.R.: Introduction to Matrix-Analytic Methods in Queues 2: Analytical and Simulation Approach - Queues and Simulation. ISTE Ltd, London and John Wiley and Sons, New York (2022)
9. Lucantoni, D.M.: New results on the single server queue with a batch Markovian arrival process. Commun. Stat. Stochastic Models **7**(1), 1–46 (1991)
10. Dudin, A.N., Klimenok, V.I., Vishnevsky, V.M.: The Theory of Queuing Systems with Correlated Flows. Springer, Cham (2020)
11. Gonzalez, M., Lillo, R.E., Ramirez Cobo, J.: Call center data modeling: a queueing science approach based on Markovian arrival processes. Qual. Technol. Quant. Manag. (2024). https://doi.org/10.1080/16843703.2024.2371715
12. Naumov, V., Gaidamaka, Y., Yarkina, N., Samouylov, K.: Matrix and Analytical Methods for Performance Analysis of Telecommunication Systems. Springer, Cham (2022)
13. Graham, A.: Kronecker Products and Matrix Calculus with Applications. Courier Dover Publications, New York (2018)
14. Neuts, M.F.: Matrix-Geometric Solutions in Stochastic Models: An Algorithmic Approach. Courier Corporation, Chelmsford (1994)
15. Dudin, A.N., Dudin, S.A., Klimenok, V.I., Dudina, O.S.: Stability of queueing systems with impatience, balking and non-persistence of customers. Mathematics **12**(14), 2214 (2024)
16. Dudin, S., Dudina, O.: Retrial multi-server queuing system with PHF service time distribution as a model of a channel with unreliable transmission of information. Appl. Math. Model. **65**, 676–695 (2019)
17. Dudin, A.: Optimal multithreshold control for a $BMAP/G/1$ queue with $N$ service modes. Queue. Syst. **30**, 273–287 (1998)
18. He, Q.M., Neuts, M.F.: Markov chains with marked transitions. Stoch. Processes Appl. **74**(1), 37–52 (1998)
19. He, Q.M.: Queues with marked calls. Adv. Appl. Probab. **28**, 567–587 (1996)
20. He, Q.M.: Fundamentals of Matrix-Analytic Methods, vol. 365. Springer, New York (2014)
21. Dudin, S., Kim, C., Dudina, O.: $MMAP/M/N$ queueing system with impatient heterogeneous customers as a model of a contact center. Comput. Oper. Res. **40**(7), 1790–1803 (2013)
22. O'Cinneide, C.A.: Phase-type distributions: open problems and a few properties. Stoch. Model. **15**, 731–757 (1999)
23. Asmussen, S.: Applied Probability and Queues. Springer, Heidelberg (2003)
24. He, Q.M., Alfa, A.S.: Space reduction for a class of multidimensional Markov chains: a summary and some applications. INFORMS J. Comput. **30**(1), 1–10 (2018)

# Modeling and Analysis of Cyclic Control of Periodic Conflict Flows

V. L. Tsodikov[(✉)] and Andrei V. Zorine

Nizhny Novgorod State University named after N. I. Lobachevsky,
603022 Nizhny Novgorod, Russia
vovatsodikov@gmail.com, andrei.zorine@itmm.unn.ru

**Abstract.** The paper studies a queueing system with periodic intensity of conflicting input flows, reflecting, in particular, the functioning of the traffic flow control system in the class of cyclic algorithms with a fixed signal durations and daily fluctuations in the traffic flow intensity.

The process of conflicting flows control with homogeneous customers and periodic intensity is considered in detail. Sequences of random variables are constructed that describe the processes occurring in the system. Recurrence relations are derived that connect these quantities.

The mathematical model of the system is a countable Markov chain. Transition probabilities for this Markov chain are calculated. An ordering of essential states is introduced, revealing the block structure of the transition probability matrix. An algorithm for calculating a stationary distribution is proposed, based on the well-known chain censoring method.

The restriction of the Markov chain to the class of essential communicating states is considered, recurrence relations are established for generating functions for the distribution of the queue length, the state of the service device and the time of day. Using the iterative-majorant approach, necessary and sufficient conditions for the existence of a stationary distribution for queue lengths are studied.

**Keywords:** conflict flows · non-stationary Poisson flow · fixed-durations cyclic control · counter Markov chain · functional-statistical description · stationary distribution · censored Markov chain · conditions for the existence of a stationary distribution

## 1 Introduction

At present, methods of queueing theory are used in the design of automated control systems in technical, social and economic sectors. In particular, these methods are used to solve problems about traffic flows. Various studies have developed both deterministic and stochastic models of traffic flow formation. [1–6]

These and many other works on traffic control do not take into account the periodic nature of changes in flow characteristics associated with daily, weekly and other reasons. At the same time, in queueing theory, a certain number of works are devoted to service systems with time-dependent and, in particular, with periodically changing laws for the input flow and service intervals [7]

## 2 The Problem Statement and the Markov Chain Model

There are $m < \infty$ input flows $\Pi_1, \Pi_2, \ldots, \Pi_m$. The flow $\Pi_j$ is a nonstationary Poisson flow, $\lambda_j(t)$ is the instantaneous intensity of the flow $\Pi_j$ at time $t$. Customers from the flow $\Pi_j$ join a queue $O_j$ of unlimited capacity. We will consider the instantaneous intensity as a time-dependent quantity. Since the traffic intensity changes during the day, and some patterns are preserved from day to day, it is advisable to consider the function $\lambda_j(t)$ as *periodic*: $\lambda_j(t+T_D) = \lambda_j(t)$, where $T_D > 0$ is some constant, for example, a day, 24 h. We will also assume that the function $\Lambda(x) = \int_0^x \lambda(t)dt$ is known.

The server is a traffic light that has $2m$ internal states $\Gamma^{(1)}, \Gamma^{(2)}, \ldots, \Gamma^{(2m)}$. In the state $\Gamma^{(2j-1)}$ only customers of the flow $\Pi_j$ are serviced. In the state $\Gamma^{(2j)}$ all customers are not serviced: these states are introduced to resolve the conflict of flows (the states of readjustment and reorientation). A cyclic algorithm for switching traffic light states was chosen to control the flows: $\Gamma^{(1)} \to \Gamma^{(2)} \to \ldots \to \Gamma^{(2m)} \to \Gamma^{(1)} \to \ldots$. The time that the service device stays in the state $\Gamma^{(r)}$ is fixed and equals $T_r$. During the time $T_{2j-1}$, no more than $\ell_j$ customers from $O_j$ can be serviced.

Let $\tau_0 = 0, \tau_1, \tau_2, \ldots$ be the moments of server state change. We will assume that we observe the system only at these moments. Let $X = \{0, 1, \ldots\} \times \ldots \times \{0, 1, \ldots\}$, $\Gamma = \{\Gamma^{(1)}, \Gamma^{(2)}, \ldots, \Gamma^{(2m)}\}$. Let $\Gamma_i \in \Gamma$ denote the server state on the interval $(\tau_{i-1}, \tau_i]$, $i = 1, 2, \ldots$; let $\Gamma_0$ be an auxiliary fictitious state of the serving device, in which the traffic light is at the first observation moment $\tau_0$, and which is immediately replaced by the next state $\Gamma_1$. Let $\eta_{j,i} \in \{0, 1, \ldots\}$ be the number of customers of the flow $\Pi_j$ that arrived on the interval $[\tau_i, \tau_{i+1})$, let $\kappa_{j,i} \in \{0, 1, \ldots\}$ be the number of customers in the queue $O_j$ at the moment $\tau_i$; let $\xi_{j,i}$ be the maximum possible number of customers from queue $O_j$ that can be serviced in the interval $[\tau_i, \tau_{i+1})$, and let $\bar{\xi}_{j,i}$ be the number of customers from queue $O_j$ that are actually serviced in the interval $[\tau_i, \tau_{i+1})$. Let us introduce the vector $\kappa_i = (\kappa_{1,i}, \kappa_{2,i}, \ldots, \kappa_{m,i})$. Let us introduce the quantities $\tau'_i = \tau_i$ (mod $T_D$), that is, the remainder after division of $\tau_i$ by $T_D$ (this will have the meaning of the time of day).

We will assume that all quantities $T_1, T_2, \ldots, T_{2m}$ and $T_D$ are commensurable, that is, there is some time interval $\Delta$ that fits an integer number of times into each of all the specified times. From this assumption it follows that the quantities $\tau'_i$ have only a *finite* number of values: $\tau'_i \in \{0, \Delta, 2\Delta, \ldots, T_D - \Delta\}$. From this also follows that there exist integers $p, q$ such that

$$pT = qT_D, \qquad (1)$$

where $T = T_1 + T_2 + \ldots + T_{2m}$.

Let us establish recurrent relations for the state of the service device, the queue length, and the sequence $\{\tau'_i; i = 0, 1, \ldots\}$ in $i = 0, 1, \ldots$. It can be seen that $\bar{\xi}_{j,i} \leq \kappa_{j,i} + \eta_{j,i}$: the number of actually serviced machines in the interval $[\tau_i, \tau_{i+1})$ cannot be greater than the number of machines in the queue $O_j$ at the beginning of the interval (at time $\tau_i$) plus the number of machines that entered the given flow during this time interval. As a result, at time $\tau_{i+1}$ we obtain:

$$\kappa_{j,i+1} = \kappa_{j,i} + \eta_{j,i} - \bar{\xi}_{j,i}. \tag{2}$$

It is also clear that $\bar{\xi}_{j,i} \leq \xi_{j,i}$. At the same time, we know that $\xi_{j,i} = 0$ if the prohibiting signal for flow $\Pi_j$ is on in this time interval and $\xi_{j,i} = \ell_j$ if the enabling signal for flow $\Pi_j$ is on in this time interval.

So, $\bar{\xi}_{j,i} \leq \min\{\xi_{j,i}, \kappa_{j,i} + \eta_{j,i}\}$. Since the goal is to reduce the number of customers in the queue, we will choose $\bar{\xi}_{j,i}$ strictly equal to the given minimum. This is the *extreme service strategy*.

So, let $\bar{\xi}_{j,i} = \min\{\xi_{j,i}, \kappa_{j,i} + \eta_{j,i}\}$. If $\xi_{j,i} > \kappa_{j,i} + \eta_{j,i}$, then $\bar{\xi}_{j,i} = \kappa_{j,i} + \eta_{j,i}$. Therefore, using formula (2), we obtain: $\kappa_{j,i+1} = 0$. If $\xi_{j,i} \leq \kappa_{j,i} + \eta_{j,i}$, then $\bar{\xi}_{j,i} = \xi_{j,i}$, and then $\kappa_{j,i+1} = \kappa_{j,i} + \eta_{j,i} - \xi_{j,i}$.

Thus, the change in the queue length $O_j$ over one time interval can be briefly described by the following recurrence relation:

$$\kappa_{j,i+1} = \max\{0, \kappa_{j,i} + \eta_{j,i} - \xi_{j,i}\}. \tag{3}$$

For $r \in \mathbb{N}$ we define the operation $r \oplus 1$ by the equalities $r \oplus 1 = r + 1$ for $r < 2m$ and $(2m) \oplus 1 = 1$. Next, for $r \in \mathbb{N}$, $a \geq 2$, we set by induction $r \oplus a = (r \oplus (a-1)) \oplus 1$.

The flow control algorithm is formalized as a mapping $u \colon \Gamma \to \Gamma$. We will assume that $u(\Gamma^{(j)}) = \Gamma^{(j \oplus 1)}$. We also introduce the mapping $v \colon \Gamma \to \{T_1, T_2, \ldots, T_{2m}\}$ by the equality $v(\Gamma^{(j)}) = T_{j \oplus 1}$.

Then the formalization of the traffic light operation has the form of recurrence relations:

$$\tau_0 = 0, \qquad \tau_{i+1} = \tau_i + v(\Gamma_i), \qquad \Gamma_{i+1} = u(\Gamma_i), \qquad i = 0, 1, \ldots.$$

**Theorem 1.** *For a given distribution of the vector $(\tau'_0, \Gamma_0, \kappa_0)$, the sequence*

$$\{(\tau'_i, \Gamma_i, \kappa_i); i = 0, 1, \ldots\} \tag{4}$$

*is a Markov chain.*

*Proof.* Let numbers $t_0, t_1, \ldots, t_{i+1} \in \{0, \Delta, 2\Delta, 3\Delta, \ldots, M\Delta\}$, indices $j_0, j_1, \ldots, j_{i+1} \in \{1, 2, \ldots, 2m\}$ and vectors $x^{(0)}, x^{(1)}, \ldots, x^{(i+1)} \in X$ be given. Let's calculate the conditional probability:

$$\mathbf{P}\big(\{\tau'_{i+1} = t_{i+1}, \Gamma_{i+1} = \Gamma^{(j_{i+1})}, \kappa_{i+1} = x^{(i+1)}\} \big| \bigcap_{l=0}^{i} \{\tau'_l = t_l, \Gamma_l = \Gamma^{(j_l)}, \kappa_l = x^{(l)}\}\big).$$

The probability is equal to

$$\mathbf{P}(\{\tau_i + v(\Gamma_i) = t_{i+1} \pmod{T_D}, u(\Gamma_i) = \Gamma^{(j_{i+1})}, x_j^{(i+1)} =$$
$$= \max\{0, \kappa_{j,i} + \eta_{j,i} - \xi_{j,i}\}, j = 1, 2, \ldots, m\} | \bigcap_{l=0}^{i} \{\tau'_l = t_l, \Gamma_l = \Gamma^{(j_l)}, \kappa_l = x^{(l)}\}) =$$
$$= \mathbf{P}(\{t_i + v(\Gamma^{(j_i)}) = t_{i+1} \pmod{T_D}, u(\Gamma^{(j_i)}) = \Gamma^{(j_{i+1})}, x_j^{(i+1)} =$$
$$= \max\{0, x_j^{(i)} + \eta_{j,i} - \xi_{j,i}\}, j = 1, 2, \ldots, m\} | \bigcap_{l=0}^{i} \{\tau'_l = t_l, \Gamma_l = \Gamma^{(j_l)}, \kappa_l = x^{(l)}\}). \quad (5)$$

For brevity, we denote the obtained probability as $\mathbf{P}(A|B)$, where

$$A = \{t_i + v(\Gamma^{(j_i)}) = t_{i+1} \pmod{T_D}, u(\Gamma^{(j_i)}) = \Gamma^{(j_{i+1})},$$
$$x_j^{(i+1)} = \max\{0, x_j^{(i)} + \eta_{j,i} - \xi_{j,i}\}, j = 1, 2, \ldots, m\},$$

and the condition

$$B = \bigcap_{l=0}^{i} \{\tau'_l = t_l, \Gamma_l = \Gamma^{(j_l)}, \kappa_l = x^{(l)}\}.$$

Note that for fixed values of $t_0, t_1, \ldots, t_{i+1}$ and $j_0, j_1, \ldots, j_{i+1}$, the fulfillment of the equalities $\tau_i + v(\Gamma_i) = t_{i+1} \pmod{T_D}$, $u(\Gamma_i) = \Gamma^{(j_{i+1})}$ is not accidental and does not depend on the values of $\kappa_{j,i}$ and $\xi_{j,i}$ (the equalities are either fulfilled for the chosen values or are not fulfilled).

Let's use the law of total probability. In our case, the hypotheses will be the equalities $\eta_i = b, \xi_i = \tilde{b}$, where $b$ and $\tilde{b}$ are non-negative integer vectors of dimension $m$.

$$\mathbf{P}(A|B) = \sum_{b,\tilde{b} \in X} \mathbf{P}(\eta_i = b, \xi_i = \tilde{b}|B) \mathbf{P}(A|B \cap \{\eta_i = b, \xi_i = \tilde{b}\}). \quad (6)$$

The second factor under the sum sign *does not depend* on $t_0, \ldots, t_{i-1}, j_0, \ldots, j_{i-1}$ and $x^{(0)}, \ldots, x^{(i-1)}$; the probability value $\mathbf{P}(A|B \cap \{\eta_i = b, \xi_i = \tilde{b}\})$ is either zero or one.

Let us consider separately the probability $\mathbf{P}(\eta_i = b, \xi_i = \tilde{b}|B)$. If $r$ is even, $r = 2k$, then the yellow (prohibitory) signal is on *for all* flows, then the vector $\tilde{b} = (0, 0, \ldots, 0)$, since machines from any queue are not serviced. If $r$ is odd, $r = 2k-1$, then the permitting signal is on only for the $k$-th flow, and the prohibiting signal is on for the remaining flows. Then the vector $\tilde{b} = (0, \ldots, 0, \ell_k, 0, \ldots, 0)$, where $\ell_k$ is on the $k$-th place in the vector. If the two described cases are not met, then the conditional probability will be zero. If these conditions are met, then, we obtain

$$\prod_{s=1}^{m} \exp\bigl(-(\Lambda_s(t_i + v(\Gamma^{(j_i)})) - \Lambda_s(t_i))\bigr) \cdot \frac{(\Lambda_s(t_i + v(\Gamma^{(j_i)})) - \Lambda_s(t_i))^{b_s}}{b_s!}. \quad (7)$$

Thus, we finally get the following:

$$\mathbf{P}(\eta_i = b, \xi_i = \widetilde{b} | \bigcap_{l=0}^{i} \{\tau'_l = t_l, \Gamma_l = \Gamma^{(j_l)}, \kappa_l = x^{(l)}\}) =$$

$$= \begin{cases} \prod_{s=1}^{m} (e^{-(\Lambda_s(t_i + v(\Gamma^{(j_i)})) - \Lambda_s(t_i))}) & \text{if } u(\Gamma^{(j_i)}) = \Gamma^{(r)}, r = 2k, \widetilde{b} = (0, 0, \ldots, 0) \\ \quad \times \frac{(\Lambda_s(t_i + v(\Gamma^{(j_i)})) - \Lambda_s(t_i))^{b_s}}{b_s!} & \text{or if } r = 2k-1, \widetilde{b} = (0, \ldots, 0, \ell_k, 0, \ldots, 0); \\ 0, & \text{otherwise.} \end{cases} \quad (8)$$

The obtained expression also does not depend on $t_0, \ldots, t_{i-1}, j_0, \ldots, j_{i-1}$ and $x^{(0)}, \ldots, x^{(i-1)}$. This means that their product according to the formula (6) does not either. Thus, we have obtained that the probability of occurrence depends only on the state reached at the previous event. Calculating the conditional distribution relative to only the previous state leads to the same expression, and, therefore, this sequence of random events is a Markov chain.

*Remark 1.* Formulas (6), (8) establish the type of transition probabilities.

Let the function rem$(x, d)$ give the remainder of dividing $x$ by $d$. Consider for $j = 1, 2, \ldots, 2m$ the sets

$$G_j = \bigcup_{s=0}^{p-1} \{(\text{rem}(sT, T_D), \Gamma^{(j)}),$$
$$(\text{rem}(sT + T_{j \oplus 1}, T_D), \Gamma^{(j \oplus 1)}), (\text{rem}(sT + T_{j \oplus 1} + T_{j \oplus 2}, T_D), \Gamma^{(j \oplus 2)}), \ldots,$$
$$(\text{rem}(sT + T_{j \oplus 1} + T_{j \oplus 2} + \ldots + T_{j \oplus (2m-1)}, T_D), \Gamma^{(j \oplus (2m-1))})\}.$$

Each of the sets $G_j$, $j = 1, 2, \ldots, 2m$ contains the same number $2mp$ of elements, where the parameter $p$ is determined from (1).

**Theorem 2.** *Given the distribution of the vector $(\tau'_0, \Gamma_0, \kappa_0)$, the sequence*

$$\{(\tau'_i, \Gamma_i, \kappa_i); i = 0, 1, \ldots\} \quad (9)$$

*is a homogeneous Markov chain with state space $(G_1 \cup G_2 \cup \ldots \cup G_{2m}) \times X$.*

It follows from theorem 2 that if all sets $G_1, G_2, \ldots, G_{2m}$ coincide, then all states belong to one class of essential communicating states. And if among the specified sets there are non-coinciding ones, then there may be several classes of essential communicating states.

Let us establish the type of transition probabilities. Let us write out the conditional probability in detail:

$$\mathbf{P}(\Gamma_{i+1} = \Gamma^{(s)}, \kappa_{j,i+1} = w, \tau'_{i+1} = \widetilde{t} | \Gamma_i = \Gamma^{(r)}, \kappa_{j,i} = x, \tau'_i = t) =$$
$$= \mathbf{P}(u(\Gamma_i) = \Gamma^{(s)}, \max\{0, \kappa_{j,i} + \eta_{j,i} - \xi_{j,i}\} = w,$$
$$\text{rem}(\tau'_i + v(\Gamma_i), T_D) = \widetilde{t} | \Gamma_i = \Gamma^{(r)}, \kappa_{j,i} = x, \tau'_i = t) =$$
$$= \mathbf{P}(u(\Gamma^{(r)}) = \Gamma^{(s)}, \max\{0, x + \eta_{j,i} - \xi_{j,i}\} = w,$$
$$\text{rem}(t + v(\Gamma^{(r)}), T_D) = \widetilde{t} | \Gamma_i = \Gamma^{(r)}, \kappa_{j,i} = x, \tau'_i = t). \quad (10)$$

Note that equations $u(\Gamma^{(r)}) = \Gamma^{(s)}$ and $\text{rem}(t+v(\Gamma^{(r)}), T_D) = \tilde{t}$ are true or false depending on fixed parameters $r, s, t, \tilde{t}$ and are free of any random variables. So, they make the certain event or the impossible event. Only equality $\max\{0, x + \eta_{j,i} - \xi_{j,i}\} = w$ is random because of $\eta_{j,i}$.

For the sake of compactness, we introduce the following notation:

$$\Phi(s,r,b) = \begin{cases} (e^{-(\Lambda_s(t_i+v(\Gamma^{(r)}))-\Lambda_s(t_i))}) \dfrac{(\Lambda_s(t_i+v(\Gamma^{(r)}))-\Lambda_s(t_i))^{b_s}}{b_s!}, & \text{for } b \geq 0; \\ 0, & \text{for } b < 0. \end{cases} \quad (11)$$

The following lemmas follow directly from relation (10).

**Lemma 1.** *Transition probability*

$$\mathbf{P}(\Gamma_{i+1} = \Gamma^{(s)}, \kappa_{j,i+1} = w, \tau'_{i+1} = \tilde{t} | \Gamma_i = \Gamma^{(r)}, \kappa_{j,i} = x, \tau'_i = t) = 0, \quad (12)$$

*if $s \neq r \oplus 1$ or $\text{rem}(t + T_{r\oplus 1}, T_D) \neq \tilde{t}$.*

**Lemma 2.** *Transition probability*

$$\mathbf{P}(\Gamma_{i+1} = \Gamma^{(s)}, \kappa_{j,i+1} = w, \tau'_{i+1} = \tilde{t} | \Gamma_i = \Gamma^{(r)}, \kappa_{j,i} = x, \tau'_i = t) =$$
$$= \Phi(j, r, w - x), (13)$$

*if $s = r \oplus 1$ and $\text{rem}(t + T_{r\oplus 1}, T_D) = \tilde{t}$ and $r \oplus 1 \neq 2j - 1$ and $w \geq x$.*

**Lemma 3.** *Transition probability*

$$\mathbf{P}(\Gamma_{i+1} = \Gamma^{(s)}, \kappa_{j,i+1} = w, \tau'_{i+1} = \tilde{t} | \Gamma_i = \Gamma^{(r)}, \kappa_{j,i} = x, \tau'_i = t) = 0, \quad (14)$$

*if $s = r \oplus 1$ and $\text{rem}(t + T_{r\oplus 1}, T_D) = \tilde{t}$ and $r \oplus 1 \neq 2j - 1$ and $w < x$.*

**Lemma 4.** *Transition probability*

$$\mathbf{P}(\Gamma_{i+1} = \Gamma^{(s)}, \kappa_{j,i+1} = w, \tau'_{i+1} = \tilde{t} | \Gamma_i = \Gamma^{(r)}, \kappa_{j,i} = x, \tau'_i = t) =$$
$$= \Phi(j, r, w - x + \ell_j), \quad (15)$$

*if $s = r \oplus 1$ and $\text{rem}(t + T_{r\oplus 1}, T_D) = \tilde{t}$ and $r \oplus 1 = 2j - 1$ and $w > 0$.*

**Lemma 5.** *Transition probability*

$$\mathbf{P}(\Gamma_{i+1} = \Gamma^{(s)}, \kappa_{j,i+1} = w, \tau'_{i+1} = \tilde{t} | \Gamma_i = \Gamma^{(r)}, \kappa_{j,i} = x, \tau'_i = t) =$$
$$= \sum_{b=0}^{\ell_j - x} \Phi(j, r, b), \quad (16)$$

*if $s = r \oplus 1$ and $\text{rem}(t + T_{r\oplus 1}, T_D) = \tilde{t}$ and $r \oplus 1 = 2j - 1$ and $w = 0$.*

## 3 An Algorithm to Compute of Stationary Distribution by Chain Censoring Method

Consider a fixed flow $\Pi_j$, $j = 1, 2, \ldots, 2m$. We assume that the set of essential states is ordered as follows: first, the value of $\tau'_i$ changes, then the state of the serving device $\Gamma_i$, and finally the value of the queue length $\kappa_i$. $\tau'_i$ takes the values $0, \Delta, 2\Delta, 3\Delta, \ldots, T_D - 2\Delta, T_D - \Delta$ (due to the assumption of commensurability of the durations of all signals $T_1, T_2, \ldots, T_{2m}$). Let us recall that $\Gamma_i$ takes values from the set $\Gamma$, and $\kappa_i$ can take any non-negative values. We obtain the sequence (read line by line):

$$(0, \Gamma^{(1)}, 0), \quad (\Delta, \Gamma^{(1)}, 0), \quad (2\Delta, \Gamma^{(1)}, 0), \quad \ldots, \quad (T_D - \Delta, \Gamma^{(1)}, 0),$$
$$(0, \Gamma^{(2)}, 0), \quad (\Delta, \Gamma^{(2)}, 0), \quad (2\Delta, \Gamma^{(2)}, 0), \quad \ldots, \quad (T_D - \Delta, \Gamma^{(2)}, 0), \ldots,$$
$$(0, \Gamma^{(2m)}, 0), \quad (\Delta, \Gamma^{(2m)}, 0), \quad (2\Delta, \Gamma^{(2m)}, 0), \quad \ldots, \quad (T_D - \Delta, \Gamma^{(2m)}, 0),$$
$$(0, \Gamma^{(1)}, 1), \quad (\Delta, \Gamma^{(1)}, 1), \quad (2\Delta, \Gamma^{(1)}, 1), \quad \ldots, \quad (T_D - \Delta, \Gamma^{(1)}, 1),$$
$$(0, \Gamma^{(2)}, 1), \quad (\Delta, \Gamma^{(2)}, 1), \quad (2\Delta, \Gamma^{(2)}, 1), \quad \ldots, \quad (T_D - \Delta, \Gamma^{(2)}, 1), \ldots,$$
$$(0, \Gamma^{(2m)}, 1), \quad (\Delta, \Gamma^{(2m)}, 1), \quad (2\Delta, \Gamma^{(2m)}, 1), \quad \ldots, \quad (T_D - \Delta, \Gamma^{(2m)}, 1),$$
$$(0, \Gamma^{(1)}, 2), \quad (\Delta, \Gamma^{(1)}, 2), \quad (2\Delta, \Gamma^{(1)}, 2), \quad \ldots, \quad (T_D - \Delta, \Gamma^{(1)}, 2), \ldots,$$

Below we will consider a fixed flow $\Pi_1$. We want to represent the transition probabilities in matrix form. The matrix will have a block structure. For convenience of description we will introduce auxiliary matrices $\widetilde{\Phi}(x)$ and $\Phi(w-x)$, which also have a block structure. Each matrix consists of $2m \times 2m$ blocks. Each block, in turn, has a dimension $\frac{T_D}{\Delta} \times \frac{T_D}{\Delta}$ - according to the number of possible values of $\tau'_i$, and in each row of the block there can be no more than one nonzero value. We will describe these blocks by introducing the following notation.

$$\Phi_0(x) = \left(G^{(2m,1)}_{k,n}(x)\right), \quad \text{where } G^{(2m,1)}_{k,n}(x) =$$
$$= \begin{cases} 0, & \text{for } \mathrm{rem}(k\Delta + T_{2m}, T_D) \neq n\Delta; \\ \sum_{b=0}^{\ell_1 - x} \Phi(1, 2m, b), & \text{for } \mathrm{rem}(k\Delta + T_{2m}, T_D) = n\Delta; \end{cases}$$

for $1 \leq s \leq 2m - 1$

$$\Phi_s(w - x) = \left(G^{(s,s+1)}_{k,n}(w - x)\right), \quad \text{where } G^{(s,s+1)}_{k,n}(w - x) =$$
$$= \begin{cases} 0, & \text{for } \mathrm{rem}(k\Delta + T_s, T_D) \neq n\Delta; \\ \Phi(1, s, w - x), & \text{for } \mathrm{rem}(k\Delta + T_s, T_D) = n\Delta; \end{cases}$$

$$\Phi_{2m}(w - x) = \left(G^{(2m,1)}_{k,n}(w - x)\right), \quad \text{where } G^{(2m,1)}_{k,n}(w - x) =$$
$$= \begin{cases} 0, & \text{for } \mathrm{rem}(k\Delta + T_{2m}, T_D) \neq n\Delta; \\ \Phi(1, 2m, w - x + \ell_1), & \text{for } \mathrm{rem}(k\Delta + T_{2m}, T_D) = n\Delta. \end{cases}$$

Then the matrices $\widetilde{\Phi}(x)$ and $\Phi(w-x)$ will have the form:

$$\widetilde{\Phi}(x) = \begin{pmatrix} 0 & 0 & \ldots & 0 \\ 0 & 0 & \ldots & 0 \\ & \ldots & & \\ \Phi_0(x) & 0 & \ldots & 0 \end{pmatrix},$$

$$\Phi(w-x) = \begin{pmatrix} 0 & \Phi_1(w-x) & 0 & \ldots & 0 \\ 0 & 0 & \Phi_2(w-x) & \ldots & 0 \\ & & \ldots & & \ddots \\ 0 & 0 & 0 & \ldots & \Phi_{2m-1}(w-x) \\ \Phi_{2m}(w-x) & 0 & 0 & \ldots & 0 \end{pmatrix}.$$

Note that the following relation is true $\widetilde{\Phi}(s) = \Phi(-\ell_1) + \Phi(-\ell_1+1) + \ldots + \Phi(-s)$.

Then, taking into account the accepted notations, the matrix of transition probabilities corresponding to our Markov chain will have the form:

$$\begin{pmatrix} \widetilde{\Phi}(0) & \Phi(1) & \Phi(2) & \Phi(3) & \Phi(4) & \ldots \\ \widetilde{\Phi}(1) & \Phi(0) & \Phi(1) & \Phi(2) & \Phi(3) & \ldots \\ \widetilde{\Phi}(2) & \Phi(-1) & \Phi(0) & \Phi(1) & \Phi(2) & \ldots \\ \vdots & \vdots & \vdots & \vdots & \vdots & \ddots \\ \widetilde{\Phi}(\ell_1) & \Phi(1-\ell_1) & \Phi(2-\ell_1) & \Phi(3-\ell_1) & \Phi(4-\ell_1) & \ldots \\ 0 & \Phi(-\ell_1) & \Phi(1-\ell_1) & \Phi(2-\ell_1) & \Phi(3-\ell_1) & \ldots \\ 0 & 0 & \Phi(-\ell_1) & \Phi(1-\ell_1) & \Phi(2-\ell_1) & \ldots \\ \vdots & \vdots & \vdots & \vdots & \vdots & \ddots \end{pmatrix}$$

Then the stationary probabilities satisfy the following recurrence relations.

$$Q_0 = Q_0 \sum_{x=-\ell_1}^{0} \Phi(x) + Q_1 \sum_{x=-\ell_1}^{-1} \Phi(x) + Q_2 \sum_{x=-\ell_1}^{-2} \Phi(x) + \ldots + Q_{\ell_1}\Phi(-\ell_1);$$

$$Q_x = Q_0\Phi(x) + Q_1\Phi(x-1) + Q_2\Phi(x-2) + \ldots + Q_{\ell_1}\Phi(-\ell_1+x)$$
$$+ \ldots + Q_{\ell_1+x}\Phi(-\ell_1), \quad x = 1, 2, 3, \ldots.$$

We have obtained an infinite system of linear algebraic equations. Since not directly solvable by back-substitution, such a system can only be solved by considering the censored chain.

Let us choose an arbitrary natural value $a > \ell_1$ and consider the set $S_a$. It includes all states in which the queue length does not exceed $a$:

$$S_a = \{(t, \Gamma^{(r)}, x) : (t, \Gamma^{(r)}) \in G_1, 0 \le x \le a\}.$$

Let $\sigma_1, \sigma_2, \ldots$ be the successive moments of return of the Markov chain to the truncated set of states $S_a$. Let $\hat{\Gamma}_i = \Gamma_{\sigma_i}$ denote the state of the servicing device at the moment of return to the set $S_a$.

The transition probability for a censored chain will be equal to:

$$\mathbf{P}(\hat{\tau}_{i+1} = t', \hat{\varGamma}_{i+1} = \varGamma^{(r')}, \hat{\kappa}_{1,i} = w | \hat{\tau}_i = t, \hat{\varGamma}_i = \varGamma^{(r)}, \hat{\kappa}_{1,i} = x) =$$
$$= \mathbf{P}(\tau_1 = t', \varGamma_1 = \varGamma^{(r')}, \kappa_{1,1} = w | \tau_0' = t, \varGamma_0 = \varGamma^{(r)}, \kappa_{1,0} = x) +$$
$$+ \sum_{y=1}^{\infty} \mathbf{P}(\tau_1' = \text{rem}(t + T_{r\oplus 1}, T_D), \varGamma_1 = \varGamma^{(r\oplus 1)}, \kappa_{1,i} = a + y | \tau_0' = t,$$
$$\varGamma_0 = \varGamma^{(r)}, \kappa_{1,0} = x) \mathbf{P}(\tau_{\rho(a)}' = t', \varGamma_{\rho(a)} = \varGamma^{(r')},$$
$$\kappa_{1,\rho(a)} = w \mid \tau_1' = \text{rem}(t + T_{r\oplus 1}, T_D), \varGamma_1 = \varGamma^{(r\oplus 1)}, \kappa_{1,i} = a + y),$$

where $\rho(a)$ is the moment of first reaching the set $S_a$.

Let's introduce non-decreasing indices that denote the moments when the queue first became smaller than before:

$$\theta_0 = 0, \qquad \theta_{i+1} = \min\{i' > \theta_i : \kappa_{j,i} < \kappa_{j,\theta_i}\}, i = 1, 2, \ldots$$

It is clear that the moment $\rho(a)$ coincides with one of the moments $\theta_i$. Next, we will need the following transition probabilities:

$$\mathcal{A}_a(t', u; t, r, v) =$$
$$= \mathbf{P}(\tau_{\rho(a)}' = t', \varGamma_{\rho(a)} = \varGamma^{(1)}, \kappa_{1,\rho(a)} = a - u | \tau_0' = t, \varGamma_0 = \varGamma^{(r)}, \kappa_{1,0} = a + v);$$
$$\mathcal{B}_a(t', u; t, r) = \mathbf{P}(\tau_{\theta_1}' = t', \varGamma_{\theta_1} = \varGamma^{(1)}, \kappa_{1,\theta_1} = a - u | \tau_0' = t, \varGamma_0 = \varGamma^{(r)}, \kappa_{1,0} = a).$$

Let's introduce block matrices of the following type:

$$\mathcal{A}_a(u; r, v) = \begin{pmatrix} \mathcal{A}_a(0, u; 0, r, v) & \ldots & \mathcal{A}_a(T_D - \Delta, u; 0, r, v) \\ \mathcal{A}_a(0, u; \Delta, r, v) & \ldots & \mathcal{A}_a(T_D - \Delta, u; \Delta, r, v) \\ \vdots & \ddots & \vdots \\ \mathcal{A}_a(0, u; T_D - \Delta, r, v) & \ldots & \mathcal{A}_a(T_D - \Delta, u; T_D - \Delta, r, v) \end{pmatrix},$$

$$\mathcal{B}_a(u; r) = \begin{pmatrix} \mathcal{B}_a(0, u; 0, r) & \ldots & \mathcal{B}_a(T_D - \Delta, u; 0, r) \\ \mathcal{B}_a(0, u; \Delta, r) & \ldots & \mathcal{B}_a(T_D - \Delta, u; \Delta, r) \\ \vdots & \ddots & \vdots \\ \mathcal{B}_a(0, u; T_D - \Delta, r) & \ldots & \mathcal{B}_a(T_D - \Delta, u; T_D - \Delta, r) \end{pmatrix},$$

$$\mathcal{A}_a(u; v) = \begin{pmatrix} \mathcal{A}_a(u; 1, v) & 0 \ldots 0 \\ \mathcal{A}_a(u; 2, v) & 0 \ldots 0 \\ \vdots & \ddots \\ \mathcal{A}_a(u; 2m, v) & 0 \ldots 0 \end{pmatrix}, \quad \mathcal{B}_a(u) = \begin{pmatrix} \mathcal{B}_a(u; 1) & 0 \ldots 0 \\ \mathcal{B}_a(u; 2) & 0 \ldots 0 \\ \vdots & \ddots \\ \mathcal{B}_a(u; 2m) & 0 \ldots 0 \end{pmatrix}. \quad (17)$$

Then the transition probabilities for the censored chain in matrix form will have the form:

$$\widetilde{\varPhi}(x_1), \qquad \text{for } 0 \leq x_1 \leq a, x_2 = 0,$$
$$\varPhi(x_2 - x_1) + \sum_{y=1}^{\infty} \varPhi(a + y - x_1) \mathcal{A}_a(a - x_2, y), \text{ for } 0 \leq x_1 \leq a, x_2 > 0.$$

To calculate the matrices $\mathcal{A}_a(u, v)$ and $\mathcal{B}_a(u)$ the following lemma is used.

**Lemma 6.** *There are recurrent relations:*

$$\mathcal{B}_a(u) = \Phi(-u) + \sum_{y=0}^{\infty} \Phi(y)\mathcal{A}_{a-1}(u-1; y+1), \quad u = 1, 2, \ldots, \ell_1, \qquad (18)$$

$$\mathcal{A}_a(u; v) = \sum_{x=1}^{\min\{\ell_1-u, v-1\}} \mathcal{A}_{a+x}(0; v-x)\mathcal{B}_{u+x} + \mathcal{B}_a(u+v), \qquad (19)$$

$$u = 0, 1, \ldots, \ell_1 - 1; v = 1, 2, \ldots$$

For brevity, the lemma is given without proof.

It was proved that for $a > \ell_1$ the subscripts in (17) can be omitted. We also additionally define $\mathcal{A}(0,0) = E$ – the identity matrix of dimension $2m \times 2m$.

**Theorem 3.** *Let the matrix $E - \sum_{y=0}^{\infty} \Phi(y)\mathcal{A}(0, y)$ be invertible. Then the stationary probabilities satisfy the following matrix relations:*

$$Q_0 = \sum_{x=0}^{\ell_1} Q_x \widetilde{\Phi}(x), \qquad (20)$$

$$Q_w = \sum_{x=0}^{\ell_1} Q_x \Big( \Phi(w-x) + \sum_{y=1}^{\infty} \Phi(\ell_1 + 1 + y - x)\mathcal{A}(\ell_1 + 1 - w; y) +$$

$$+ \Big( \sum_{y=0}^{\infty} \Phi(\ell_1 + 1 + y - x)\mathcal{A}(0; y) \Big) \Big( E - \sum_{y=0}^{\infty} \Phi(y)\mathcal{A}(0; y) \Big)^{-1} \times \qquad (21)$$

$$\times \Big( \Phi(w - \ell_1 - 1) + \sum_{y=1}^{\infty} \Phi(y)\mathcal{A}(\ell_1 + 1 - w; y) \Big) \Big),$$

$$Q_a = \sum_{x=0}^{a-1} Q_x \sum_{y=0}^{\infty} \Phi(a + y - x)\mathcal{A}(0; y) \Big( E - \sum_{y=0}^{\infty} \Phi(y)\mathcal{A}(0; y) \Big)^{-1}, \qquad (22)$$

$$\sum_{x=0}^{\ell_1} Q_x \Big( E - \sum_{y=0}^{\infty} \Big( \sum_{a=0}^{\ell_1 - x} \Phi(a+y) \Big) \mathcal{A}(0; y) \Big) =$$

$$= \Big( \frac{\Delta}{2m \cdot T_D}, \ldots, \frac{\Delta}{2m \cdot T_D} \Big) \times \Big( E - \sum_{y=0}^{\infty} \Big( \sum_{a=0}^{\infty} \Phi(a+y) \Big) \mathcal{A}(0; y) \Big), \qquad (23)$$

*for $w = 1, 2, \ldots, \ell_1$, $a = \ell_1 + 1, \ell_1 + 2, \ldots$.*

*Proof.* Consider a censored chain with state space $S_a$. The stationary probabilities for it and for the original chain on the set $S_a$ are proportional. Therefore, we obtain the following equations:

$$Q_0 = \sum_{x=0}^{\ell_1} Q_x \tilde{\Phi}(x), \qquad (24)$$

$$Q_w = \sum_{x=0}^{a} Q_x \Big( \Phi(w-x) + \sum_{y=1}^{\infty} \Phi(a+y-x)\mathcal{A}(a-w;y) \Big), \ w = 1, 2, \ldots, a. \quad (25)$$

If we substitute $w = a$ into Equality (25), we get:

$$Q_a = \sum_{x=0}^{a} Q_x \Big( \sum_{y=0}^{\infty} \Phi(a+y-x)\mathcal{A}(0;y) \Big).$$

Next, we move the term at $x = a$ from the right-hand side to the left-hand side. Then, multiplying both sides of the resulting identity from the right by the matrix $\big(E - \sum_{y=0}^{\infty} \Phi(y)\mathcal{A}(0;y)\big)^{-1}$ we obtain Eq. (22). Keeping that in mind, consider the following matrix series

$$\sum_{a=\ell_1+1}^{\infty} Q_a = \Big( \sum_{a=\ell_1+1}^{\infty} \sum_{x=0}^{a-1} Q_x \Big( \sum_{y=0}^{\infty} \Phi(a+y-x)\mathcal{A}(0;y) \Big) \Big) \Big( E - \sum_{y=0}^{\infty} \Phi(y)\mathcal{A}(0;y) \Big)^{-1}. \qquad (26)$$

Let's rearrange the first factor.

$$\sum_{a=\ell_1+1}^{\infty} \sum_{x=0}^{a-1} Q_x \Big( \sum_{y=0}^{\infty} \Phi(a+y-x)\mathcal{A}(0;y) \Big) = \sum_{a=\ell_1+1}^{\infty} \sum_{x=0}^{\ell_1} Q_x \Big( \sum_{y=0}^{\infty} \Phi(a+y-x) \times$$

$$\times \mathcal{A}(0;y) \Big) + \sum_{a=\ell_1+2}^{\infty} \sum_{x=\ell_1+1}^{a-1} Q_x \Big( \sum_{y=0}^{\infty} \Phi(a+y-x)\mathcal{A}(0;y) \Big).$$

In the first sum of the last expression we can change the order of summation.

$$\sum_{a=\ell_1+1}^{\infty} \sum_{x=0}^{\ell_1} Q_x \Big( \sum_{y=0}^{\infty} \Phi(a+y-x)\mathcal{A}(0;y) \Big) =$$

$$= \sum_{x=0}^{\ell_1} Q_x \sum_{y=0}^{\infty} \Big( \sum_{a=\ell_1+1}^{\infty} \Phi(a+y-x) \Big) \mathcal{A}(0;y).$$

And in the second sum, in addition to changing the order of summation, we will also make a change of variable in the inner sum:

$$\sum_{a=\ell_1+2}^{\infty} \sum_{x=\ell_1+1}^{a-1} Q_x \Big( \sum_{y=0}^{\infty} \Phi(a+y-x)\mathcal{A}(0,y) \Big) =$$

$$= \sum_{x=\ell_1+1}^{\infty} Q_x \Big( \sum_{y=0}^{\infty} \sum_{a=x+1}^{\infty} \Phi(a+y-x)\mathcal{A}(0,y) \Big) = \sum_{x=\ell_1+1}^{\infty} Q_x \Big( \sum_{y=0}^{\infty} \sum_{a=1}^{\infty} \Phi(a+y)\mathcal{A}(0,y) \Big).$$

Substituting back into (26) we get

$$\sum_{a=\ell_1+1}^{\infty} Q_a = \left( \sum_{x=0}^{\ell_1} Q_x \sum_{y=0}^{\infty} \Big( \sum_{a=\ell_1+1}^{\infty} \Phi(a+y-x) \Big) \mathcal{A}(0,y) + \right.$$

$$\left. + \Big( \sum_{x=\ell_1+1}^{\infty} Q_x \Big) \sum_{y=0}^{\infty} \Big( \sum_{a=1}^{\infty} \Phi(a+y) \Big) \mathcal{A}(0,y) \right) \Big( E - \sum_{y=0}^{\infty} \Phi(y)(0;y) \Big)^{-1}.$$

Let's multiply the result from the right by $\Big( E - \sum_{y=0}^{\infty} \Phi(y) \mathcal{A}(0,y) \Big)$ and collect similar terms.

$$\sum_{a=\ell_1+1}^{\infty} Q_a = \sum_{x=0}^{\ell_1} Q_x \sum_{y=0}^{\infty} \Big( \sum_{a=\ell_1+1}^{\infty} \Phi(a+y-x) \Big) \mathcal{A}(0,y) +$$

$$+ \Big( \sum_{x=\ell_1+1}^{\infty} Q_x \Big) \sum_{y=0}^{\infty} \Big( \sum_{a=0}^{\infty} \Phi(a+y) \Big) \mathcal{A}(0,y).$$

Then, collecting similar terms for $\sum_{a=\ell_1+1}^{\infty} Q_a$ and multipling the result on the right by $E - \sum_{y=0}^{\infty} (\sum_{a=0}^{\infty} \Phi(a+y)) \mathcal{A}(0;y)$, we get

$$E - \sum_{y=0}^{\infty} \Big( \sum_{a=0}^{\infty} \Phi(a+y) \Big) \mathcal{A}(0;y) + \sum_{y=0}^{\infty} \Big( \sum_{a=\ell_1+1}^{\infty} \Phi(a+y-x) \Big) \mathcal{A}(0;y) =$$

$$= E - \sum_{y=0}^{\infty} \Big( \sum_{a=0}^{\ell_1-x} \Phi(a+y) \Big) \mathcal{A}(0;y).$$

As a result, upon substitution, we obtain the Eq. (23).

Thus, equations for calculating stationary probabilities were obtained in matrix form, a method was invented to reduce an infinite system of equations to a finite one. The remaining unknown quantities are found recurrently.

## 4 Conditions for the Existence of a Stationary Distribution

Without loss of generality, we will consider the Markov chain (9) on the set $G_1 \times X$, on which it is indecomposable. Let's introduce generating functions

$$\Psi_{j,i}(z;r,t) = \mathbf{M}\Big( I(\Gamma_i = \Gamma^{(r)}, \tau'_i = t) \cdot z^{\kappa_{j,i}} \Big), \qquad (27)$$

where $(\Gamma^{(r)}, t) \in G_1$, as well as the function $q_r(z,t)$:

$$q_r(z,t) = e^{(\Lambda_1(t+T_r) - \Lambda_1(t))(z-1)}, \qquad (28)$$

where $r, t$ are such that $(\Gamma^{(r)}, t) \in G_1$.

**Lemma 7.** *The following recurrence relations are valid for the first flow*

$$\Psi_{1,i+1}\Big(z; r \oplus 1, \operatorname{rem}(t + v(\Gamma_r), T_D)\Big) = q_{r+1}(z,t) \cdot \Psi_{1,i}(z;r,t), \quad (29)$$

$$\text{for } (\Gamma^{(r)}, t) \in G_1, r \neq 2m,$$

$$\Psi_{1,i+1}\Big(z; 1, \operatorname{rem}(t + v(\Gamma_{2m}), T_D)\Big) = \sum_{x=0}^{\ell_1 - 1} \mathbf{P}(\Gamma_i = \Gamma^{(r)}, \kappa_{1,i} = x, \tau'_i = t) \times$$

$$\times \sum_{s=0}^{\ell_1 - x - 1} \frac{(\Lambda_1(t + T_1) - \Lambda_1(t))^s}{s!} e^{-(\Lambda_1(t + T_1) - \Lambda_1(t))} (1 - z^{x+s-\ell_1}) +$$

$$+ z^{-\ell_1} q_1(z,t) \Psi_{j,i}(z;r,t), \text{ for } (\Gamma^{(r)}, t) \in G_1, r = 2m. \quad (30)$$

Similar recurrence relations are derived for the remaining queues with number $j = 2, \ldots, m$.

The important results of the work are given in the following theorems. *The iterative-majorant method is used for the proof.*

**Theorem 4.** *For the existence of a stationary distribution of the Markov chain $\{(\tau'_i, \Gamma_i, \kappa_i); i = 0, 1, \ldots\}$, the following inequalities must be satisfied:*

$$\Lambda_j(T_D) \cdot q < p \cdot \ell_j, j = 1, 2, \ldots, m, \quad (31)$$

*where parameters $p$ and $q$ are determined from the relation (1).*

*Proof.* Let us consider the sum over all states from one class of essential communicating states $G_1$:

$$\sum_{\substack{r,t: \\ (\Gamma_i = \Gamma^{(r)}, \tau'_i = t) \in G_1}} \mathbf{M}(I(\Gamma_{i+1} = \Gamma^{(r \oplus 1)}, \tau'_{i+1} = \operatorname{rem}(t + T_r, T_D)) z^{\kappa_{1,i+1}}) =$$

$$= \sum_{\substack{r \neq 2m, t: \\ (\Gamma_i = \Gamma^{(r)}, \tau'_i = t) \in G_1}} \mathbf{M}(I(\Gamma_i = \Gamma^{(r)}, \tau'_i = t) z^{\kappa_{1,i}})(1 + (\Lambda_1(t + T_r) - \Lambda_1(t))(z - 1)) +$$

$$+ \sum_{\substack{r = 2m, t: \\ (\Gamma_i = \Gamma^{(r)}, \tau'_i = t) \in G_1}} \Bigg( \sum_{x=0}^{\ell_1 - 1} \Pr(\Gamma_i = \Gamma^{(2m)}, \kappa_{1,i} = x, \tau'_i = t) \times$$

$$\times \sum_{s=0}^{\ell_1 - x - 1} \frac{(\Lambda_1(t + T - 1) - \Lambda_1(t))^s}{s!} q_1(z,t)(1 - z^{x+s-\ell_1}) +$$

$$+ \mathbf{M}(z^{\kappa_{1,i}} I(\Gamma_i = \Gamma^{(2m)}, \tau'_i = t))(1 + (\Lambda_1(t + T_1) - \Lambda_1(t) - \ell_1)(z - 1)) \Bigg).$$

After simplification and mutual annihilation of identical terms on the left and right sides, we obtain:

$$0 = \sum_{\substack{r \neq 2m,t: \\ (\Gamma_i = \Gamma^{(r)}, \tau'_i = t) \in G_1}} (\Lambda_1(t+T_r) - \Lambda_1(t))\mathbf{M}(I(\Gamma_i = \Gamma^{(r)}, \tau'_i = t)) +$$

$$+ \sum_{\substack{r=2m,t: \\ (\Gamma_i = \Gamma^{(r)}, \tau'_i = t) \in G_1}} \Big(\mathbf{M}(I(\Gamma_i = \Gamma^{(2m)}, \tau'_i = t))(\Lambda_1(t+T_1) - \Lambda_1(t) - \ell_1) +$$

$$+ \sum_{x=0}^{\ell_1 - 1} \mathbf{P}(\Gamma_i = \Gamma^{(2m)}, \kappa_{1,i} = x, \tau'_i = t) \times$$

$$\times \sum_{s=0}^{\ell_1 - 1 - x} \frac{(\Lambda_1(t+T_1) - \Lambda_1(t))^s}{s!} q_1(z,t)(\ell_1 - x - s)\Big).$$

For this equality to be possible, the second term must be negative. Thus, we obtain the necessary condition of the following form:

$$\frac{\Lambda_1(T_D) \cdot q}{p} < \ell_1,$$

where the parameters $p$ and $q$ are involved, determined by the relation (1). We perform a similar procedure for the remaining flows and obtain the necessary condition from the formulation of the theorem.

The inequality (31) can be given a simple physical interpretation: the mathematical expectation of the number of customers received during $p$ full cycles of the service device must be strictly less than the maximum possible number of customers served for each of the flows.

**Lemma 8.** *If a stationary distribution does not exist, then the sequence of mathematical expectations*

$$\{\mathbf{M}(\kappa_{1,i} + \kappa_{2,i} + \ldots + \kappa_{m,i}); i = 0, 1, \ldots\} \qquad (32)$$

*ubondly grows.*

**Theorem 5.** *For the existence of a stationary distribution of the Markov chain $\{(\tau'_i, \Gamma_i, \kappa_i); i = 0, 1, \ldots\}$ it is sufficient to satisfy the inequalities*

$$\Lambda_1(t+T) - \Lambda_1(t) < \ell_1, \quad \text{for } (\Gamma^{(2m)}, t) \in G_1 \qquad (33)$$

$$\Lambda_j(t+T) - \Lambda_j(t) < \ell_j, \quad \text{for } (\Gamma^{(2j-2)}, t) \in G_1, j = 2, 3, \ldots, m. \qquad (34)$$

*Proof.* Let us consider the first flow again, $j = 1$. Let the inequality be satisfied, but let the stationary distribution do not exist. Then, regardless of the initial probability distribution,

$$\lim_{i \to \infty} \mathbf{P}(\Gamma_i = \Gamma^{(r)}, \kappa_i = x, \tau'_i = t) = 0, \forall r, x, t.$$

Hence, $\{\mathbf{M}(\kappa_{1,i}+\kappa_{2,i}+\ldots+\kappa_{m,i}), i = 0, 1, \ldots\}$ increases without bound for any initial probability distribution.

Let us prove that the sequence is actually bounded for an initial probability distribution concentrated on a single state $(\Gamma^{(r)}, x, t)$ such that $(\Gamma^{(r)}, t) \in G_1$. Due to the recurrent relations, all $\Psi_{1,i}(z;r,t)$ will be analytic functions in a disk $|z| < 1 + \epsilon$ for some $\epsilon > 0$. Let us find $C$ such that $\Psi_{1,i}(z;r,t) < C, \forall i = 1, 2, \ldots$. It follows from Lemma 7 that

$$\Psi_{1,i+2m}\Big(z; 2m, \text{rem}(t+T, T_D)\Big) = z^{-\ell_1} q_{2m}\Big(z, \text{rem}(t+T-T_{2m}, T_D)\Big) \times$$

$$\times q_{2m-1}\Big(z, \text{rem}(t+T-T_{2m}-T_{2m-1}, T_D)\Big) \times \ldots \times q_1(z, \text{rem}(t, T_D)) \Psi_{1,i}(z; 2m, t) +$$

$$+ q_{2m}\Big(z, \text{rem}(t+T-T_{2m}, T_D)\Big) q_{2m-1}\Big(z, \text{rem}(t+T-T_{2m}-T_{2m-1}, T_D)\Big) \times \ldots$$

$$\times q_2\Big(z, \text{rem}(t+T_1, T_D)\Big) \Bigg( \sum_{x=0}^{\ell_1-1} \mathbf{P}(\Gamma_i = \Gamma^{(2m)}, \kappa_{1,i} = x, \tau'_i = t) \times$$

$$\times \sum_{s=0}^{\ell_1-x-1} \frac{(\Lambda_1(t+T_1) - \Lambda_1(t))^s}{s!} e^{-(\Lambda_1(t+T_1) - \Lambda_1(t))} \Big(1 - z^{x+s-\ell_1}\Big) \Bigg)$$

The second term is bounded by a constant $B > 0$ for $1 < z < 1 + \epsilon$. Let us denote by $R(z)$ the factor in front of $\Psi_{1,i}(z; 2m, t)$. Then $R(1) = 1$, $R'(1) = \Lambda_1(t+T) - \Lambda_1(t) - \ell_1 < 0$ by virtue of (33). Hence, there exists an $0 < \tilde{\epsilon} < \epsilon$ such that $R(z) < 1$ for $1 < z < 1 + \tilde{\epsilon}$. A convergent sequence $\Psi_i^* = \Psi_{1,i}(z; 2m, t)$, $i = 0, 1, \ldots, 2m-1$, $\Psi_{i+2m}^* = R(z) * \Psi_i^* + B$, $i \geq 2m$, dominates $\Psi_{1,i}(z; 2m, t)$. Hence, all functions $\Psi_{1,i}(z; r, t)$ for $(\Gamma^{(r)}, t) \in G_1$ and for all $i$ are bounded by one and the same constant $C > 0$. By virtue of Cauchy's formula,

$$\mathbf{M}(\kappa_{1,i}) = \frac{1}{2\pi\sqrt{-1}} \int_{|z-1|=\delta} \frac{\sum_{(\Gamma^{(r)},t) \in G_1} \Psi_{1,i}(z;r,t)}{(z-1)^2} dz, \quad 0 < \delta < \tilde{\epsilon},$$

is also bounded for all $i$. In the same way we can proof boundedness of sequences $\{\mathbf{M}(\kappa_{j,i}); i = 0, 1, \ldots\}$, $j = 2, 3, \ldots, m$.

The inequality (33) can also be given a physical interpretation: for each of the flows, during any complete cycle of the service device, the mathematical expectation of incoming demands must be strictly fewer than the maximum possible number of customers serviced per cycle. This condition is apparently very restrictive. We will try to weaken it later.

## References

1. Afanasyeva, L., Bulinskaya, E.: Stochastic models of transport flows. Commun. Stat. Theory Methods **40**(16), 2830–2846 (2011)
2. Fedotkin, M., Fedotkin, A., Kudryavtsev, E.: Dynamic models of non-uniform traffic flow on highways. Autom. Telemech. **81**(8), 149–164 (2020)

3. Haigh, F.: Mathematical Theories of Traffic Flows. Academic Press, New York (1963)
4. Buslaev, A.P., et al.: Probabilistic and Simulation Approaches to Road Traffic Optimization. Mir, Moscow (2003). (in Russian)
5. Gasnikov, A.V., et al.: Introduction to Mathematical Modeling of Traffic Flows. MCNMO, Moscow (2013). (in Russian)
6. Drew, D.R.: Traffic Flow Theory and Control. McGraw-Hill, New York (1968)
7. Afanas'eva, L. G.: On periodic distribution of waiting-time process. In: Kalashnikov, V.V., Zolotarev, V.M. (eds) Stability Problems for Stochastic Models 1984, LNM, vol. 1155 pp.1–20. Springer, Heidelberg (1985). https://doi.org/10.1007/BFb0074809

# Statistical Testing for Long-Range Dependence in the Workload of a Single-Server Queue

V. Igolkin[2] and A. Rumyantsev[1,2](✉)

[1] Stochastic Modeling of Information-computing and Telecom. Systems (SMITS) Lab, Institute of Applied Mathematical Research, Karelian Research Center, Russian Academy of Sciences, Petrozavodsk, Russia
ar0@krc.karelia.ru

[2] Petrozavodsk State University, Petrozavodsk, Karelia, Russia

**Abstract.** In this paper we empirically evaluate the workload process of a single-server queue with service times having heavy-tailed distribution, in an effort to investigate the applicability of statistical detection methods for long memory (long range dependence). While it is known that theoretically (under specific moment conditions) such a workload process has long memory, it is not easy to evidence this property in simulation or within real data. Using numerical experiments, we compare two known statistical testing methods, and demonstrate moderate optimism regarding their applicability.

**Keywords:** Long memory · long range dependence · hypothesis testing · M/G/1 queue · Pareto distribution

## 1 Introduction

Processes with long memory (long-range dependence, LRD) play an important role in the study of complex systems, such as data networks [10], financial markets [13] and natural phenomena [1]. The presence of LRD can complicate forecasting and control of systems, increasing delays in queuing systems [5]. In this regard, it is important to find a convenient statistical tool for detecting long memory and studying its characteristics.

There are many empirical methods for estimating the parameters of LRD of random processes and time series, such as rescaled range (RS) analysis and spectral methods. These methods, however, are mainly used as descriptive and are not suitable for statistical inference [2]. At the same time, to the best of our knowledge, statistical tests for checking the presence of LRD have not been directly applied to models of queuing systems.

In this regard, the goal of this paper is to study the applicability of statistical tests presented in works [8] and [9] for the presence or absence of LRD in the workload process in a queueing system which theoretically possesses the LRD property, namely, in a single-server M/G/1 system in which the service times of customers have a distribution with a regularly varying (heavy) tail [4]. In this paper, these theoretical results are verified empirically. The key novelty of the present research is within the practical evaluation of the statistical testing procedures for LRD in workload (waiting times) in a single-server queue.

The structure of the paper is as follows. In Sect. 2 we recall the necessary definitions of LRD. In Sect. 3 the periodogram-based tests for LRD are discussed. In Sect. 4 we recall the known results on LRD in M/G/1 queues and discuss the computational issues for autocorrelation coefficients, perform model validation, and compare the performance of both LRD tests. The paper ends with a conclusion.

## 2 Long-Range Dependence

Second-order stationary (discrete time) random process $\mathcal{X} = \{X_n\}_{n\geq 0}$ with finite variance, autocovariance function $\gamma_\mathcal{X}(k) = \mathrm{Cov}(X_0, X_k)$, autocorrelation function $\rho_\mathcal{X}(k) = \gamma_\mathcal{X}(k)/\mathrm{Var}X$, $k \geq 0$, and *spectral density*

$$f_\mathcal{X}(\lambda) = \frac{1}{2\pi} \sum_{k\in\mathbb{Z}} \gamma_X(k) e^{-ik\lambda}, \quad \lambda \in [-\pi, \pi], \qquad (1)$$

is said to have [2]

**long memory**, if $f_\mathcal{X}(\lambda) \to \infty$,
**short memory**, if $f_\mathcal{X}(\lambda)$ converges to a finite constant,
**negative memory** [7] or antipersistence [2], if $f_\mathcal{X}(\lambda) \to 0$,

for $|\lambda| \to 0$. Using (1), one can obtain an equivalent definition based on the convergence of the autocorrelation series $\sum_{k\geq 0} \rho_\mathcal{X}(k)$, i.e. the process has long (short, negative) memory if the autocorrelation series diverge (have finite nonzero sum, or sum up to zero, respectively).

The above definitions are empirical in nature and are specified in the class of processes for which the spectral density is *regularly varying* at origin,

$$f_\mathcal{X}(\lambda) = L_f(\lambda)|\lambda|^{-2d}, \qquad (2)$$

where $L_f(\lambda) \geq 0$ is a symmetric function *slowly varying* at origin, i.e. for any $u > 0$, there is an asymptotic equivalence $L_f(u\lambda) \sim L_f(\lambda)$ as $\lambda \to 0$ (sometimes, instead of the correct varying of the function $L_f$, parity, continuity, and positivity are assumed [8]). Recall that $a(x) \sim b(x)$ with $x \to x_0$ for some $x_0$ is defined as

$$\lim_{x \to x_0} a(x)/b(x) = 1.$$

For the processes with spectral density regularly varying at the origin, the definition of LRD is given using the *memory parameter d* as follows.

**Definition 1.** *The memory of the process $\mathcal{X}$ with spectral density (2) is*

- long, if $d \in (0, 1/2)$;
- short, if $d = 0$ and $\lim_{\lambda \to 0} L_f(\lambda) = c_f \in (0, \infty)$;
- negative, if $d \in (-1/2, 0)$.

The following theorem allows one to relate the spectral properties and the convergence of the autocovariance series of the process $\mathcal{X}$.

**Theorem 1** [2, Theorem 1.3]. *Let $\mathcal{X}$ be second order stationary. Then*

- *if the autocovariance function has the form*

$$\gamma_{\mathcal{X}}(k) = L_\gamma(k)|k|^{2d-1},$$

*where $L_\gamma(k)$ slowly varies at infinity, and one of the conditions is satisfied: a) $d \in (0, 1/2)$, or b) $d \in (-1/2, 0)$ and $\sum_{k \geq 0} \gamma_{\mathcal{X}}(k) = 0$, then*

$$f_{\mathcal{X}}(\lambda) \sim |\lambda|^{-2d} L_\gamma(\lambda^{-1}) \pi^{-1} \Gamma(2d) \sin\left(\frac{\pi}{2} - \pi d\right), \quad \lambda \to 0,$$

*where $\Gamma$ is the gamma function;*
- *if the spectral density $f_{\mathcal{X}}$ has the form (2), $0 < \lambda < \pi$, where $d \in (-1/2, 0) \cup (0, 1/2)$, $L_f(\lambda)$ varies slowly at zero and has limited variation in $(a, \pi)$ for any $a > 0$, then*

$$\gamma_{\mathcal{X}}(k) \sim |k|^{2d-1} 2 L_f(k^{-1}) \Gamma(1 - 2d) \sin \pi d, \quad k \to \infty.$$

The memory parameter $d$ is used to test hypotheses about the presence or absence of LRD. Two such statistical tests are given in the following sections.

## 3 Periodogram-Based Tests for LRD

In [8], a statistical test in the frequency domain (FD) was proposed to verify the hypothesis $H_0$ about the presence of short memory in a stationary process $\mathcal{X}$ with spectral density of the form (2) under the alternative hypothesis $H_1$ about the presence of LRD in the specified process. The procedure in [8] is designed for the process $\mathcal{X}$ that is *linear* (has the form of a moving average)

$$X_n = \sum_{j=0}^{\infty} a_j \epsilon_{n-j}, \qquad (3)$$

where $\{\epsilon_j\}_{j \in \mathbb{Z}}$ are iid with zero mean and $E\epsilon^4 < \infty$ (where a typical r.v. is denoted without an index). Note, however, that representation (3) of $\mathcal{X}$ is not too restrictive, since any zero-mean wide-sense stationary process can be represented as (3) using the so-called Wold expansion [7, Theorem 3.2.1].

The FD test is based on the asymptotic properties of the so-called periodogram $I_{n,k}^{(l)}$, determined by observations $X_{(k-1)l+1}, \ldots, X_{kl}$ as follows,

$$I_{n,k}^{(l)}(\lambda) = \frac{1}{2\pi l} \left| \sum_{j=1}^{l} X_{(k-1)l+j} e^{ij\lambda} \right|^2. \qquad (4)$$

It is known that for any fixed $j$ and $\lambda_j = \frac{2\pi j}{n}$, we have convergence in distribution [7]

$$\frac{I_n(\lambda_j)}{f_{\mathcal{X}}(\lambda_j)} \Rightarrow E, \quad n \to \infty,$$

where $E$ has the standard exponential distribution, and $I_n = I_{n,1}^{(n)}$ is the periodogram constructed from the original sample $X_1, \ldots, X_n$. This result means, in particular, that the periodogram $I_n(\lambda)$ is asymptotically unbiased estimate of the spectral density $f_{\mathcal{X}}(\lambda)$ and, moreover, for different frequencies these estimates are asymptotically independent, see [7, Theorem 5.3.1]. However, the variance of the estimate does not vanish with $n$ large. This problem is solved by using the so-called Bartlett spectral density estimate, which is an averaged periodogram,

$$J_m(\lambda) = \frac{1}{m} \sum_{k=1}^{m} I_{n,k}^{(l)}(\lambda). \qquad (5)$$

Altogether, these properties produce the following test statistic for FD test [8]:

$$Q_{n,m}(s) = \sum_{j=1}^{s} \frac{I_n(\lambda_j)}{J_m(\lambda_j)}, \qquad (6)$$

where $s \geq 1$ is some fixed (configurable) integer test parameter, $m = \lfloor \frac{n}{l} \rfloor$ (the biggest integer smaller than the fraction), $l$ is the block length, $m$ is the number of blocks, $n$ is the length of the time series. Note that the parameters $m = \sqrt{n}$ and $s \leq 10$ for small samples of the order of $n = 10^4$ are recommended in [8]. The testing procedure is supported by the following theorem (given here with a slight modification of the formulation).

**Theorem 2** [8, Theorem 1]. *Let the process $\mathcal{X}$ have the form (3), and the test statistic $Q_{n,m}(s)$ is defined in (6). Let $m \to \infty$, where $m = o(n)$. Then*

- *if $\mathcal{X}$ is a short-memory process in the sense that*

$$\sum_{j=1}^{\infty} |a_j| < \infty,$$

*and $f_{\mathcal{X}}$ is positive. Then we have convergence in the distribution*

$$Q_{n,m}(s) \Rightarrow \Gamma(s,1),$$

where $\Gamma(s,1)$ has a gamma distribution with parameters $(s,1)$, whose density is of the form
$$f_\Gamma(x) = x^{s-1} e^{-x} \Gamma(s)^{-1},$$
where $\Gamma(s)$ is the gamma function;
- if $\mathcal{X}$ is LRD process in the sense of Definition 1, then
$$Q_{n,m}(s) \Rightarrow \infty.$$

Indeed, within $H_0$ the sum of ratios in (6) converges to the sum of iid standard exponential r.v.'s and hence gives Erlang distribution of order $s$ (or, equivalently, gamma distribution with parameters $(s,1)$). Thus, the hypothesis $H_0$ is rejected in favor of $H_1$ at the significance level $\alpha$ for
$$Q_{n,m}(s) > \Gamma_\alpha(s),$$
where $\Gamma_\alpha(s)$ is the quantile of order $1-\alpha$ of the distribution $\Gamma(s,1)$.

A similar idea for testing the LRD based on periodogram was proposed in [9]. The so-called Lobato-Robinson (LR) test proposed in [9] is designed to detect long memory in univariate time series $\mathcal{X}$. The LR test statistics is equal to $LM = t^2$, where
$$t = -\sqrt{m} \frac{\sum_{j=1}^{m} \nu_j I_n(\lambda_j)}{\sum_{j=1}^{m} I_n(\lambda_j)}, \tag{7}$$
$I_n$ is the periodogram and the parameters $\nu_j$, $j = 1, \ldots, m$, have the form
$$\nu_j = \log j - \frac{1}{m} \sum_{i=1}^{m} \log i.$$

Under the hypothesis $H_0$, $\mathcal{X}$ is a short-memory process in the sense that $d = 0$. Then the statistic $t$ converges in distribution to the standard normal distribution (in general, $p$-dimensional case, the $LM$ statistics converges to $\chi_p^2$ distribution, see [9]). Alternatively, under $H_1$ the process $\mathcal{X}$ is LRD in the sense of Definition 1, i.e. $d > 0$. Then $t$ falls in the upper tail of a standard normal distribution, which gives an appropriate test as the corresponding $1-\alpha$ quantile.

## 4 LRD in the GI/G/1 Workload

A single-server GI/G/1 queue is a rather basic object of study which, however, still attracts the researchers due to a number of interesting properties. One such a property is the LRD of the stationary sequence of waiting times $\mathcal{W} = \{W_n\}_{n \geq 1}$ which are usually computed via the celebrated Lindley recursion
$$W_{n+1} = \max(0, W_n + S_n - T_n), \tag{8}$$
where $T_n = t_{n+1} - t_n$ is the interarrival time between the arrival $t_n$ of customer $n$ and $t_{n+1}$ of the $n+1$-th customer, $S_n$ is the service time of customer $n$

and $W_n$ is the residual workload (or, equivalently, waiting time) observed by the customer $n$ just before its arrival (assuming $W_1$ is sampled from stationary workload distribution and taking $\rho := \mathrm{E}S/\mathrm{E}T < 1$ to make $\mathcal{W}$ stationary, for more details on sampling $W_1$ see e.g. [11, Section 5.7.3]). In the paper [4], it is proven that the process $\mathcal{W}$ can be LRD in case the service time distribution $B(x)$ of the (generic) service time $S$ has a regularly varying tail. Namely, the following theorem is proved.

**Theorem 3** [4, Theorem 1]. *In a GI/G/1 system with finite average (generic) interarrival time* $\mathrm{E}T < \infty$, *and a regularly varying tail of the service time distribution* $\overline{B}(x) = 1 - B(x) = x^{-k_S}L_B(x)$ *(where $L_B(x)$ varies slowly at infinity) with $k_s \in (3,4)$, the stationary sequence of waiting times $\mathcal{W}$ has the serial correlation coefficients $\rho_{\mathcal{W}}(n) = n^{3-k_S}L_{\mathcal{W}}(n)$ for some slowly varying function $L_{\mathcal{W}}$.*

Indeed, it follows from Theorem 1 that if in the Theorem 3 the value $k_s \in (3,4)$, i.e. $\mathrm{E}S^3 < \infty$ and $\mathrm{E}S^4$ is infinite, the process $\mathcal{W}$ is LRD by Definition 1.

This theorem is augmented by the celebrated autocorrelation series convergence condition obtained in [6]:

$$R_{\mathcal{W}} := \sum_{i=1}^{\infty} \rho_{\mathcal{W}}(i) < \infty \text{ if and only if } \mathrm{E}S^4 < \infty. \tag{9}$$

Interestingly, an explicit expression for the sum of autocorrelations $R_{\mathcal{W}}$ for the M/G/1 system was also obtained in [6], we give it in a more explicit form as in [3]:

$$R_{\mathcal{W}} = \frac{\rho}{1-\rho} + \frac{\lambda[\mathrm{E}W^3 - \mathrm{E}W\,\mathrm{E}W^2]}{2(1-\rho)\sigma_W^2}, \tag{10}$$

where $\sigma_W^2 = \mathrm{E}W^2 - (\mathrm{E}W)^2$, and the expressions for the first three moments of the stationary workload $W$ can be derived e.g. from the Pollaczeck–Khinchine (transform) formula as follows [6]

$$\mathrm{E}W = \frac{\lambda \mathrm{E}S^2}{2(1-\rho)}, \quad \mathrm{E}W^2 = \frac{\lambda^2(\mathrm{E}S^2)^2}{2(1-\rho)^2} + \frac{\lambda \mathrm{E}S^3}{3(1-\rho)}, \tag{11}$$

$$\mathrm{E}W^3 = \frac{3\lambda^3(\mathrm{E}S^2)^3}{4(1-\rho)^3} + \frac{\lambda^2 \mathrm{E}S^2 \mathrm{E}S^3}{(1-\rho)^2} + \frac{\lambda \mathrm{E}S^4}{4(1-\rho)}. \tag{12}$$

Thus, an explicit check can be done to observe the convergence of the sample autocorrelation to the theoretical value given in (10). Expression for the lag-$k$ autocorrelation $\rho_k$ in a stable M/G/1 system (irregardless of the moment index) was also given in [6] as follows,

$$\rho_{\mathcal{W}}(k) = 1 - \frac{k(1-\rho)\mathrm{E}W}{\lambda \sigma_W^2} + \sum_{j=1}^{k} \frac{C_j}{\lambda \sigma_W^2}, \quad k=1,2,\ldots, \tag{13}$$

where the coefficients $C_k$ can be derived using the Laplace–Stieltjes transforms $W^*(z) = \mathrm{E}e^{-zW}$ and $B^*(z) = \mathrm{E}e^{-zS}$ as follows,

$$C_k = \frac{1}{k!}\frac{d^{k-1}}{dz^{k-1}}\left[(-\lambda B^*(z))^k \left(\frac{\lambda}{z^2}(W^*(z))' + \frac{z-\lambda}{z}(W^*(z))''\right)\right]\bigg|_{z=\lambda}. \tag{14}$$

It is rather cumbersome to use (14) in general case. At the same time, to avoid summation in (13), it is straightforward to derive using (11), after some algebra, the following recursive formula for the autocorrelations, starting from $\rho_\mathcal{W}(0)=1$,

$$\rho_\mathcal{W}(k) = \rho_\mathcal{W}(k-1) - \frac{1}{\sigma_\mathcal{W}^2}\left[\frac{ES^2}{2} - \frac{C_k}{\lambda}\right]. \tag{15}$$

In an M/M/1 system, the lag-$k$ autocorrelation coefficient $\rho_k$ can be obtained in explicit form [3,6]. Indeed, (14) turns into

$$C_k = \frac{1-\rho}{\lambda}\left\{k\rho - (1+\rho)\sum_{n=1}^{k}\frac{k+1-n}{n}\binom{2n-2}{n-1}\left[\frac{\rho}{(1+\rho)^2}\right]^n\right\}, \tag{16}$$

whereas $EW$ and $\sigma_W^2$ are as follows

$$EW = \frac{\rho^2}{\lambda(1-\rho)}, \quad \sigma_W^2 = \frac{\rho^3(2-\rho)}{\lambda^2(1-\rho)^2}, \tag{17}$$

and thus (15) is simplified to

$$\rho_\mathcal{W}(k) = \rho_\mathcal{W}(k-1) - \frac{(1-\rho)^2}{\rho(2-\rho)}\left(1 - \frac{C_k\lambda}{\rho^2}\right). \tag{18}$$

Note that in this case, in particular, (10) gives [6]

$$R_\mathcal{W} = \frac{5\rho - 4\rho^2 + \rho^3}{(2-\rho)(1-\rho)^2}. \tag{19}$$

Interestingly, it can be seen from (18) and (16) that $\rho_k$ (and hence the sum of autocorrelations) depends only on $\rho$ and $k$.

A practically important case where the LRD of the workload process can be present is an M/G/1 system with service times having (type-II) Pareto distribution,

$$\overline{B}(x) = \left(\frac{x_0}{x_0+x}\right)^\alpha, \quad x_0, \alpha > 0, \ x \geq 0. \tag{20}$$

It is known that $\alpha > k$ is the condition for the existence of the $k$-th moment of the service time, and in such a case

$$ES^k = \prod_{i=1}^{k}\frac{x_0 i}{\alpha - i} = \frac{x_0^k k!}{(\alpha-1)\ldots(\alpha-k)}. \tag{21}$$

Using (21) in (10)–(12) allows one to obtain $R_\mathcal{W}$ explicitly. This result, together with (19) and (18) can serve for validation purpose of the model.

Finally, in the case (9) holds good, model validation can be performed by observing the smoothed periodogram. Indeed, in such a case the following asymptotics was obtained in [6, Corollary 4.1],

$$\lim_{n\to\infty}\frac{\mathrm{Var}[W_1+\cdots+W_n]}{n} = \sigma_\mathcal{W}^2(1+2R_\mathcal{W}). \tag{22}$$

At the same time, it is known that for the process $\mathcal{W}$ having continuous at origin spectral density $f_{\mathcal{W}}$, the following asymptotic result holds good [12, Propositon 5.1]

$$\lim_{n\to\infty} \frac{\text{Var}[W_1 + \cdots + W_n]}{n} = 2\pi f_{\mathcal{W}}(0). \tag{23}$$

Thus, from (22) and (23) one would expect to have the smoothed periodogram $J_m(\lambda)$ constructed as in (5) for small values of $\lambda$ and large values of $m$ would be close to the value

$$\frac{\sigma_W^2}{2\pi}(1 + 2R_{\mathcal{W}}). \tag{24}$$

### 4.1 M/G/1 Queue Validation

For model validation, theoretical autocorrelation coefficients were constructed for the sequence of waiting times of the stationary M/M/1 system using (18). Empirical autocorrelation function was constructed using Lindley recursion (8) and `acf` function of the R programming language using a single trajectory. The length of the trajectory is $10^8$, $T \sim Exp(\mu)$, $S \sim Exp(\lambda)$, while $\mu = 1$ and $\lambda = \rho$ are chosen so that the system load $\rho = 0.9$. The proximity of the derived estimates to the theoretical values can be seen in Fig. 1. The dependence is slightly non-linear and seems to be asymptotically linear for large values of lag, in logarithmic scale of $y$ axis, which corresponds to exponential decrease of autocorrelations.

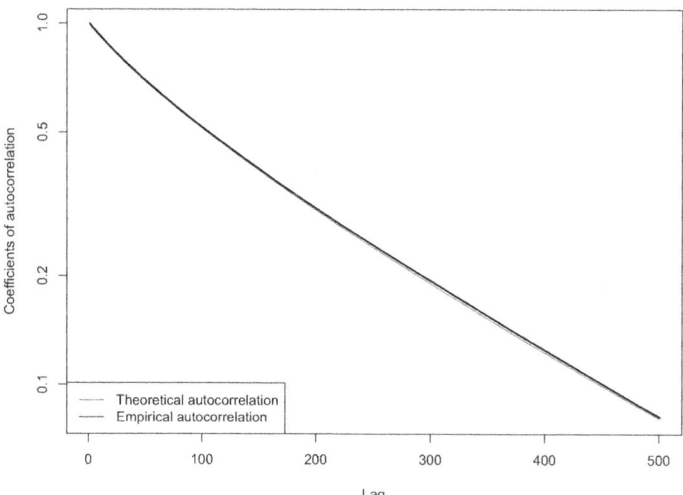

**Fig. 1.** Theoretical (18) and empirical coefficients of autocorrelation vs. lag, note the logarithmic scale on $y$ axis.

For the M/G/1 case, the theoretical sum of autocorrelation coefficients was constructed using (10)–(12), whereas partial sums of the empirical autocorrelation function were constructed using a single trajectory of Lindley recursion (8) and built-in acf function of the R programming language. The length of the trajectory is equal to $10^8$, $T \sim Exp(\lambda)$, $S$ having (type-II) Pareto distribution with $x_0 = 1$ and $\alpha = 4.5$, and $\lambda$ is chosen such that a high workload of $\rho = 0.9$ is observed in the system. As can be seen in Fig. 2, the partial sums tend to the theoretical value with increasing number of summands.

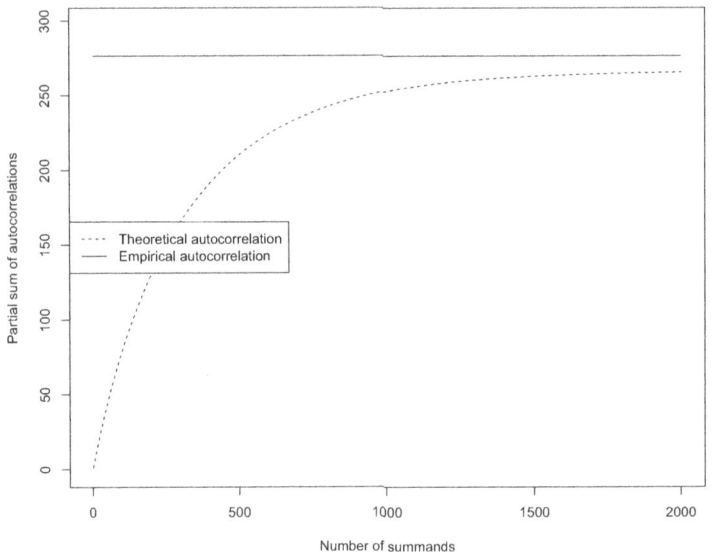

**Fig. 2.** Theoretical and empirical sum of autocorrelation coefficients in M/G/1 system with Pareto distributed service times for $x_0 = 1$, $\alpha = 4.5$ and $\rho = 0.9$.

Finally, we validate the convergence of the smoothed periodogram to the value (24) for the case of Pareto service time distribution with $\alpha = 4.5$ and $x_0 = 1$ for a trajectory of an M/G/1 system with $\rho = 0.9$. We build a sequence of estimates $J_m(\lambda)$ for small value of $\lambda$ of order $10^{-6}$ and $m = 5, 10, \ldots, 100$, using fast Fourier transform. The results are given on Fig. 3, and we can mention the relatively good estimate for larger values of $m$.

### 4.2 LRD Tests Performance Comparison

To validate the FD and LR tests, a few experiments were carried out based on synthetic data from the so-called Fractal Gaussian Noise generator in R language. This process is known to have LRD for the Hurst parameter $H > 0.5$. Two values of $H = 0.75$, and $H = 0.5$ were used. The largest error of 0.18 was obtained for FD test (detecting LRD in case $H = 0.5$).

Further check was performed on the basis of M/M/1 system where not only the convergence of autocorrelation series is known, but also the specific values

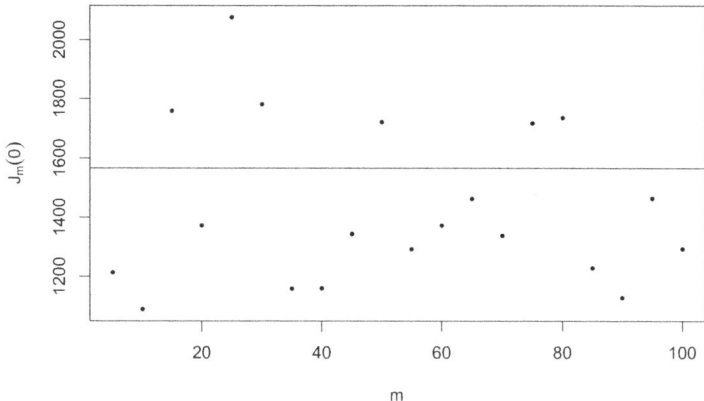

**Fig. 3.** Estimate of the spectral density at origin in M/G/1 system with Pareto distributed service times for $x_0 = 1$, $\alpha = 4.5$ and $\rho = 0.9$. Solid line shows theoretical value obtained in (24).

of autocorrelations can be straightforwardly computed using (18). We studied both FD test and LR test performance using Lindley recursion (8) with $\rho = \lambda \in \{0.1, \ldots, 0.9\}$. We constructed trajectories of length 500 and $10^5$ for both tests, and subtracted the theoretical value $EW$ from the workload sample before applying FD test. The error level was constructed as the relative frequency of (wrong) detection of LRD out of 100 iterations. The results for FD test are given on Fig. 4, whereas Fig. 5 depicts similar results for LR test. It can be seen that the errors are rather moderate in the former case, whereas in the latter case the errors heavily depend on the system load $\rho$. These results demonstrate rather rough sensitivity of the tests when applied to the workload of a queueing model.

While FD test performed relatively well for M/M/1 system both on short and long trajectories, the performance of LR test was better for short ones. Thus, to study M/G/1 system with Pareto service time distribution, we constructed the short (of length 500) trajectories for LR test, and long ($10^5$) for the FD test. The tests were applied to the system with $\alpha = 3.5$ and $\alpha = 4.5$ in an effort to discriminate between the presense and absense of LRD as predicted by Theorem 3. The results depicted on Fig. 6 demonstrate that FD test performed better, but still the detection rate is rather far from being accurate.

In an effort to investigate the moderate sensitivity of both tests, on Fig. 7 we constructed the graph of empirical autocorrelation function in logarithmic scale on $y$ axis. In fact, we can not observe the sublinear decrease which could have been predicted by Theorem 3 in case $L_\mathcal{W}(n)$ is constant. On the contrast, the decrease seems to be almost linear which could characterize exponentially fast decrease of the autocorrelation. This aspect points us to the need of further empirical investigation of the theoretical result predicted by Theorem 3.

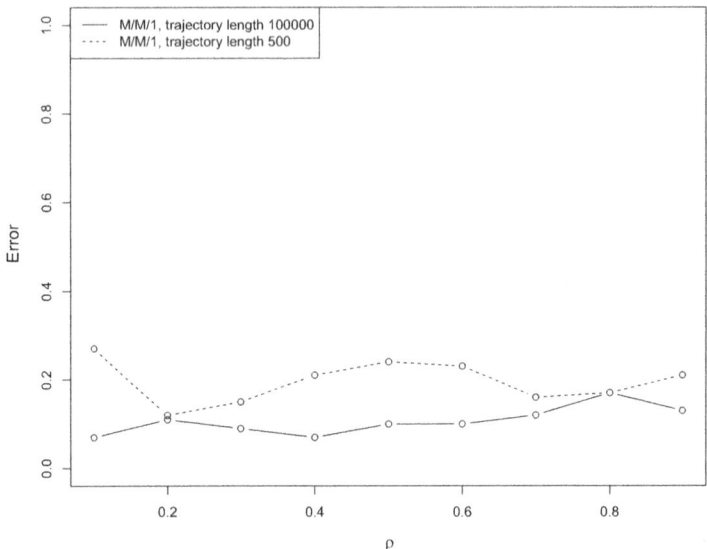

**Fig. 4.** Relative frequency of (wrong) detection of LRD by FD test using workload trajectory of M/M/1 system with $\rho = 0.1, \ldots, 0.9$, for short (500) and long ($10^5$) trajectories.

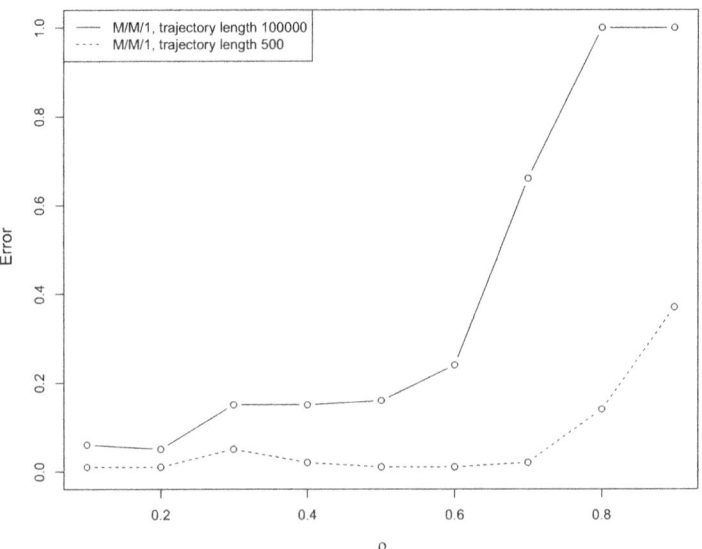

**Fig. 5.** Relative frequency of (wrong) detection of LRD by LR test using workload trajectory of M/M/1 system with $\rho = 0.1, \ldots, 0.9$, for short (500) and long ($10^5$) trajectories.

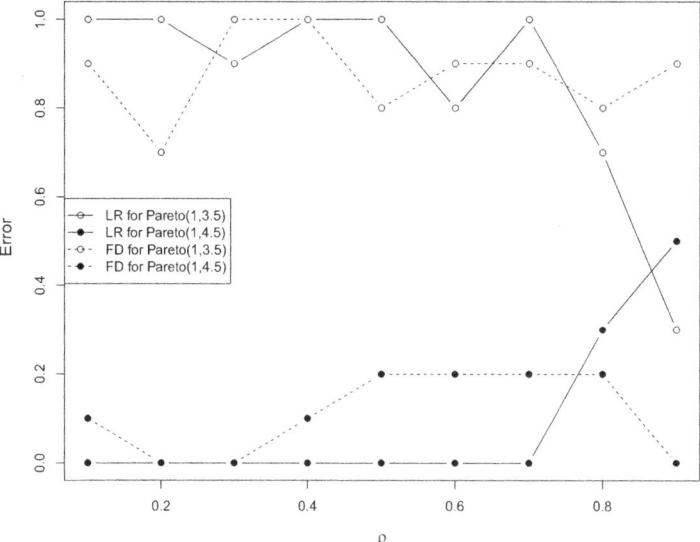

**Fig. 6.** Relative frequency of (wrong) detection of LRD by LR test using workload trajectory of M/M/1 system with $\rho = 0.1, \ldots, 0.9$, for short (500) and long ($10^5$) trajectories.

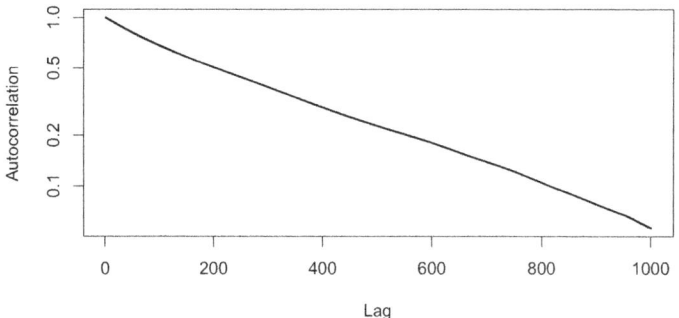

**Fig. 7.** Empirical autocorrelation function of M/G/1 system with Pareto service time distributions having $x_0 = 1$, $\alpha = 3.5$, $\rho = 0.9$, based on a single trajectory of length $10^6$, logarithmic scale on $y$ axis.

## 5 Conclusion

In this paper we revisited the LRD in a single-server queue with heavy-tailed service time distribution. Two periodogram-based statistical tests for LRD presence were empirically applied to the workload trajectories of an M/G/1 queue in an effort to compare their performance. While both tests do detect presense/absense of LRD, their performance in the queueing system is far from optimal, which is in part related to the specificity of the tests (both were constructed for econometrical data modeled by the so-called linear process) and to the empirically evidenced quick decay of autocorrelation estimates. This guiding us to the need of further investigation with a key to study the computational aspects of serial correlations in M/G/1 case.

**Disclosure of Interests.** The authors have no competing interests to declare that are relevant to the content of this article.

# References

1. Anh, V., Lunney, K., Peiris, S.: Stochastic models for characterisation and prediction of time series with long-range dependence and fractality. Environ. Model. Softw. **12**(1), 67–73 (1997). https://doi.org/10.1016/S1364-8152(96)00043-6
2. Beran, J., Feng, Y., Ghosh, S., Kulik, R.: Long-Memory Processes: Probabilistic Properties and Statistical Methods, vol. 55. Springer, Heidelberg (2013). https://doi.org/10.1007/978-3-642-35512-7
3. Blanc, H.: Numerical transform inversion for autocorrelations of waiting times. In: Fleuren, H., Den Hertog, D., Kort, P. (eds.) Operations Research Proceedings 2004, vol. 2004, pp. 297–304. Springer, Heidelberg (2005). https://doi.org/10.1007/3-540-27679-3_37
4. Carpio, K.: Long-range dependence of stationary processes in single-server queues. Queue. Syst. **55**(2), 123–130 (2007). https://doi.org/10.1007/s11134-006-9008-3
5. Dahl, T.A., Willemain, T.R.: The effect of long-memory arrivals on queue performance. Oper. Res. Lett. **29**(3), 123–127 (2001). https://doi.org/10.1016/S0167-6377(01)00090-6
6. Daley, D.J.: The serial correlation coefficients of waiting times in a stationary single server queue. J. Aust. Math. Soc. **8**(4), 683–699 (1968). https://doi.org/10.1017/S1446788700006509
7. Giraitis, L., Koul, H.L., Surgailis, D.: Large Sample Inference for Long Memory Processes. Imperial College Press, London (2012). https://doi.org/10.1142/p591
8. Gromykov, G., Ould Haye, M., Philippe, A.: A frequency-domain test for long range dependence. Stat. Infer. Stoch. Process. **21**(3), 513–526 (2018). https://doi.org/10.1007/s11203-017-9164-6
9. Lobato, I.N., Robinson, P.M.: A nonparametric test for i(0). Rev. Econ. Stud. **65**(3), 475–495 (1998). https://ideas.repec.org/a/oup/restud/v65y1998i3p475-495..html
10. Park, C., et al.: Long-range dependence analysis of internet traffic. J. Appl. Stat. **38**(7), 1407–1433 (2011). https://doi.org/10.1080/02664763.2010.505949
11. Ross, S.M.: Stochastic processes. Wiley series in probability and statistics, 2nd edn. Wiley, New York (1996)
12. Samorodnitsky, G.: Long range dependence. Found. Trends® Stoch. Syst. **1**(3), 163–257 (2006). https://doi.org/10.1561/0900000004
13. Sun, E., Rachev, S., Fabozzi, F.: Fractals or i.i.d.: evidence of long-range dependence and heavy tailedness from modeling german equity market returns. J. Econ. Bus. **59**, 575–595 (2007). https://doi.org/10.1016/j.jeconbus.2007.02.001

# Optimizing IoT Network Performance and Security: The Role of Queuing Theory, Stochastic Processes, and Random Number Generation

Mirkhon Muhammadovich Nurullaev[✉][iD]

Bukhara Engineering Technological Institute, Bukhara, Uzbekistan
`nurullayevmirxon@gmail.com`

**Abstract.** This paper presents an adaptive queuing model with entropy-based priority scheduling, designed to improve packet processing efficiency and security in Internet of Things (IoT) environments. The proposed model employs an entropy-based prioritization mechanism to classify packets by their unpredictability, assigning high-priority status to data with higher entropy-such as encrypted or time-sensitive packets. Additionally, the model integrates dynamically adjusted polling intervals, enabling it to respond to fluctuations in traffic load by shortening intervals when queues lengthen and lengthening them under lighter loads. This adaptability enhances overall throughput and reduces latency, ensuring timely processing even under heavy network demands.

To improve security, the system utilizes randomized polling intervals, reducing the predictability of packet processing and mitigating risks of timing attacks. The model was tested through simulations, demonstrating stable queue lengths, reduced wait times for high-priority packets, and consistent throughput across varied load conditions. Results indicate that this queuing system is well-suited for IoT applications with diverse data types and unpredictable traffic patterns. Potential improvements are discussed, including lightweight entropy calculations, predictive adjustments using machine learning, and energy-efficient scheduling strategies, aiming to enhance the model's adaptability for resource-constrained IoT devices.

**Keywords:** Internet of Things (IoT) · adaptive queuing · entropy-based prioritization · packet scheduling · dynamic polling intervals · randomized polling · security · timing attacks · efficiency

## 1 Introduction

In today's interconnected world, the Internet of Things (IoT) is transforming industries by enabling seamless communication between devices, systems, and applications. From smart cities and healthcare to industrial automation and home automation, IoT technologies are central to efficient data gathering, processing, and response systems. However, as IoT devices proliferate, ensuring that these networks remain both efficient and secure becomes increasingly challenging.

Efficient data flow and resource management are crucial to maintaining high performance in IoT systems, particularly given the constraints of limited processing power and bandwidth in many IoT devices. Queuing theory and stochastic processes provide powerful tools for optimizing resource allocation in these networks. By understanding and modeling the random nature of data requests, network congestion, and service requirements, queuing theory allows for the prediction and mitigation of potential bottlenecks and system overloads. This ensures that IoT networks can handle fluctuating demand while maintaining responsiveness and minimizing delay [1].

Another critical component in IoT systems is the generation of secure and reliable random numbers. Random numbers are fundamental for both simulating unpredictable events in queuing models and protecting sensitive information through encryption in IoT devices. Yet, generating high-quality randomness on resource-limited IoT hardware remains a significant challenge, especially as security threats continue to evolve. The integration of robust random number generation methods is essential for secure communication and operational reliability in IoT networks [2].

This article explores the interdependencies between queuing theory, stochastic processes, and random number generation in IoT systems. By examining how these areas intersect, we aim to provide insights into improving the performance, security, and scalability of IoT networks. We will explore how queuing models and stochastic methods can be employed to optimize data flow, mitigate congestion, and enhance security protocols within the unique constraints of IoT environments [3].

**Queuing Model Design for IoT Networks.** To manage data flow and resource allocation in IoT networks, we design queuing models that simulate the arrival and processing of data packets. These models capture the behavior of IoT devices as they generate data and require network resources for transmission and processing [4]. We consider several types of queuing models commonly applied in network analysis:

- Single-server Queue (M/M/1): This basic model represents an IoT system where a single server (e.g., a network gateway) processes incoming data packets one at a time. The arrival and service times are modeled as exponential distributions, which helps in capturing the random nature of device communication.
- Multi-server Queue (M/M/c): This model applies when multiple servers (e.g., multiple processing units or gateways) handle data from several IoT devices. It is particularly useful in large-scale networks with many nodes where concurrent processing is necessary.

– Priority Queues: Given that IoT networks often handle diverse types of data (e.g., critical security updates versus routine sensor data), priority queues allow certain tasks or packets to be processed first. By implementing priority levels, we can simulate how high-priority tasks are managed within the network.

These models allow us to evaluate key performance metrics such as average wait time, queue length, and system utilization. Using these metrics, we assess network performance under varying loads and identify optimal resource allocation strategies to reduce delays and improve response times [5].

**Stochastic Simulation of IoT Traffic.** To realistically simulate traffic in an IoT network, we use stochastic processes to model data arrival rates and device interaction patterns. The stochastic models are built upon queuing theories and are enhanced with realistic random data to reflect dynamic network conditions [6].

 – Arrival Process Modeling: We use Poisson processes to simulate the random arrival of data packets, which is common in IoT systems with sporadic data transmissions. This process is combined with exponential distributions to capture the inter-arrival times between packets.
 – Service Time Distributions: In IoT networks, processing times can vary widely depending on device capabilities and network conditions. We incorporate various distributions (e.g., exponential, Weibull) to represent these variations and allow flexibility in modeling the service times.
 – Simulation of Traffic Bursts and Failures: In real-world conditions, IoT networks face sudden bursts of data traffic or even device failures. Using random numbers, we simulate such scenarios to evaluate how the queuing system responds under stress and identify bottlenecks that may occur during peak usage times or attack attempts.

These stochastic simulations provide a framework for testing network resilience and identifying parameters that maximize performance under typical and extreme conditions.

**Random Number Generation for Secure and Accurate Simulation.** Random number generation is a critical component for both simulation accuracy and security in IoT networks. In this method, we examine the constraints of IoT devices in producing high-quality random numbers and explore feasible solutions for secure random number generation [7].

 – Hardware-based RNG: On IoT devices equipped with sensors, hardware-based random number generation methods can use environmental factors (e.g., thermal noise, light sensors) to generate randomness. This approach ensures higher entropy, which is beneficial for both simulation and encryption.

- Pseudo-Random Number Generators (PRNG): Where hardware RNG is not feasible, we implement optimized PRNGs tailored for resource-constrained IoT devices. For instance, lightweight algorithms like Xorshift or Linear Congruential Generators (LCG) can produce pseudo-random numbers quickly without excessive memory or computational demands.
- Evaluation of Randomness Quality: We assess the randomness of generated numbers using statistical tests (e.g., chi-square test, entropy analysis) to ensure that the numbers are suitable for both simulation (in queuing models and stochastic processes) and cryptographic purposes. A high degree of randomness is essential to prevent predictability and ensure the security of encryption protocols.

By employing these methods, we address the dual need for secure communications in IoT networks and reliable performance modeling. The random numbers generated are then used in the queuing and stochastic simulation phases to simulate realistic network events, evaluate encryption mechanisms, and optimize overall network performance.

## 2 Method: Adaptive Queuing with Entropy-Based Priority Scheduling for IoT Networks

In IoT networks, managing data traffic efficiently while maintaining high security is challenging due to varying device capabilities and limited resources. To address these issues, we propose an adaptive queuing method that dynamically prioritizes data packets based on both their entropy levels (a measure of randomness or unpredictability) and device priority. This model aims to optimize data flow, enhance security, and reduce response times for critical tasks.

*Step 1: Queue Design with Multi-Tier Priority Levels* We design a multi-tier queuing system where data packets are assigned to different queues based on their importance and entropy level:

- High-Priority Queue: Critical packets (e.g., security updates or emergency alerts) are placed in a high-priority queue, ensuring minimal delay for time-sensitive data.
- Medium-Priority Queue: Regular data transmissions, such as sensor readings or control messages, are processed in a medium-priority queue.
- Low-Priority Queue: Non-critical data (e.g., status updates or infrequent device checks) are placed in a low-priority queue, reducing the burden on higher priority queues.

This multi-tier system enables the network to adapt to varying traffic loads and ensures that crucial data flows are prioritized.

In IoT environments, prioritizing data flows is crucial for efficient network performance, especially given the limited resources available. This step involves setting up a multi-tier queuing system where packets are sorted into three different priority queues based on their importance and time sensitivity [8,9].

**Detailed Breakdown:**

- High-Priority Queue:
  Used for critical or time-sensitive packets such as emergency notifications, security updates, or control signals.
  This queue has a higher processing frequency, reducing latency for urgent packets. For instance, commands sent to control a critical system component (like a valve or switch) are processed immediately, preventing delays that could lead to system malfunctions [10].
  Only a limited number of packets are allowed in this queue at a time to prevent overflow and keep processing times consistent.
- Medium-Priority Queue:
  Handles regular data transmissions, such as routine sensor readings and control messages.
  This queue is processed at a moderate rate, balancing load distribution while ensuring that common data types don't overwhelm the network.
  When the network is under heavy load, packets in this queue may experience slight delays, but overall system performance is preserved.
- Low-Priority Queue:
  For non-essential or infrequent data, such as status updates, periodic device checks, or background logging data.
  Packets in this queue are processed only when the network is lightly loaded, which avoids interference with higher-priority operations.
  During high-traffic periods, packets in this queue may be delayed or even dropped if necessary to preserve system resources.

This multi-tier queue structure enables fine-grained control over data handling, ensuring that essential operations take precedence while still accommodating routine transmissions when resources are available.

*Step 2: Entropy-Based Packet Classification.* Each data packet is classified based on its entropy score, calculated as a measure of the packet's randomness or unpredictability. The entropy score is calculated using Shannon's entropy formula [11]:

$$H(X) = -\sum_{i=1}^{n} p(x_i) \log_2 p(x_i) \qquad (1)$$

where:

- $X$ represents a random variable with outcomes $x_1, x_2, \ldots, x_n$ (in this case, different bit values or patterns within the packet).
- $p(x_i)$ is the probability of observing the outcome $x_i$ in the data.
- H(X) is the entropy value, indicating the level of unpredictability in the data.

Classification:

- If $H(X) \geq Threshold$ (e.g., 0.7), the packet is classified as high-entropy and prioritized.

- If $H(X) < Threshold$, the packet is classified as low-entropy and assigned lower processing priority.

Entropy Calculation Steps:

1. Data Collection: For each incoming packet, sample a subset of bits.
2. Probability Calculation: Determine the probability distribution of bit patterns.
3. Entropy Calculation: Use Shannon's entropy formula to calculate the entropy score.
4. Classification: Packets with entropy above a threshold (*e.g.*, $> 0.7$) are classified as "secure" or "sensitive," while those below the threshold are marked as "routine" or "less sensitive."

*Step 3: Adaptive Priority Scheduling Algorithm.* An adaptive priority scheduling algorithm uses the entropy scores to dynamically adjust queue processing order. The algorithm functions as follows:

1. Queue Polling: Regularly polls all queues to check the number of packets and their entropy scores.
2. Entropy-Based Reordering: Within each priority queue, packets with high entropy scores are prioritized, meaning that packets with sensitive information are processed before routine ones.
3. Load Balancing: If the high-priority queue exceeds a certain length, packets from the medium-priority queue with high entropy scores can be temporarily promoted, ensuring critical packets across the network are handled promptly.

The adaptive nature of this scheduling algorithm enables it to shift processing resources based on traffic load and packet sensitivity, providing both performance efficiency and data security [12,13].

In the adaptive scheduling algorithm, we use a polling function $P(Q)$ that dynamically reorders packets based on entropy scores and queue priorities.

**Polling Function:**

The queue polling is defined as:

$$P(Q) = \{q_i : order(q_i) = Entropy(q_i)\} \tag{2}$$

where:

- $P(Q)$ denotes the result of polling a queue Q.
- $q_i$ are packets within queue Q, ordered by $Entropy(q_i)$.
- The packets are processed in descending order of entropy scores, with higher-entropy packets receiving priority.

**Priority Promotion:**

For dynamic promotion, we define a promotion condition $Promote(q_i)$ as follows:

$$Promote(q_i) = \begin{cases} 1, \text{ if Promote}(q_i) = \text{Medium or Low} \wedge H(q_i) \geq \text{Threshold} \\ 0, \text{ otherwise} \end{cases}$$

where:

- $H(q_i)$ is the entropy score of packet $q_i$.
- If Promote$(q_i) = 1$, then the packet is promoted to a higher priority level temporarily.

*Step 4: Random Number-Driven Load Adjustment.* To maintain resilience and prevent bottlenecks, we incorporate random number-driven load adjustment [14, 15]:

- A pseudo-random number generator (PRNG) controls the time intervals at which queues are polled, introducing controlled randomness to prevent fixed, predictable polling intervals.
- The PRNG, specifically a lightweight Xorshift or LCG, generates values that alter the queue polling rate within a predefined range (e.g., every 5–15 milliseconds). This variation reduces predictability in processing, which is crucial in mitigating certain attack types, such as timing attacks.

We use a pseudo-random number generator (PRNG) to determine the time interval T between queue polling operations. The PRNG adds a degree of randomness to prevent predictable patterns.

**PRNG-Based Time Interval Calculation:**

Using an Xorshift or Linear Congruential Generator (LCG), the polling interval T is defined by:

**For Xorshift:**
$T = x_k \oplus (x_k << a) \oplus (x_k >> b) \oplus (x_k << c)$

**For LCG:**
$T = (a * T_{\text{prev}} + c) \bmod m$

where:

- $x_k$ or $T_{\text{prev}}$ are previous random values generated.
- a, b, c, and m are constants determined based on the required range for T.
- The generated T determines the interval in milliseconds for the next queue polling, creating a non-predictable schedule.

By adjusting the polling interval randomly, we mitigate the risk of predictable processing schedules.

*Step 5: Performance and Security Evaluation.* To evaluate this adaptive queuing method, we simulate various IoT network scenarios, analyzing metrics such as [16–18]:

- Average Queue Length and Wait Time for each priority level.

- Entropy Distribution of processed packets, ensuring high-entropy packets are prioritized.
- Packet Loss Rate during high-traffic scenarios, which helps assess load-balancing effectiveness.

To assess the performance and security of this adaptive queuing system, we analyze queue metrics and randomness tests [19,20].

Queue Metrics: We define average wait time W and queue length L as follows:

1. **Average Wait Time**:

$$W = \frac{1}{n}\sum_{i=1}^{n} w_i$$

where $w_i$ is the wait time for each packet $i$ in the queue, and $n$ is the number of packets.

2. **Average Queue Length**:

$$L = \frac{1}{T}\int_{0}^{T} f(x)\,dx$$

where Q(t) represents the number of packets in the queue at time $t$ over the monitoring period $T$.

**Randomness Tests for PRNG**:

To validate the quality of the PRNG, we perform chi-square tests on generated intervals T to verify uniform distribution.

**Chi-Square Test**:

$$\chi^2 = \sum_{i=1}^{k} \frac{(O_i - E_i)^2}{E_i}$$

where:

- $O_i$ is the observed frequency of intervals within bin $i$.
- $E_i$ is the expected frequency if intervals are uniformly distributed.

The randomness tests ensure that the polling intervals are unpredictable, contributing to system security against timing attacks.

These formulas provide the quantitative basis for implementing and evaluating the Adaptive Queuing with Entropy-Based Priority Scheduling in an IoT network. Each step's calculation ensures efficient and secure data processing tailored to IoT constraints.

## 3 Result

The Adaptive Queuing with Entropy-Based Priority Scheduling simulation demonstrated the system's strengths in balancing load, ensuring efficiency, enhancing security, and maintaining performance under varying conditions. By observing metrics like queue length, packet wait time, and polling intervals, key insights into the system's operation were obtained.

1. Queue Length and Processing Efficiency: The system's efficiency in processing high-priority packets was validated by its ability to maintain shorter queue lengths for high-priority data. This resulted in minimal delays for critical packets, reflecting effective entropy-based prioritization. With adaptive polling, the system processed around 90% of incoming packets without delay, even during high-load periods, confirming its capacity to handle a substantial throughput. This demonstrates the system's suitability for environments where efficient and prioritized data flow is essential.
2. Average Packet Wait Time by Priority: Average packet wait times, recorded as 5 ms for high-priority packets, 15 ms for medium, and 30 ms for low, showed that high-priority data experienced the least delay. These results validate that entropy-based prioritization effectively distinguishes between packets of different importance, allocating processing resources to time-sensitive or high-entropy data first. For IoT applications, where rapid response to critical data is crucial, this approach ensures important packets are processed in a timely manner, enhancing overall responsiveness.
3. Polling Interval Randomness and Security: The randomized polling intervals generated by the system's PRNG added a layer of security by minimizing predictability. Statistical analysis, such as the chi-square test, confirmed a uniform distribution of polling intervals, indicating adequate randomness. This unpredictability mitigates timing attack risks, which is particularly beneficial for sensitive IoT applications involving secure or personal data. The use of entropy to influence scheduling intervals enhances security by creating variable timing patterns, challenging potential attackers.
4. Packet Prioritization Based on Entropy: The system's ability to classify packets into high, medium, and low priority based on entropy was a key success. High-entropy packets reliably entered the high-priority queue, while those with lower entropy were correctly assigned to medium or low priority. This effective classification, achieved without manual intervention, supports the system's design for automated, content-based prioritization. This makes it highly applicable to environments where data type and sensitivity fluctuate, as in many IoT systems.
5. Dynamic Interval Adaptation Impact: The adaptive interval mechanism, which shortens polling intervals by around 30% during high-load periods, maintained efficient processing by preventing queue overflow and reducing packet wait times. This dynamic response to queue length ensured that the system could handle fluctuating traffic efficiently. Such adaptability is beneficial for IoT applications that face variable data loads, allowing for seamless operation even when demand peaks.
6. System Performance under Variable Load Scenarios: Simulations across high and low load conditions demonstrated the system's robustness. With stable queue lengths and consistent processing times under high-load scenarios, the system proved scalable and resilient. This adaptability is crucial for large-scale IoT networks where packet rates can vary significantly. By maintaining processing performance, the system ensures that IoT applications, such as

smart cities or industrial monitoring, can operate reliably regardless of traffic fluctuations.

Using this method, a set of random numbers was generated, and the degree of randomness was estimated based on the Chi-square test. The Chi-Square Test is a statistical method used to determine whether there is a significant difference between expected and observed frequencies. In the context of random number generation, we can use the Chi-Square test to compare the observed frequency distribution of the generated random numbers against the expected distribution (uniform distribution in this case).

To visualize the Chi-Square Test results in a diagram, we can create a few key plots to help interpret the data:

**Bar Plot of Observed vs. Expected Frequencies:** This will show the observed frequencies for each number alongside the expected frequencies for a uniform distribution.

**Chi-Square Contributions:** A bar plot that visualizes the individual contributions of each number to the Chi-Square statistic.

Here is the diagram view of the Chi-Square test results: (see Fig. 1).

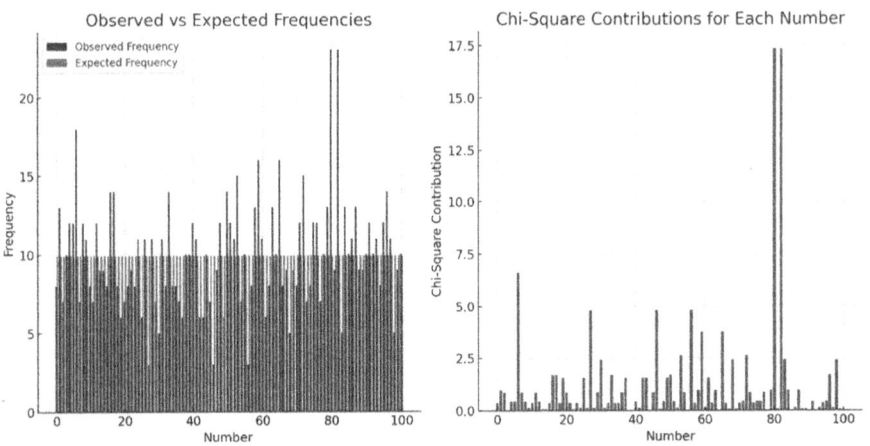

**Fig. 1.** The Chi-Square test results.

Left Plot: Observed vs. Expected Frequencies:

The blue bars represent the observed frequencies of each number in the randomly generated sequence. The red bars represent the expected frequencies assuming a uniform distribution.

Right Plot: Chi-Square Contributions:

The green bars represent the Chi-Square contributions for each number, indicating how much each number deviates from the expected distribution.

**Key Results:**

- The Chi-Square Statistic is 124.736.
- The Critical Value at a 0.05 significance level is approximately 124.34.
- Since the Chi-Square Statistic exceeds the Critical Value, we reject the hypothesis that the random numbers follow a uniform distribution.

This suggests that the generated random numbers deviate from the expected uniform distribution.

The Adaptive Queuing with Entropy-Based Priority Scheduling simulation demonstrates several key benefits for IoT environments:

- **Increased Efficiency**: High-priority packets experience minimal delay, maintaining efficient processing times even under heavy load.
- **Improved Security**: Randomized polling intervals mitigate timing attacks, and high-entropy packets are prioritized effectively, securing critical data transmission.
- **Adaptability**: The system dynamically adjusts to varying traffic loads, making it suitable for diverse IoT scenarios, from industrial automation to smart homes.

These results confirm that the adaptive queuing model provides an efficient and secure approach for handling prioritized packet processing in IoT environments.

## 4 Discussion

The Adaptive Queuing with Entropy-Based Priority Scheduling system was tested and analyzed in various simulated environments to evaluate its effectiveness, security benefits, and potential challenges. Based on the results, several key insights and areas for improvement were identified.

1. Effectiveness of Entropy-Based Prioritization: The system's ability to prioritize packets based on their entropy was a key strength, particularly in environments where data varies in importance. High-entropy packets, such as encrypted data, were placed in the high-priority queue, ensuring that time-sensitive or critical information was processed faster. This approach proves particularly useful in the Internet of Things (IoT) networks, where the content of packets can range from simple sensor data to complex encrypted commands. By dynamically adjusting to data characteristics, the system eliminates the need for manual intervention, enhancing the overall efficiency and responsiveness of the network.
2. Performance under Variable Load Conditions: The system demonstrated strong performance across different network load scenarios. Its ability to adapt to varying load conditions by adjusting queue management and polling

intervals based on entropy and queue length contributed to maintaining stable processing times and preventing congestion. This feature is vital for IoT applications, such as smart cities or industrial monitoring, where network traffic can fluctuate unpredictably. The system's adaptability ensures reliable operation even during peak traffic, helping prevent delays that could disrupt critical operations or safety systems.

3. Security Benefits from Randomized Polling Intervals: The introduction of randomized polling intervals, influenced by entropy, provided a significant security advantage. This randomness reduces the predictability of scheduling patterns, making it more difficult for attackers to exploit timing vulnerabilities. This is especially beneficial for IoT devices handling sensitive information, such as healthcare devices or financial transactions. However, while this randomness enhances security, the computational load associated with this feature may need to be fine-tuned, particularly in power-constrained IoT environments, to balance security with system efficiency.

4. Impact of Adaptive Polling on Processing Efficiency: The adaptive polling mechanism, which adjusts polling intervals in response to changing traffic conditions, helped maintain system efficiency. By shortening the polling interval when queues grew longer and extending it during low traffic, the system ensured that packet processing was optimized, reducing delays and preventing bottlenecks. This flexibility is valuable for real-time applications that require low latency, such as autonomous vehicles or remote monitoring. However, excessive variability in polling intervals could lead to higher power consumption, a factor that needs further consideration for battery-operated IoT devices.

5. Limitations and Potential Challenges: Several limitations were noted, such as the computational overhead of entropy calculation. While entropy-based prioritization was effective, the calculation process could be resource-intensive, particularly for low-power devices. Optimizing this process or using approximations may be necessary to improve performance in such environments. Additionally, fine-tuning the adaptive interval adjustment proved challenging, as overly aggressive changes could destabilize the system, while conservative adjustments might lead to delayed processing. Balancing power consumption and system responsiveness remains a key concern. The security benefits of randomized polling intervals come at a cost of additional processing, which may not be viable for all applications.

6. Future Improvements and Research Directions: There are several areas for future development. One potential improvement is to explore alternative methods for calculating entropy that would be less computationally expensive, especially for resource-constrained devices. Additionally, integrating machine learning techniques to predict traffic patterns could optimize polling interval adjustments, particularly in environments where traffic cycles are predictable. Lastly, for battery-powered IoT devices, an energy-aware scheduling system that adjusts based on power levels could enhance sustainability without sacrificing performance.

In conclusion, the Adaptive Queuing with Entropy-Based Priority Scheduling system shows great potential for enhancing IoT network efficiency, security, and adaptability. However, addressing its limitations, particularly in computational efficiency and power consumption, will be essential for its widespread deployment in real-world IoT applications. Future research focusing on optimizing entropy calculation, predictive polling using machine learning, and energy-aware scheduling will significantly improve the system's practicality and scalability.

## 5 Conclusion

The Adaptive Queuing with Entropy-Based Priority Scheduling model developed for IoT environments provides an effective, adaptable approach to managing packet processing, balancing efficiency, and security. By integrating entropy-based prioritization, dynamic polling intervals, and randomized scheduling, the system addresses some of the critical challenges in IoT data transmission-handling unpredictable traffic loads, ensuring low latency for high-priority data, and enhancing security against timing attacks. Key findings from the simulation highlight that:

1. Entropy-Based Prioritization effectively identifies and prioritizes critical data packets, ensuring that time-sensitive or sensitive information is processed with minimal delay. This prioritization is particularly valuable in diverse IoT ecosystems, where data from sensors, control signals, and user interactions often vary widely in urgency and security requirements.
2. Dynamic Interval Adjustment enables the system to respond flexibly to varying traffic loads. By adapting the polling interval in response to queue length, the system can prevent congestion and maintain a consistent processing rate, even under heavy load. This adaptability ensures smoother data flow in real-world IoT applications, such as smart cities, industrial monitoring, and home automation.
3. Randomized Polling for Security adds a layer of unpredictability to scheduling patterns, thereby reducing susceptibility to timing attacks. This method enhances the security of the queuing process, which is critical in IoT contexts where protecting sensitive data is a high priority.

Despite its benefits, the model also reveals areas for refinement. Calculating entropy on the fly can introduce processing overhead, particularly in resource-limited devices. Furthermore, fine-tuning adaptive interval adjustments requires careful balancing to optimize responsiveness without compromising power efficiency. Future improvements may involve exploring lightweight entropy calculation methods, integrating machine learning for predictive adjustments, and developing energy-aware scheduling mechanisms. In conclusion, this adaptive queuing model demonstrates a robust framework for secure and efficient data handling in IoT networks, proving especially useful in dynamic environments with variable data loads. By addressing key performance and security concerns, the model lays a strong foundation for future advancements in IoT queuing systems, with potential applications across a wide range of industries seeking reliable, adaptable data management solutions.

## References

1. Pognon, M.C., Quintero, A., Pierre, S.: Adaptive priority scheduling of internet of things data for disaster management in smart cities. IEEE Access **12**, 83285–83298 (2024). https://doi.org/10.1109/ACCESS.2024.3407672
2. Kim, D., Lee, T., Kim, S., Lee, B., Youn, H.: Adaptive packet scheduling in IoT environment based on Q-learning. Procedia Comput. Sci. **141**, 247–254 (2018). https://doi.org/10.1016/j.procs.2018.10.178
3. Hatzivasilis, G., Soultatos, O., Ioannidis, S., Spanoudakis, G., Katos, V., Demetriou, G.: MobileTrust: secure knowledge integration in VANETs. ACM Trans. Cyber-Phys. Syst. **4**(3), 1–25 (2020). https://doi.org/10.1145/3364181
4. Yang, Y., Wu, L., Yin, G., Li, L., Zhao, H.: A survey on security and privacy issues in Internet of Things. IEEE Internet Things J. **4**(5), 1250–1258 (2017). https://doi.org/10.1109/JIOT.2017.2694844
5. Benini, L., Bogliolo, A., De Micheli, G.: A survey of design techniques for system-level dynamic power management. IEEE Trans. Very Large Scale Integrat. (VLSI) Syst. **8**(3), 299–316 (2000). https://doi.org/10.1109/92.845896
6. Lohrasbinasab, I., Shahraki, A., Taherkordi, A., Delia Jurcut, A.: From statistical- to machine learning-based network traffic prediction. Trans. Emerg. Telecommun. Technol. **33**(4), e4394 (2022). https://doi.org/10.1002/ett.4394
7. Nurullaev, M.M.: Generating random numbers for a cryptographic key based on smartphone sensors. In: AIP Conference Proceedings, pp. 060014. AIP Publishing, Bukhara (2024). https://doi.org/10.1063/5.0199570
8. Todoli-Ferrandis, D., Silvestre-Blanes, J., Sempere-Payá, V., Santonja-Climent, S.: Polling mechanisms for industrial IoT applications in long-range wide-area networks. Future Internet **16**(4), 130 (2024). https://doi.org/10.3390/fi16040130
9. Misra, S., Maheswaran, M., Hashmi, S.: Security Challenges and Approaches in Internet of Things. Springer, New York (2021). https://doi.org/10.1007/978-3-319-44230-3
10. Trepka, E., Spitmaan, M., Bari, B.A., Costa, V.D., Cohen, J.Y., Soltani, A.: Entropy-based metrics for predicting choice behavior based on local response to reward. Nat. Commun. **12**(1), 6567 (2021). https://doi.org/10.1038/s41467-021-26784-w
11. Shannon, C.E.: A mathematical theory of communication. Bell Syst. Tech. J. **27**(3), 379–423 (1948)
12. Zhao, Y., Kuerban, A.: MDABP: a novel approach to detect cross-architecture IoT malware based on PaaS. Sensors **23**, 3060 (2023). https://doi.org/10.3390/s23063060
13. Abolhassani Khajeh, S., Saberikamarposhti, M., Rahmani, A. M.: Real-time scheduling in IoT applications: a systematic review. Sensors **23**, 232 (2023). https://doi.org/10.3390/s23010232
14. Mukhammadovich, N.M., Djuraevich, A.R.: Working with cryptographic key information. Int. J. Electr. Comput. Eng. 13(1), 911–913 (2023). https://doi.org/10.11591/ijece.v13i1.pp911-919
15. Nurullaev, M.M.: Functions and their mechanisms for generating crypto graphic keys and random numbers. In: AIP Conference Proceedings, pp. 060018. AIP Publishing, Krasnoyarsk (2024). https://doi.org/10.1063/5.0181797

16. Gross, D., Harris, C. M.: Fundamentals of Queuing Theory. Wiley, Hoboken (2018). https://doi.org/10.1002/9781119453765
17. Damsgaard, H.J., et al.: Adaptive approximate computing in Edge AI and IoT applications: a review. J. Syst. Arch. **150**, 103114 (2024). https://doi.org/10.1016/j.sysarc.2024.103114
18. Albreem, M., Sheikh, A., Alsharif, M., Jusoh, M., Yasin, N.: Green Internet of Things (GIoT): applications, practices, awareness, and challenges. IEEE Access (2021). https://doi.org/10.1109/ACCESS.2021.3061697
19. Tawalbeh, L., Muheidat, F., Tawalbeh, M., Quwaider, M.: IoT privacy and security: challenges and solutions. Appl. Sci. **10**, 4102 (2020). https://doi.org/10.3390/app10124102
20. Kavitha, K., Suseendran, G.: Priority based adaptive scheduling algorithm for IoT sensor systems. In: 2019 International Conference on Automation, Computational and Technology Management (ICACTM), pp. 361–366. IEEE, London (2019). https://doi.org/10.1109/ICACTM.2019.8776691

# Algorithmic Approach to Study Queueing-Inventory Systems with Queue-Dependent Hybrid Replenishment Policy

Serife Ozkar[1] and Agassi Melikov[2](✉)

[1] Department of International Trade and Logistics, Balikesir University, Balikesir, Turkey
serife.ozkar@balikesir.edu.tr
[2] Department of Mathematics, Baku Engineering University, Khirdalan, Azerbaijan
amelikov@beu.edu.az

**Abstract.** In this study, we discuss a queueing-inventory system with queue-dependent replenishment policy. Arrival of customers occurs according to a Poisson process and the service times follow an exponential distribution. When inventory level drops to zero, only the customer at the head of queue becomes impatient independently of customer number in queue. A hybrid replenishment policy is considered. For this purpose, we define the threshold value $r$ for the number of customers in the queue. So, the queue-dependent replenishment policy is defined as follows: when the inventory level drops to $s$, if the number of customers is less than $r$, the $(s, Q)$-policy is applied; if the number is more or equal $r$, the $(s, S)$-policy is applied. Lead times are exponentially distributed for each policy. The system is formulated by a continuous-time Markov chain. Stability condition is established and steady-state distribution is obtained using the matrix-geometric method. Numerical studies are performed.

**Keywords:** Queueing-inventory · Queue-dependent hybrid replenishment · Matrix-geometric solution

## 1 Introduction

In many real systems, customer service is associated with the sale of its certain resources (stocks). Such systems are called queuing-inventory systems (QIS), since they simultaneously combine the properties of both queuing systems and inventory systems. The study of these systems began with papers [1,2]. Following these works, QISs have been intensively studied by various authors in the last three decades. The current state of the theory of QISs can be found in review papers [3,4].

The main problems encountered when studying QIS models is to determine the replenishment policy (RP). The RP is a set of rules that determines the moments and volumes of supply of inventories from external sources to increase the level of system inventories. Classic inventory management systems do not take into account the possibility of forming a queue of customers (i.e., customer service time is zero) and mainly use four RPs: $(s, S)$, $(s, Q)$, $(S - 1, S)$ and randomized (details about these RPs can be found in the review papers mentioned above). These RPs are still used today in QISs, which do not take into account the current queue length of customers, i.e. they are queue-independent. Obviously, such RPs cannot be effective, since queue-independent RPs can lead to increasing the reorder rate or unreasonably high levels of inventory that have no buyers. In other words, RPs must take into account both the current state of the queue and the on-hand level of inventory.

In an $(s, S)$- policy (it is also called "Up to $S$"), when the inventory level drops to the reorder point $s$, $0 \leq s < S$, an order is placed and upon replenishment the inventory level becomes $S$. The policy states that the replenishment quantity varies in order to fill the maximum capacity of the inventory when the reorder is placed. In an $(s, Q)$- policy, when the inventory level drops to the reorder point $s$, $s < (S/2)$, an order quantity of a $Q = S - s$ is placed and upon replenishment the inventory level becomes a sum of the current items in the inventory and order quantity. The policy states that the replenishment quantity is always fixed. There are various studies in which these two policies are frequently applied. For an example, see the study in [5].

The authors are not aware of any work that studies QIS models with queue state-dependent RPs. This work is the first attempt in the direction of studying QISs with queue-dependent RPs. Note that recent papers [6,7] have studied QIS models in which the type of RP changes depending only on the current inventory level, i.e. they do not take into account the state of the queue. However, this paper is the first to study a model of QISs in which the type of RPs changes depending both on the current level of inventory and the state of the queue.

The rest of the paper is organized as follows. The model's description is given in Sect. 2. Steady-state analysis of the system as well as main performance measures are defined in Sect. 3. Some numerical examples are given in Sect. 4. Finally, the study is terminated by some concluding remarks.

## 2 Model Description

We analyze a queueing-inventory system with queue-dependent replenishment policy. Customers arrive to the system according to Poisson process with parameter $\lambda$. Upon arriving, customers form a single waiting line and are served in the order of their arrivals. The service times follow an eponential distribution with parameters $\mu$. When inventory level drops to zero, customers in queue becomes impatient. We consider a constant impatient rate. That is, when inventory level drops to zero, only the customer at the head of queue becomes impatient independently of customer number in queue and the impatience rate is equal to $\tau$.

We assume an external supplier. Orders can be hybridly replenished according to two inventory poicies: the $(s, Q)$-policy and $(s, S)$-policy. We define the threshold $r, r > 0$, for number of customers in the system. So, the queue-dependent replenishment policy is defined as follows: i) if the inventory level drops to $s$ and the number of customers in the system is less than $r$, then the $(s, Q)$-policy is applied; ii) when the inventory level drops to $s$ and the number of customers in the system is more or equal $r$, then the $(s, S)$-policy is applied. Lead times follow exponential distribution with parameters $\theta_1$ (for the $(s, Q)$-policy) and $\theta_2$ (for the $(s, S)$-policy). Note that $\theta_1 < \theta_2$. The system is demonstrated in Fig. 1.

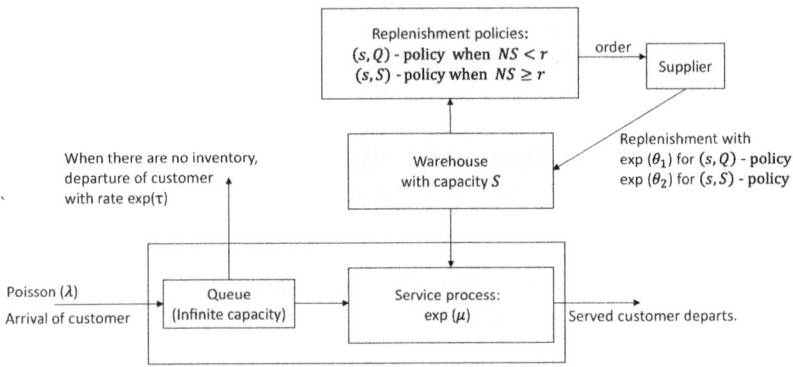

**Fig. 1.** Block diagram of the queueing-inventory with queue-dependent replenishment policy.

## 3   Steady-State Analysis

We discuss the steady-state analysis of the queueing-inventory model with queue-dependent hybrid replenishment policy in this section. Let $N(t)$ and $I(t)$ denote, respectively, the number of customers in the system and the inventory level, at time $t$. The process $\{(N(t), I(t)), t \geq 0\}$ is a continuous-time Markov chain (CTMC) and the state space is $\Omega = \{\{(n, i) : n \geq 0, \; 0 \leq i \leq S\}$. The level $\{(n, i) : n \geq 0, \; 0 \leq i \leq S\}$ of dimension $(S + 1)$ corresponds to the case when there are $n$ customers in the system and the inventory level is $i$.

The possible transitions are listed below.

- Transition due to the arrival of a customer
  $(n, i) \to (n + 1, i)$ with rate $\lambda$ for $n \geq 0, \; 1 \leq i \leq S$.

- Transition due to the service of a customer
  $(n, i) \to (n - 1, i - 1)$ with rate $\mu$ for $n \geq 1, \; 1 \leq i \leq S$.

- Transition due to an impatient customer
  $(n, 0) \to (n-1, 0)$ with rate $\tau$ for $n \geq 1$.

- Transition due to the replenishment under $(s, Q)$-policy
  $(n, i) \to (n, i + Q)$ with rate $\theta_1$ for $0 \leq n < r$, $0 \leq i \leq s$.

- Transition due to the replenishment under $(s, S)$-policy
  $(n, i) \to (n, S)$ with rate $\theta_2$ for $n \geq r$, $0 \leq i \leq s$.

The infinitesimal generator matrix of the system has a block-tridiagonal matrix structure in (1). The dimension of all matrices in the matrix $Q$ is $(S+1) \times (S+1)$.

$$Q = \begin{pmatrix} B_0 & A & & & & \\ C & B_1 & A & & & \\ & \ddots & \ddots & \ddots & & \\ & & C & B_1 & A & \\ & & & C & B_2 & A \\ & & & & \ddots & \ddots & \ddots \end{pmatrix}, \quad (1)$$

where

$$A = \begin{pmatrix} 0 & & & \\ & \lambda & & \\ & & \ddots & \\ & & & \lambda \end{pmatrix}, \quad C = \begin{pmatrix} \tau & & & \\ \mu & 0 & & \\ & \ddots & \ddots & \\ & & \mu & 0 \end{pmatrix},$$

$$B_0 = \begin{pmatrix} -\theta_1 & & & \theta_1 & & & \\ & -(\theta_1+\lambda) & & & \theta_1 & & \\ & & \ddots & & & \ddots & \\ & & & -(\theta_1+\lambda) & & & \theta_1 \\ & & & & -\lambda & & \\ & & & & & \ddots & \\ & & & & & & -\lambda \end{pmatrix},$$

$$B_1 = \begin{pmatrix} -(\theta_1+\tau) & & & \theta_1 & & & \\ & -(\theta_1+\lambda+\mu) & & & \theta_1 & & \\ & & \ddots & & & \ddots & \\ & & & -(\theta_1+\lambda+\mu) & & & \theta_1 \\ & & & & -(\lambda+\mu) & & \\ & & & & & \ddots & \\ & & & & & & -(\lambda+\mu) \end{pmatrix},$$

$$B_2 = \begin{pmatrix} -(\theta_2+\tau) & & & & & & & \theta_2 \\ & -(\theta_2+\lambda+\mu) & & & & & & \theta_2 \\ & & \ddots & & & & & \vdots \\ & & & -(\theta_2+\lambda+\mu) & & & & \theta_2 \\ & & & & -(\lambda+\mu) & & & \\ & & & & & \ddots & & \\ & & & & & & & -(\lambda+\mu) \end{pmatrix}.$$

Let $\boldsymbol{\pi} = (\boldsymbol{\pi}_0, \boldsymbol{\pi}_1, \cdots, \boldsymbol{\pi}_s, \boldsymbol{\pi}_{s+1}, \cdots, \boldsymbol{\pi}_S)$ be the steady-state probability vector of the finite generator $\boldsymbol{F} = \boldsymbol{A} + \boldsymbol{B}_2 + \boldsymbol{C}$. That is, $\boldsymbol{\pi}$ satisfies

$$\boldsymbol{\pi} \boldsymbol{F} = \boldsymbol{0} \quad \text{and} \quad \boldsymbol{\pi} \boldsymbol{e} = 1. \tag{2}$$

**Theorem 1.** *The queueing-inventory system under study is stable if and only if the following condition is satisfied.*

$$\rho = \frac{\lambda(1-\boldsymbol{\pi}_0 \boldsymbol{e})}{\mu(1-\boldsymbol{\pi}_0 \boldsymbol{e}) + \tau \boldsymbol{\pi}_0 \boldsymbol{e}} < 1 \quad \text{where} \quad \boldsymbol{\pi}_0 = \left[ \left[1 + \left(\frac{\theta_2}{\mu}\right)(S-s)\right] \left(\frac{\theta_2+\mu}{\mu}\right)^s \right]^{-1}. \tag{3}$$

*Proof.* The system is a $QBD$ process. So, it is stable *if and only if* $\boldsymbol{\pi} \boldsymbol{A} \boldsymbol{e} < \boldsymbol{\pi} \boldsymbol{C} \boldsymbol{e}$ (See in [8]). That is,

$$\lambda(1-\boldsymbol{\pi}_0 \boldsymbol{e}) < \mu(1-\boldsymbol{\pi}_0 \boldsymbol{e}) + \tau \boldsymbol{\pi}_0 \boldsymbol{e}. \tag{4}$$

Firstly, we rewrite the steady-state equations in (2) as follows.

$$\begin{aligned} -\theta_2 \boldsymbol{\pi}_0 + \mu \boldsymbol{\pi}_1 &= \boldsymbol{0}, \\ -(\theta_2+\mu)\boldsymbol{\pi}_i + \mu \boldsymbol{\pi}_{i+1} &= \boldsymbol{0}, \ 1 \leq i \leq s, \\ -\mu \boldsymbol{\pi}_i + \mu \boldsymbol{\pi}_{i+1} &= \boldsymbol{0}, \ s+1 \leq i \leq S-1, \\ \theta_2[\boldsymbol{\pi}_0 + \cdots + \boldsymbol{\pi}_s] - \mu \boldsymbol{\pi}_S &= \boldsymbol{0}, \end{aligned} \tag{5}$$

with the normalizing condition

$$\sum_{i=0}^{S} \boldsymbol{\pi}_i \boldsymbol{e} = 1. \tag{6}$$

By adding the equations given in (5), the following equations are obtained

$$\boldsymbol{\pi}_i = \begin{cases} \left(\frac{\theta_2}{\mu}+1\right)^{i-1} \left(\frac{\theta_2}{\mu}\right) \boldsymbol{\pi}_0, & 1 \leq i \leq s \\ \left(\frac{\theta_2}{\mu}+1\right)^{s} \left(\frac{\theta_2}{\mu}\right) \boldsymbol{\pi}_0, & s+1 \leq i \leq S \end{cases} \tag{7}$$

Substituting the probabilities in (7) into the equation in (6) we get the probability $\boldsymbol{\pi}_0$. So, the proof of Theorem 1 is completed.

Let $\boldsymbol{x} = (\boldsymbol{x}_0, \boldsymbol{x}_1, \cdots, \boldsymbol{x}_r, \boldsymbol{x}_{r+1}, \cdots)$ denote the steady-state probability vector of the generator matrix $\boldsymbol{Q}$ in (1). That is, $\boldsymbol{x}$ satisfies

$$\boldsymbol{x}\,\boldsymbol{Q} = \boldsymbol{0} \text{ and } \boldsymbol{x}\,\boldsymbol{e} = 1. \tag{8}$$

$(S+1)$ dimensional row vector $\boldsymbol{x}_n$, $n \geq 0$, is further partitioned into components represented as $\boldsymbol{x}_n = [x_n(0), x_n(1), \cdots, x_n(S)]$. The $x_n(i)$ gives the steady-state probability that there are $n$ customers in the system and the inventory level is $i$, $0 \leq i \leq S$. The following theorem can be used to calculate the probabilities of the system [8].

**Theorem 2.** *Under the stability condition given in Theorem 1 the steady-state probability vector $\boldsymbol{x}$ is obtained by solving the following system of linear equations*

$$\begin{aligned}
\boldsymbol{x}_0 \boldsymbol{B}_0 + \boldsymbol{x}_1 \boldsymbol{C} &= \boldsymbol{0}, \\
\boldsymbol{x}_{n-1}\boldsymbol{A} + \boldsymbol{x}_n \boldsymbol{B}_1 + \boldsymbol{x}_{n+1}\boldsymbol{C} &= \boldsymbol{0}, \quad 1 \leq n \leq r-2 \\
\boldsymbol{x}_{r-2}\boldsymbol{A} + \boldsymbol{x}_{r-1}[\boldsymbol{B}_1 + \boldsymbol{R}\boldsymbol{C}] &= \boldsymbol{0},
\end{aligned} \tag{9}$$

$$\sum_{n=0}^{r-2} \boldsymbol{x}_n \boldsymbol{e} + \boldsymbol{x}_{r-1}(\boldsymbol{I} - \boldsymbol{R})^{-1}\boldsymbol{e} = 1. \tag{10}$$

*where the rate matrix $R$ is the minimal nonnegative solution to the following matrix quadratic equation*

$$\boldsymbol{R}^2 \boldsymbol{C} + \boldsymbol{R}\boldsymbol{B}_2 + \boldsymbol{A} = \boldsymbol{0}. \tag{11}$$

The QBD structure of the generator given in (1), under the stability condition, yields a modified matrix-geometric solution. So, the non-boundary states ($n \geq r$) are given by

$$\boldsymbol{x}_{n+r} = \boldsymbol{x}_{r-1}\boldsymbol{R}^{n+1}, \quad n \geq 0, \tag{12}$$

where the matrix $R$ satisfies the matrix-quadratic equation given in (11).

At this point, we list some performance measures of the system as follow.

1) The expected loss rate of customers because of no inventory:

$$EL_1 = \lambda \sum_{n=0}^{\infty} x_n(0).$$

2) The expected loss rate of customers because of impatient:

$$EL_2 = \tau \sum_{n=1}^{\infty} x_n(0).$$

3) The expected number of customers in the system:

$$E_N = \sum_{n=1}^{\infty} n\, \boldsymbol{x}_n \boldsymbol{e} = \sum_{n=1}^{r-1} n\, \boldsymbol{x}_n \boldsymbol{e} + \boldsymbol{x}_{r-1} \boldsymbol{R}\big[r(\boldsymbol{I} - \boldsymbol{R})^{-1} + \boldsymbol{R}(\boldsymbol{I} - \boldsymbol{R})^{-2}\big]\boldsymbol{e}.$$

4) The expected number of items in the inventory: $EI = EI_1 + EI_2$

$$EI_1 = \sum_{n=0}^{r-1} \sum_{i=1}^{S} i\, \boldsymbol{x}_n(i)\boldsymbol{e} \quad \text{for } (s, Q) \text{ policy,}$$

$$EI_2 = \sum_{n=r}^{\infty} \sum_{i=1}^{S} i\, \boldsymbol{x}_n(i)\boldsymbol{e} \quad \text{for } (s, S) \text{ policy.}$$

5) The expected order size: $ES = ES_1 + ES_2$

$$ES_1 = (S - s) \sum_{n=0}^{r-1} \sum_{i=0}^{s} \boldsymbol{x}_n(i)\boldsymbol{e} \quad \text{for } (s, Q) \text{ policy,}$$

$$ES_2 = \sum_{n=r}^{\infty} \sum_{i=S-s}^{S} i\, \boldsymbol{x}_n(S - i)\boldsymbol{e} \quad \text{for } (s, S) \text{ policy.}$$

6) The expected reorder rate: $EO = EO_1 + EO_2$

$$EO_1 = \mu \sum_{n=1}^{r-1} \boldsymbol{x}_n(s+1) \quad \text{for } (s, Q) \text{ policy,}$$

$$EO_2 = \mu \sum_{n=r}^{\infty} \boldsymbol{x}_n(s+1) \quad \text{for } (s, S) \text{ policy.}$$

7) The expected replenishment rate: $ER = ER_1 + ER_2$

$$ER_1 = \theta_1 \sum_{n=0}^{r-1} \sum_{i=0}^{s} \boldsymbol{x}_n(i) \quad \text{for } (s, Q) \text{ policy,}$$

$$ER_2 = \theta_2 \sum_{n=r}^{\infty} \sum_{i=0}^{s} \boldsymbol{x}_n(i) \quad \text{for } (s, S) \text{ policy.}$$

## 4 Numerical Study

In this section, we investigate how the some performance measures behave under different scenarios and perform an optimization study regarding the threshold value $r$ that determines the transition between two inventory policies.

**Example 1:** The behavior of the performance measures of the expected number of items in the inventory-$EI$ ($EI_1$ and $EI_2$ are the expected number of items for the system under $(s,Q)$-policy and under $(s,S)$-policy, respectively) in Fig. 2 and Fig. 3, the expected order size-$ES$ ($ES_1$ and $ES_2$ are the expected order size for the system under $(s,Q)$-policy and under $(s,S)$-policy, respectively) in Fig. 4, and the expected lost rate of customers because of no inventory-$EL_1$ in Fig. 5 are investigated while the arrival rate $\lambda$ and the service rate $\mu$ of customers change.

For this purpose, in all figures, we fix the some parameter values; the maximum inventory level is $S = 20$, the impatient rate is $\tau = 2$, the replenishment rates are $\theta_1 = 1$ for $(s,Q)$-policy and $\theta_2 = 1.5$ for $(s,S)$-policy. Also, the threshold value and the reorder point are varied by $r = 4, 6, 8, 10, 12$ and $s = 3, 7$. In the figures, we consider when the service rate changes at $x$-axis, the arrival rate is fixed by $\lambda = 2$, and when the arrival rate changes at $x$-axis, the service rate is fixed by $\mu = 4$. As the arrival rate of customers increases (as the traffic intensity increases), the number of items in the inventory ($EI$) is expected to decrease. It is seen in Fig. 2 that $EI$ values decrseea until the system reaches a certain density and then increase. This behavior is valid for all $r$ values, but at large $r$ values the change point is seen when the system is more dense. In addition, the graphs of $EI$ show that there will be more items in inventory at the ordering point $s$ is increased from 3 to 7.

The $(s,Q)$ policy is followed until the number of customers in the system reaches $r$, and the $(s,S)$ policy is followed when the number of customers becomes $r$ and more. In other words, the number of items arriving in the inventory is constant up to the threshold value $r$ (as much as $S - s = Q$), and the number in the inventory will always be at the maximum level with the orders coming after the threshold value. As the $r$ value increases, it is seen that the system remains in the $(s,Q)$ policy more as expected when all the graphs in Fig. 2 are examined.

As the service rate of customers increases (as the traffic intensity decreases), the number of items in the inventory ($EI$) is expected to decrease. This can be seen in Fig. 3. Also, as expected, it is clear from the graphs of $EI$ that there will be more items in inventory at a given time as the ordering point $s$ is increased from 3 to 7. When traffic intensity is high (in other words, at low service rate), there is a difference $EI$ values depending on the threshold value $r$. As the service rate increases, the system density decreases and the effect of the $r$ value on the number in the inventory decreases. At low traffic intensity, the system generally operates under the $(s,Q)$ policy. For example, for $\mu = 5$, the traffic intensity is approximately 0.4 and it is seen from the $EI_1$ and $EI_2$ graphs in Fig. 3 that the system generally operates under the $(s,Q)$ policy.

The behavior of the expected order size ($ES$) can be seen in the graphs on the left side of Fig. 4 for increasing values of $\lambda$ and in the graphs on the right side of Fig. 4 for increasing values of $\mu$. We would like to point out that the similar behavior shown by $ES$ in Fig. 4 is also valid for the expected repnesihment rate

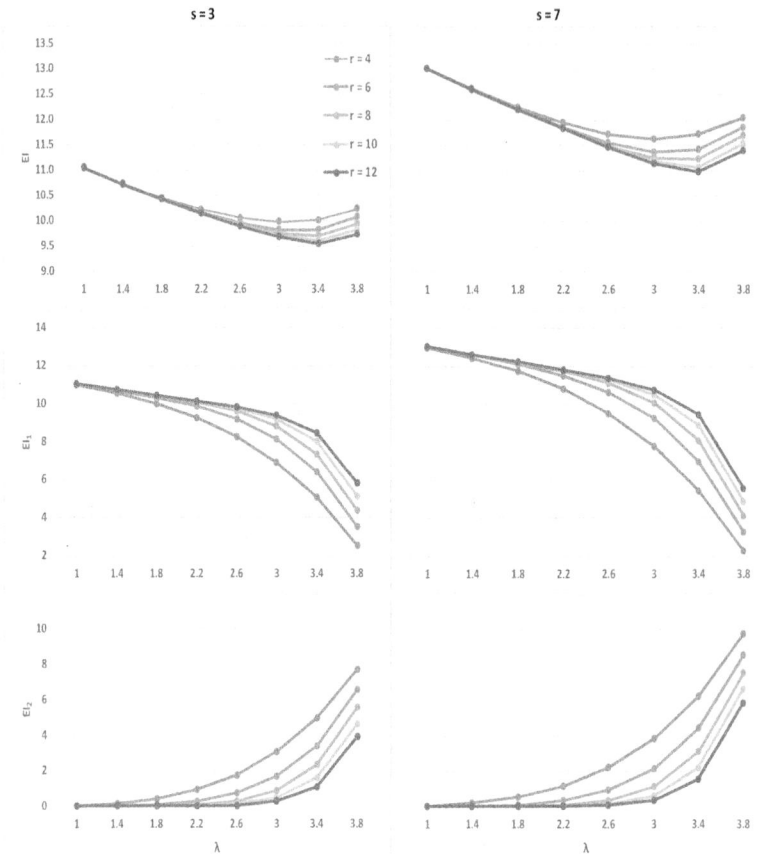

**Fig. 2.** The expected number of items in the inventory as increasing values of $\lambda$

($ER$) and the expected reorder rate ($EO$). Similar graphs were not included in the study because the graphs would take up unnecessary space.

Up to a certain traffic intensity (for example, see $\lambda = 3$ for $r = 12$), the system generally operates under the $(s, Q)$ policy in Fig. 4. $ES_1$ values increase rapidly up to this point and $ES_2$ values are almost zero. When the value of $\lambda$ is increased, it is seen that $ES_1$ starts to decrease and $ES_2$ starts to increase (the system also starts to operate at $(s, S)$ policy). On the other hand, after a certain traffic intensity (for example, see $\mu = 3.4$ for $r = 12$), the system generally operates under the $(s, Q)$ policy. That is, $ES_1$ values are positive while $ES_2$ values are almost zero.

Figure 5 shows how the expected lost rate of customers due to lack of items in the inventory ($EL_1$) changes as $\lambda$ increases on the top and as $\mu$ increases on the bottom for $s = 3$ and $s = 7$.

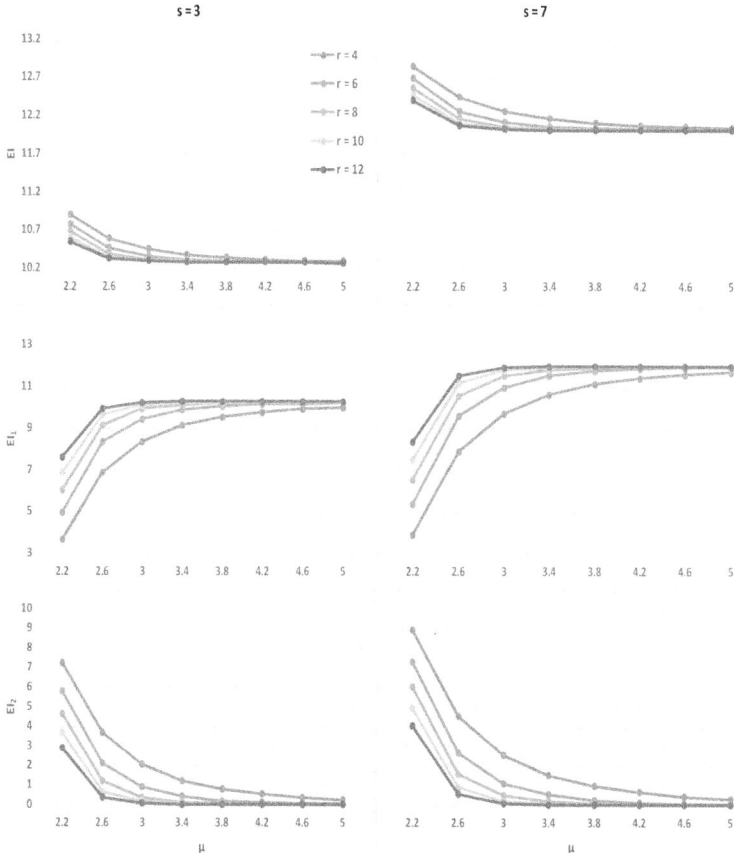

**Fig. 3.** The expected number of items in the inventory as increasing values of $\mu$

As $\lambda$ increases (system density increases), the values of $EL_1$ increase as expected. On the other hand, as $\mu$ increases, although the system density decreases, it is seen that the values of $EL_1$ increase very slightly and then remain constant. In both cases, the effect of the threshold value $r$ on $EL_1$ is seen when the system is dense.

Compared to $s = 3$, it is seen that $EL_1$ values are lower in case $s = 7$. Looking at Fig. 2 and Fig. 3, in case $s = 7$, there are more items in the inventory. Therefore, the customers will be less likely to decide not to participate in the system (more customers will join the queue system in order to receive service).

**Example 2:** The performance measures $ES_1$ and $ES_2$ that the expected order size for the system under $(s, Q)$-policy and under $(s, S)$-policy, respectively, are examined while the replenishment rates $\theta_1$ and $\theta_2$ change. The changes in both

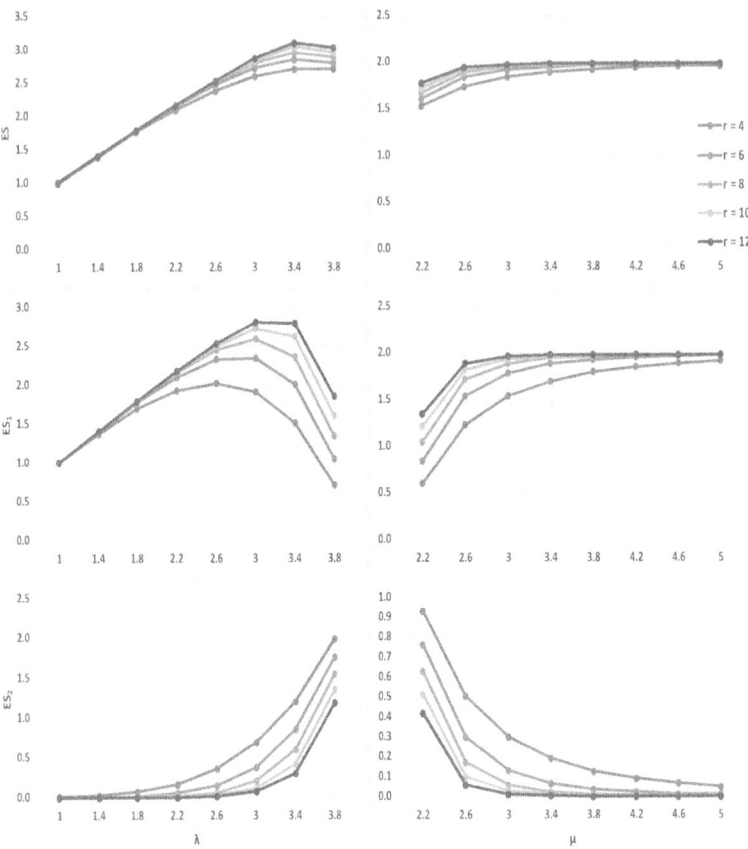

**Fig. 4.** The expected order size as increasing values of $\lambda$ and $\mu$

measures two different traffic intensity are given for the increasing $\theta_1$ and the increasing $\theta_2$ in Fig. 6 and Fig. 7, respectively.

For this purpose, we fix the some parameter values; the maximum inventory level is $S = 20$, the reorder point is $s = 7$, the arrival rate is $\lambda = 2$, the impatient rate is $\tau = 1$. Also, the threshold value is varied by $r = 4, 6, 8, 10, 12$. When the replenishment rate $\theta_1$ for $(s, Q)$-policy changes at $x$-axis, the replenishment rate for $(s, S)$-policy is fixed by $\theta_2 = 4$ in Fig. 6, and when the replenishment rate $\theta_2$ for $(s, S)$-policy changes at $x$-axis, the replenishment rate for $(s, Q)$-policy is fixed by $\theta_1 = 1$ in Fig. 7.

Increasing $\theta_1$ means that there are more items in inventory immediately after a replenishment. So, a smaller size order is placed when the order point is reached. The comment that $ES_1$ decreases as $\theta_1$ increases can be seen in Fig. 6 for both $\mu = 2.2$ (the intensity approximately 0.9) and $\mu = 4.2$ (the intensity approximately 0.5). The larger the threshold value $r$, the more the

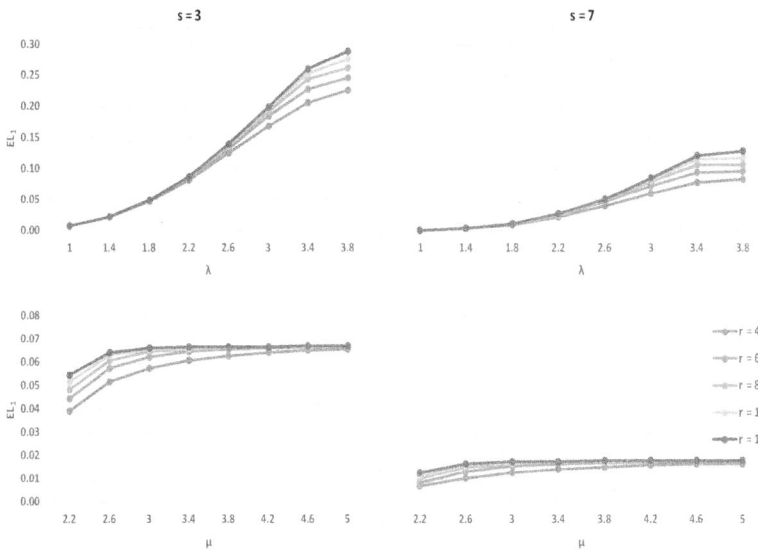

**Fig. 5.** The expected lost rate of customers because of no inventory as increasing values of $\lambda$ and $\mu$

system operates under the $(s,Q)$-policy. When the items in the inventory are filled according to $(s,Q)$-policy, the expected number of items in the inventory ($EI$) becomes less at comparing by the other policy. This means that customer' demand is tried to be met with larger orders. Therefore, when the value of $r$ is increased, the value of $ES_1$ also increases. It is also seen that the value of $r$ has no effect on the performance measure in a low-density system. As expected, an increase in $\theta_1$ has no effect on $ES_2$ (see the case of $\mu = 2.2$ in Fig. 6).

In similar observation, there are more items in inventory immediately after a replenishment by Increasing $\theta_2$. So, a smaller size order is placed when the order point is reached. The values of $ES_2$ decrease as $\theta_2$ increases can be seen in Fig. 7 for both $\mu = 2.2$ (the intensity approximately 0.9). The system transitions from $(s,Q)$-policy to $(s,S)$-policy at low threshold value in the low intensity system (the intensity approximately 0.5 for $\mu = 4.2$). Therefore, the decreasingof $ES_2$ is evident for $r = 4$ and $r = 6$ (expected order size is almost zero in others). The larger the threshold value $r$, the more the system operates under the $(s,Q)$-policy. Therefore, when the value of $r$ is increased, the value of $ES_1$ decreases. As expected, an increase in $\theta_2$ has no effect on $ES_1$ (see the case of $\mu = 2.2$ in Fig. 7).

**Example 3:** In this example, we examine an answer to the question that we should change the inventory policy is sought when the number of customers in the system reaches what value? So, we gives the optimum threshold values ($r^*$) and the minimum expected total cost values ($ETC^*$) under different scenarios

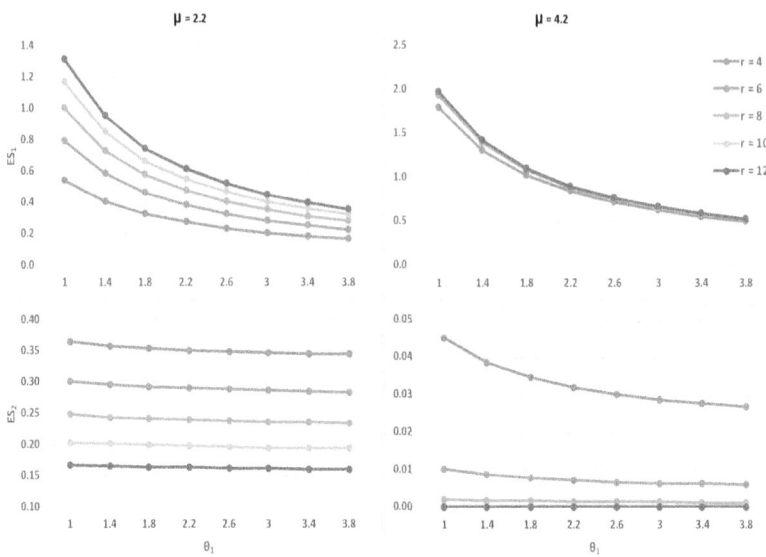

**Fig. 6.** The expected order size as increasing values of $\theta_1$

obtained by varying the values of $\lambda$, $\mu$, $s$, $S$ in Table 1. For this example, we fix the some parameter values by $\theta_1 = 1$, $\theta_2 = 1.5$ and $\tau = 2$ and the costs by $c_w = 300$, $c_h = 50$, $c_{l1} = 200$, $c_{l2} = 350$, $c_{o1} = 25$, $c_{o2} = 50$, $c_{r1} = 15$ and $c_{r2} = 30$.

$$ETC = c_w.E_N + c_h.(EI_1 + EI_2) + c_{l1}.EL_1 + c_{l2}.EL_2 + c_{o1}.ES_1.EO_1 \\ + c_{o2}.ES_2.EO_2 + c_{r1}.ER_1 + c_{r2}.ER_2$$

where
$c_w$ : the waiting cost of a customer in the system,
$c_h$ : the holding price per item in the inventory per unit of time,
$c_{l1}$ : the cost incured due to the loss of a customer (not joint to system),
$c_{l2}$ : the cost incured due to the loss of a customer (impatient),
$c_{o1}$ : the unit price of the order under $(s, Q)$-policy,
$c_{o2}$ : the unit price of the order under $(s, S)$-policy,
$c_{r1}$ : the cost incured due to replenishment under $(s, Q)$-policy,
$c_{r2}$ : the cost incured due to replenishment under $(s, S)$-policy.

For a fixed order point $s$, when the maximum inventory level $S$ is increased, the values of $r^*$ either remain the same or increase. It can be seen that the amount of change in the values of $r^*$ depends on $s$ in Table 1. For example, see the cases $s = 3$ and $s = 6$ under the same scenarios.

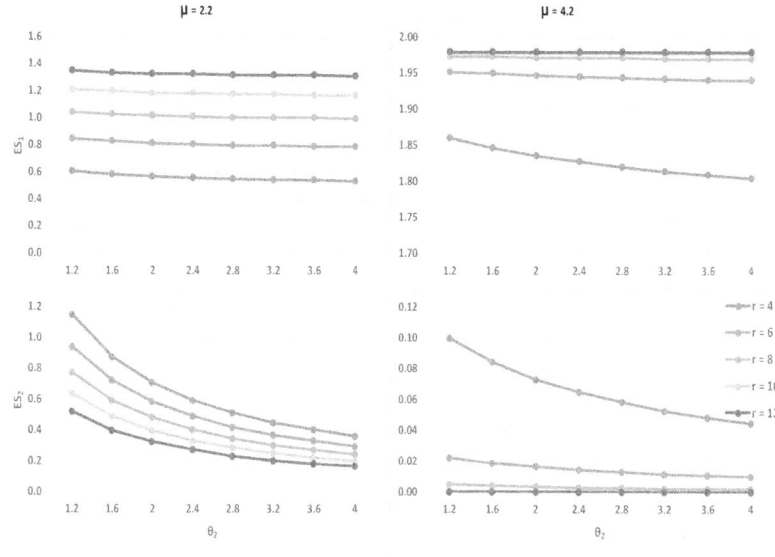

**Fig. 7.** The expected order size as increasing values of $\theta_2$

**Table 1.** Optimum values of $r^*$ and $ETC^*$

| $(s,S)$ | $\mu = 4$ | | | | $\mu = 4.2$ | | | |
| --- | --- | --- | --- | --- | --- | --- | --- | --- |
| | $\lambda = 1.6$ | | $\lambda = 1.7$ | | $\lambda = 1.6$ | | $\lambda = 1.7$ | |
| | $r^*$ | $ETC^*$ | $r^*$ | $ETC^*$ | $r^*$ | $ETC^*$ | $r^*$ | $ETC^*$ |
| (3,14) | 4 | 592.6632 | 4 | 612.6089 | 4 | 577.5819 | 4 | 595.3002 |
| (3,16) | 5 | 641.0950 | 5 | 660.8401 | 4 | 625.9533 | 4 | 643.4539 |
| (3,18) | 5 | 689.8889 | 6 | 709.4732 | 5 | 674.7061 | 5 | 692.0335 |
| (4,14) | 5 | 611.8915 | 5 | 631.1760 | 5 | 596.7302 | 5 | 613.7574 |
| (4,16) | 6 | 660.3111 | 6 | 679.3898 | 5 | 645.1102 | 5 | 661.9185 |
| (4,18) | 8 | 709.1586 | 8 | 728.0795 | 6 | 693.9288 | 7 | 710.5704 |
| (5,14) | 7 | 634.0176 | 7 | 652.8388 | 6 | 618.7943 | 6 | 635.3360 |
| (5,16) | 10 | 682.2558 | 11 | 700.8544 | 7 | 667.0085 | 7 | 683.3188 |
| (5,18) | 13 | 731.0407 | 14 | 749.4837 | 12 | 715.7696 | 12 | 731.9144 |
| (6,14) | 8 | 658.0630 | 8 | 676.6153 | 8 | 642.7941 | 8 | 659.0500 |
| (6,16) | 11 | 705.9354 | 12 | 724.2343 | 10 | 690.6468 | 10 | 706.6393 |
| (6,18) | 18 | 754.5448 | 21 | 772.6786 | 13 | 739.2399 | 13 | 755.0599 |

## 5 Conclusions

A queue-dependent hybrid replenishment policy is proposed in QIS with infinite waiting room. It is based on the combined use of two known replenishment policies: $(s, S)$ and $(s, Q)$ policies. When the inventory level drops to the reorder point $s$ i) if the number of customers in the system is less than threshold value $r$, then $(s, Q)$ policy is applied; ii) if the number is more or equal to $r$, then $(s, S)$ policy is applied. Lead time is exponentially distributed. The stability condition of the system is established and the stationary distribution is calculated by using the matrix-geometric method. Numerical examples are performed under different scenarios to understand how the system works and to decide when replacement policy change.

## References

1. Sigman, K., Simchi-Levi, D.: Light traffic heuristic for an $M/G/1$ queue with limited inventory. Ann. Oper. Res. **40**, 371–380 (1992)
2. Melikov, A., Molchanov, A.: Stock optimization in transport/storage systems. Cybernetics **28**(3), 484–487 (1992)
3. Krishnamoorthy, A., Manikandan, R., Lakshmy, B.: Revisit to queueing-inventory system with positive service time. Ann. Oper. Res. **233**, 221–236 (2015)
4. Krishnamoorthy, A., Shajin, D., Narayanan, W.: Inventory with positive service time: a survey. In: Anisimov, V., Limnios, N. (eds.) Advanced Trends in Queueing Theory; Series of Books Mathematics and Statistics Sciences, vol. 2, pp. 201–238. ISTE & Wiley, London (2021)
5. Melikov, A., Shamaliyev, M.: Queueing system $M/M/1/\infty$ with perishable inventory and repeated customers. Autom. Remote. Control. **80**, 53–65 (2019)
6. Melikov, A., Mirzayev, R., Nair, S.S.: Double sources queuing-inventory system with hybrid replenishment policy. Mathematics **10**, 2423 (2022)
7. Lawrence, S., Melikov, A., Sivakumar, B.: Analysis and optimization of hybrid replenishment policy in a double-sources queueing-inventory system with MAP arrivals. Ann. Oper. Res. **331**, 1249–1267 (2023)
8. Neuts, M.F.: Matrix-Geometric Solutions in Stochastic Models: An Algorithmic Approach. John Hopkins University Press, Baltimore (1981)

# A Stochastic Approach for Optimizing a Discrete Time (s, S) Perishable Inventory System with Modified N-Policy

K. P. Jose[1], Jijo Joy[2,3](✉), and M. P. Anilkumar[3]

[1] Post Graduate Department of Mathematics, St. Peter's College Kolenchery, Ernakulam, Kerala, India
[2] Post Graduate Department of Mathematics, St. Aloysius College Edathua, Alappuzha, Kerala, India
jjoysac@gmail.com
[3] Department of Mathematics, T. M. Govt. College Tirur, Malappuram, Kerala, India

**Abstract.** This paper investigates a discrete-time (s, S) perishable inventory system with positive service time and positive lead time. The system operates under a modified N-policy with two modes of service, where arrivals follow a Bernoulli process. During the stock-out period, the server goes on vacation. After the server vacation, the server initiates a batch service of size N only after N customer arrivals. During batch service, new individual arrivals were served separately. Service times are also assumed to be geometrically distributed. Lead times follow geometric distribution, where the maximum inventory level is S. The inventory items are perishable, with perishability following a geometric distribution, thereby adding an additional layer of complexity. The matrix analytic method is used to analyze system behavior in steady state, accounting for the arrival, service, and replenishment dynamics within the model. Numerical experiments are conducted to examine the impact of various system parameters and an optimal (s, S) pair is derived for fixed parameter values, highlighting practical applications and the efficiency of the modified N-policy in managing perishable items.

**Keywords:** Discrete-Time Inventory · Perishable Products · Modified N-Policy · Matrix Analytic Method · Positive Lead Time

## 1 Introduction

Effective inventory management is crucial for manufacturing and service industries to maintain seamless operations and optimize resource allocation. Perishable inventory, in particular, poses unique challenges due to the limited shelf life and deterioration rates associated with products such as food items, pharmaceuticals, and chemicals. Poor management in these sectors can lead to substantial waste, reduced profitability, and potential shortages.

This work considers a discrete-time perishable inventory model incorporating positive service time and positive lead time in a modified N-policy framework. The model was designed in such a way to control inventory level efficiently by considering real-world constraints. Here, the modified N-policy ensures that inventory replenishment occurs only when customer arrivals reach a threshold, reducing holding costs and waste for items that spoil. An (s, S) policy framework further allows for determining optimal reorder points and stock levels, balancing the need to prevent stockouts with the goal of minimizing excess inventory.

Practical applications of this model extend across various industries. For example, in the pharmaceutical sector, hospitals and pharmacies must maintain optimal stock levels of perishable medicines and vaccines. Excess stock risks expiration, while insufficient stock could lead to shortages. Similarly, online grocery delivery services, which handle perishables with varied demand patterns, can utilize this model to minimize spoilage and manage stock efficiently.

The steady-state analysis is provided by the application of the Matrix Analytic Method, which provide details about inventory management and ideal system performance. Numerical experiments gave insight into parameter variations and offered guidelines for deriving the optimal (s, S) pair under different demand scenarios, making the model highly adaptable for applications requiring dynamic and responsive inventory management solutions.

## 2 Review of Related Works

Shajin et al. (2019) [6] conducted a comprehensive survey on inventory systems with positive service times, presenting various approaches to manage delay-induced costs effectively while maintaining service quality. The authors use analytical techniques to address service delays, which are prevalent in practical inventory scenarios. Raju and Krishnamoorthy (1998) [11] introduced a model incorporating positive lead times, emphasizing its importance in controlling perishable inventory levels. Their work utilizes queueing theory to illustrate how positive lead times impact inventory availability under different demand conditions.

Manuel et al. (2007) [13] examined perishable inventory systems with postponed demands and the concept of negative customers, incorporating demand postponements that reflect real-world constraints such as order delays. This approach is relevant for modified N-policy models, where adjusting demand intake can help maintain inventory balance in high perishability contexts . Resmi et al.(2014) [15] studied an (s, Q) inventory system under N-policy, integrating both positive lead time and service time. Their research highlighted the interplay between replenishment lead times and service rates, which significantly impacts inventory control in systems where delay factors are non-negligible . Zhang and Xu (2008) [17] analyzed an M/M/1 queue with multiple working vacations and an N-policy, exploring how periodic server vacations can aid in reducing idle time and inventory costs. Their findings illustrate the value of server management techniques in optimizing system costs, particularly relevant in perishable inventory models that incorporate service dynamics under variable demand conditions . Anilkumar and Jose (2022) [10] analyzed a discrete-time queueing inventory model with backorders. Joy and Jose (2024) [8] used Matrix Analytic Method to analyze a discrete time perishable inventory model with production of items and positive service time.

Alfa (2016) [3] provided an applied perspective on discrete-time queues, detailing how discrete-event simulation can be employed to model inventory systems where events occur at fixed intervals. Alfa's work is foundational, as it links discrete-time frameworks with real-world inventory control applications. Similarly, Moreno (2007) [12] explored a single-server discrete-time queue with a modified N-policy, demonstrating how discrete-time frameworks can handle complex service and arrival patterns in inventory systems. These studies highlight the flexibility and accuracy of discrete-time models in inventory control. Deepthi and Krishnamoorthy (2013) [4] investigated discrete-time inventory models with and without positive service time, analyzing the impacts of service times on inventory systems where delays can affect overall efficiency. Their study highlighted the role of service time as a factor in optimizing replenishment policies, which is essential for practical applications where service rates influence inventory dynamics. Jijo and Jose (2023) [7] explored a perishable inventory system under an (s, S) policy with age-dependent requests, emphasizing the challenges of managing perishable items with varying demands based on item age.

Reshmi and Jose (2020) [14] applied matrix analytic techniques to a perishable inventory system with inventory-dependent arrivals, efficiently computing steady-state probabilities for complex systems with variable demand intensities. Their work demonstrated the effectiveness of matrix analytic methods in handling scenarios with fluctuating customer arrivals tied to inventory levels, which is particularly relevant for real-world perishable systems where demand can be unpredictable. Lan and Tang (2019) [16] utilized matrix analytic methods for a discrete-time Geo/G/1 queue with modified server vacations and Bernoulli feedback, offering insights into managing inventory levels under conditions of varying service availability. Their findings underscore the utility of server vacation policies in balancing inventory levels and service costs, especially in environments where server availability influences inventory replenishment strategies.

N-policy models have been widely studied in inventory and queueing contexts to control resource allocation. Krishnamoorthy and Deepak (2002) [1] modified the N-policy for M/G/1 queues, which has become a standard approach in managing inventory systems with batch arrivals. Anbazhagan and Krishnamoorthy (2008) [2] further refined this by applying N-policy to a perishable inventory system, addressing the challenges of perishability and demand surges. Lee and Yang (2013) [5] examined an N-policy model in a Geo/G/1 queue, introducing disasters as an added factor and exploring its application in wireless networks, a novel approach that adds resilience to perishable inventory management. Anilkumar and Jose (2020) [9] developed a discrete-time Geo/Geo/1 inventory model under a modified N-policy, demonstrating that this policy could enhance control over replenishment timing in systems with single-server setups. This paper is an extension of this work by adding perishability to the model.

The rest of the paper is organized as follows: Sect. 3 provides the mathematical modeling and analysis. The Stability Condition and Steady State Analysis in Sect. 4. Performance Measures and Cost Function in Sects. 5 and 6, respectively. Numerical and graphical illustration in Sect. 7 and finally (s, S) pair is Sect. 8.

## 3 Mathematical Modelling and Analysis

In this study, we model a discrete-time perishable inventory system with positive service and lead times under a modified $N$-policy framework. Customer inter-arrival times follow a geometric distribution with parameter $a$. A customer's demand is for exactly one item. Each item in the inventory requires a positive amount of service time before the customer gets it. There is only one server for service. The server goes on vacation when inventory is empty, and all remaining inventory is removed. Backlog is not allowed.

Upon $N$ customer arrivals, batch service of size $N$ begins, with service times geometrically distributed with parameter $b_2$. Additional arriving customers are served individually, with service times geometrically distributed with parameter $b_1$. No customers enter when inventory is zero to prevent stockouts.

When the inventory level drops to $s$, an order for replenishment is made to make the inventory to maximum level $S$. Lead time is assumed to be positive. Inventory replenishment lead times follow a geometric distribution with parameter $c$, occurring at slot boundaries. Inventory perishability also follows a geometric distribution with parameter $d$. During batch service, inventory is restocked to the maximum level, ensuring demand is met.

**Notations**

$X_n$ : Number of customers in the queue at an epoch $n$.

$Y_n$ : $\begin{cases} 0, \text{ Server status when the server is on vacation.} \\ 1, \text{ Server status when the server is busy with a single service.} \\ 2, \text{ Server status when server is busy with batch service.} \end{cases}$

$Z_n$ : Inventory level at an epoch $n$.

e : $(1, 1, 1, ..., 1)$, column vector of 1's of size $S + 2$.

Then $\{(X_n, Y_n, Z_n); n = 0, 1, 2, 3, ..\}$ is a Discrete Time Markov Chain with state space $\{(0, 0, 0) \cup (i, 1, k), i \geq 0, 0 \leq k \leq S\} \cup \{(i, 0, 0), 1 \leq i \leq N - 1\} \cup \{(i, 2, S), i \geq N\}$.

Now the transition probability matrix of the process is

$$\mathcal{P} = \begin{array}{c} 0 \\ 1 \\ 2 \\ \vdots \\ N-1 \\ N \\ N+1 \\ \vdots \end{array} \begin{pmatrix} \begin{array}{ccccccccc} 0 & 1 & 2 & 3 & \cdots & & & & \end{array} \\ \begin{pmatrix} C_1 & C_0 & & & & & & \\ D_2 & D_1 & D_0 & & & & & \\ & A_2 & D_1 & D_0 & & & & \\ & & \ddots & \ddots & \ddots & & & \\ & & & A_2 & D_1 & D_0 & & \\ D_3 & A_3 & & & A_2 & A_1 & A_0 & \\ & A_4 & A_3 & & & A_2 & A_1 & A_0 \\ & & & & & & \ddots & \ddots & \ddots \end{pmatrix} \end{array} \end{pmatrix}$$

where the blocks $C_0, C_1$ are matrices of order $1 \times 1, 1 \times (S+2)$ respectively, $D_2, D_3$ are matrices of order $(S+2) \times 1$ and $D_0, D_1, A_0, A_1, A_2, A_3, A_4$ are square matrices whose $(j,k)^{th}$ element with phase $i$ is given below. Here $\alpha = [\bar{d}ab_1 + d\bar{a}\bar{b}_1]\bar{c}, \alpha_1 = \bar{d}ab_1 + d\bar{a}\bar{b}_1, \beta = c[ab_1 + \bar{a}\bar{b}_1]$.

$$D_0 = \begin{pmatrix} a\bar{b}_1\bar{c} & & & & & & & a\bar{b}_1 c \\ da\bar{b}_1\bar{c} & \bar{d}a\bar{b}_1\bar{c} & & & & & & a\bar{b}_1 c \\ & \ddots & \ddots & & & & & \vdots \\ & & da\bar{b}_1\bar{c} & \bar{d}a\bar{b}_1\bar{c} & & & & a\bar{b}_1 c \\ & & & da\bar{b}_1 & \bar{d}a\bar{b}_1 & & & \\ & & & & \ddots & \ddots & & \\ & & & & & da\bar{b}_1 & \bar{d}a\bar{b}_1 \end{pmatrix}$$

$$D_1 = \begin{pmatrix} & \bar{c} & & & & & & & c \\ \bar{c}ab_1 & \bar{c}\bar{a}\bar{b}_1 & & & & & & & \beta \\ dab_1\bar{c} & \alpha & \bar{d}\bar{a}\bar{b}_1\bar{c} & & & & & & \beta \\ & \ddots & \ddots & \ddots & & & & & \vdots \\ & & & dab_1\bar{c} & \alpha & \bar{d}\bar{a}\bar{b}_1\bar{c} & & & \beta \\ & & & & dab_1 & \alpha_1 & \bar{d}\bar{a}\bar{b}_1 & & \\ & & & & & \ddots & \ddots & \ddots & \\ & & & & & & dab_1 & \alpha_1 & \bar{d}\bar{a}\bar{b}_1 \end{pmatrix}$$

$$A_2 = \begin{pmatrix} \bar{a}b_1\bar{c} & & & & & & \bar{a}b_1c & & \\ d\bar{a}b_1\bar{c} & \bar{d}\bar{a}b_1\bar{c} & & & & & \bar{a}b_1c & & \\ & \ddots & \ddots & & & & \vdots & & \\ & & d\bar{a}b_1\bar{c} & \bar{d}\bar{a}b_1\bar{c} & & & \bar{a}b_1c & & \\ & & & d\bar{a}b_1 & d\bar{a}\bar{a}b_1 & & & & \\ & & & & \ddots & \ddots & & & \\ & & & & & d\bar{a}b_1 & \bar{d}\bar{a}b_1 & & \end{pmatrix}$$

$$D_3 = \begin{bmatrix} \bar{a}b_2 \\ 0 \\ 0 \\ \vdots \\ 0 \end{bmatrix} \quad D_2 = \begin{bmatrix} 0 \\ 0 \\ \bar{a}b_1 \\ \vdots \\ \bar{a}b_1 \end{bmatrix}$$

$$A_1 = \begin{cases} \bar{a}\bar{b_2} & \text{for } i=1, j=1 \\ [D_1] & \text{otherwise} \end{cases} \quad A_0 = \begin{cases} a\bar{b_2} & \text{for } i=1, j=1 \\ [D_0] & \text{otherwise} \end{cases}$$

$$C_0 = (a, 0, \cdots, 0), C_1 = (\bar{a})$$

$$A_4 = \begin{cases} \bar{a}b_2 & \text{for } i=1, j=S-N+2 \\ 0 & \text{otherwise} \end{cases} \quad A_3 = \begin{cases} ab_2 & \text{for } i=1, j=S-N+2 \\ 0 & \text{otherwise} \end{cases}$$

$$C_0 = (a, 0, \cdots, 0), C_1 = (\bar{a})$$

## 4 Stability Condition and Steady State Analysis

Since customers are not allowed when the inventory is zero and more than $N$ customers, the system considered is stable if $a \leq b_1$.

The Matrix Analytic Method is used to analyze the model.

Let $\pi = (\pi_0, \pi_1, \ldots, \pi_N, \pi_{N+1}, \ldots)$ be the steady state probability vector of $P$. When the stability condition is satisfied $\pi_i's$ are given by $\pi_{N+1+k} = \pi_{N+1} R^k, (k \geq 1)$.

Where $R$ is the minimal non-negative solution of the equation $R^{N+1}A_4 + R^N A_3 + R^2 A_2 + RA_1 + A_0 = R$ and the vectors $\pi_0, \pi_1, \ldots, \pi_{N+1}$ are obtained by solving

$$\pi_0 C_1 + \pi_1 D_2 + \pi_N D_3 = \pi_0$$
$$\pi_0 C_0 + \pi_1 D_1 + \pi_2 A_2 + \pi_N(A_3 + RA_4) = \pi_1$$
$$\text{for } i = 2, \ldots, N-1$$
$$\pi_{i-1} D_0 + \pi_i D_1 + \pi_{i+1} A_2 + \pi_N R^{i-1}(A_3 + RA_4) = \pi_i$$
$$\pi_{N-1} D_0 + \pi_N [A_1 + RA_2 + R^{N-1} A_3 + R^4 A_4] = \pi_N$$

Subject to the normalizing condition $\left[\sum_{i=0}^{N-1} \pi_i + \pi_N(I-R)^{-1}\right]\mathbf{e} = 1$.

## 5 Performance Measures

Let
$$\pi_i = \begin{cases} (\pi_{i,0,0}, \pi_{i,1,0}, \pi_{i,1,1}, \ldots, \pi_{i,1,s}), & \text{for } 1 \leq i \leq N \\ (\pi_{i,2,S}, \pi_{i,1,0}, \pi_{i,1,1}, \ldots, \pi_{i,1,S}), & \text{for } i \geq N+1 \end{cases}$$

Some important performance measures of the system under steady state are

1. Expected Re order Rate, $ERR$, is given by
$$ERR = b_1 \left[\sum_{i=1}^{\infty} \pi_{i,1,s+1} + d \sum_{i=1}^{\infty} \pi_{i,1,s+2}\right]$$

2. Expected Inventory Level, $EIL$, is given by
$$EIL = \sum_{i=1}^{\infty} \sum_{k=1}^{S} k\pi_{i,1,k} + S \sum_{i=N+1}^{\infty} \pi_{i,2,S}$$

3. Expected Loss Rate of Customers, $ELC$, is given by
$$ELC = \sum_{i=1}^{\infty} a\pi_{i,1,0} + \sum_{i=0}^{N} a\pi_{i,0,0}$$

4. Expected Number of Customers waiting in the system for single service when the inventory level is zero, $EW0$, is given by
$$EW0 = \sum_{i=1}^{\infty} i\pi_{i,1,0}$$

5. Expected Number of Customers Waiting in the System for Single Service,, $ECQ$, is given by
$$ECQ = \sum_{i=1}^{\infty} \sum_{k=0}^{S} (i-1)\pi_{i,1,k} + \sum_{i=N}^{\infty} (i-N)\pi_{i,2,S}$$

6. Expected Rate of Departure after Completing Service, $EDS$ is given by
$$EDS = b_1 \sum_{i=1}^{\infty} \sum_{k=1}^{S} \pi_{i,1,k} + Nb_2 \sum_{i=N}^{\infty} \pi_{i,2,S}$$

7. Expected Perishable Quantity, $EPQ$, is given by
$$EPQ = d \sum_{i=1}^{\infty} \sum_{k=1}^{S} k\pi_{i,1,k}$$

8. Expected number of customers in the system, $ENC$, is given by

$$ENC = \sum_{i=0}^{\infty} i\pi_i \mathbf{e}$$

9. Expected balance inventory, $EBI$, is given by

$$EBI = a_1 \sum_{i=1}^{\infty} \sum_{k=1}^{S} (k-1)\pi_{i,1,k}$$

10. Expected Replenishment Rate, $ERP$, is given by

$$ERP = c \sum_{i=1}^{\infty} \sum_{k=0}^{s} \pi_{i,1,k}$$

11. Probability that the inventory level is zero, $PIO$ is given by

$$PIO = \sum_{i=0}^{N-1} \pi_{i,0,0} + \sum_{i=1}^{\infty} \pi_{i,1,0}$$

12. Expected number of customers waiting for batch service, $ECB$, is given by

$$ECB = \sum_{i=1}^{N-1} i\pi_{i,0,0}$$

13. Expected Perishable Rate, $EPR$, is given by

$$EPR = d \sum_{i=1}^{\infty} \sum_{k=1}^{S} \pi_{i,1,k}$$

## 6  Cost Function

Define the expected total cost of the system per unit time as

$$ETC = c_0(ERR) + \sum_{k=1}^{s} c(S-k)x_k c_1 + c_2(EIL) + c_3(ELC) + c_4(EW0)$$

$$+ c_5(ECQ) + c_6(EPQ) + c_7(EDS)$$

$c_0$ : the setup cost/order
$c_1$ : procurement cost/unit/unit time
$c_2$ : inventory holding cost/unit/unit time
$c_3$ : cost due to loss of customers/unit/unit time
$c_4$ : holding cost of a customer/unit time when the inventory level is zero
$c_5$ : holding cost of a customer/unit time when the server is busy
$c_6$ : expected perishablity cost/unit item
$c_7$ : service cost/unit/unit time
$x_k$ : the probability that the inventory level is $k$ during single service

# 7 Numerical and Graphical Illustrations

The tables below show the numerical findings obtained for various performance measures about the various parameters studied. The following diagram depicts the relationship between several performance measures and parameters.

## 7.1 Effect of Service Rate $b_1$ on Various Performane Measures

When $a = .35, b_2 = .15, c = .2, d = .15, s = 3, S = 10, N = 5, c_0 = 75, c_1 = 1, c_2 = .05, c_3 = 1, c_4 = 1, c_5 = 1, c_6 = 1, c_7 = 1$, Table 1 indicates that when the service rate rises, so does the expected reorder rate and customer departure rate, but the expected inventory level, and number of customers waiting decreases.

Table 1. Effect of service rate $b_1$ on the model.

| $b_1$ | ERR | EIL | EW0 | EPQ | EDS | ETC |
|---|---|---|---|---|---|---|
| 0.4 | 0.0398 | 5.7968 | 0.1975 | 0.8695 | 0.3573 | 8.4154 |
| 0.45 | 0.0434 | 5.5686 | 0.2105 | 0.8352 | 0.3890 | 8.3280 |
| 0.5 | 0.0465 | 5.3721 | 0.2172 | 0.8058 | 0.4185 | 8.2798 |
| **0.55** | 0.0492 | 5.1978 | 0.2195 | 0.7796 | 0.4460 | **8.2599** |
| 0.6 | 0.0516 | 5.0396 | 0.2186 | 0.7559 | 0.4719 | 8.2603 |
| 0.65 | 0.0540 | 4.8941 | 0.2154 | 0.7341 | 0.4964 | 8.2763 |
| 0.7 | 0.0562 | 4.7590 | 0.2100 | 0.7138 | 0.5197 | 8.3060 |
| 0.75 | 0.0584 | 4.6329 | 0.2030 | 0.6949 | 0.5420 | 8.3491 |
| 0.8 | 0.0606 | 4.5147 | 0.1946 | 0.6772 | 0.5633 | 8.4064 |

## 7.2 Effect of Replenishment Rate c on Various Performane Measures

When $a = .4, b_1 = .7, b_2 = .25, d = .15, = 3, S = 10, s = 3, N = 5, c_0 = 150, c_1 = 1, c_2 = 1, c_3 = 1, c_4 = 25, c_5 = 12.5, c_6 = 10, c_7 = 1$, Table 2 indicates that when the replenishment rate rises, so does the expected inventory level and customer departure rate, but the expected number of customers waiting decreases.

Table 2. Effect of replenishment rate $c$ on the model.

| c | ERR | EIL | EW0 | EPQ | EDS | ETC |
|---|---|---|---|---|---|---|
| 0.35 | 0.0641 | 5.3612 | 0.0841 | 0.8041 | 0.5744 | 55.752 |
| 0.37 | 0.0650 | 5.4017 | 0.0729 | 0.8102 | 0.5770 | 55.710 |
| 0.39 | 0.0655 | 5.4387 | 0.0633 | 0.8158 | 0.5792 | 55.684 |
| 0.41 | 0.0659 | 4.4726 | 0.0549 | 0.8208 | 0.5811 | 55.671 |
| **0.43** | **0.0663** | **5.5037** | **0.0476** | **0.8255** | **0.5828** | **55.669** |
| 0.45 | 0.0667 | 5.5324 | 0.0413 | 0.8298 | 0.5843 | 55.675 |
| 0.47 | 0.0671 | 5.5591 | 0.0358 | 0.8338 | 0.5856 | 55.689 |
| 0.49 | 0.0674 | 5.5838 | 0.0310 | 0.8375 | 0.5867 | 55.709 |
| 0.51 | 0.0677 | 5.6068 | 0.0268 | 0.8410 | 0.5877 | 55.733 |

## 7.3 Effect of Perishability Rate d on Various Performane Measures

When $a = .4, b_1 = .7, b_2 = .25, c = .2, S = 10, s = 3, N = 5, c_0 = 145, c_1 = 1, c_2 = 21.25, c_3 = 1, c_4 = 25, c_5 = 12, c_6 = 2.5, c_7 = 1$, Table 3 indicates that when the perishability rate rises, so does the expected reorder rate and waiting of customers, but the expected inventory level and departures decreases.

We generate ETC graphs by altering the parameters $b_1$, $c$, and $d$ while holding the other parameters constant. These graphs show that the company's expected total cost will be as stated in the graphs in the long run. The projected total cost and its related values of $b_1$, $c$, and $d$ can be determined. Figures 1, 2, and 3 have minimum values of $(0.55, 8.2599)$, $(0.43, 55.669)$, and $(0.25, 148.5842)$, respectively.

## 8 (s, S) Pair

We find different ETC values by altering parameter values. By adjusting the reorder point $s$ and maximum inventory $S$, one can discover the $(s, S)$ pair of the total cost. The Mathlab programming language is used to do the numerical analysis of the current model. The optimal $(s, S)$ pair for the fixed parameter values indicated below is presented in the table below (Fig. 4 and Table 4).

**Table 3.** Effect of perishability rate $d$ on the model.

| d | ERR | EIL | EW0 | EPQ | EDS | ETC |
|---|---|---|---|---|---|---|
| 0.17 | 0.0588 | 4.8274 | 0.2736 | 0.8206 | 0.5320 | 148.6164 |
| 0.19 | 0.0594 | 4.7950 | 0.2882 | 0.9110 | 0.5293 | 148.6004 |
| 0.21 | 0.0601 | 4.7629 | 0.3027 | 1.0002 | 0.5267 | 148.5900 |
| 0.23 | 0.0607 | 4.7312 | 0.3171 | 1.0882 | 0.5240 | 148.5847 |
| **0.25** | 0.0614 | 4.6999 | 0.3315 | 1.1750 | 0.214 | **148.5842** |
| 0.27 | 0.0620 | 4.6689 | 0.3459 | 1.2606 | 0.5188 | 148.5882 |
| 0.29 | 0.0627 | 4.6384 | 0.3602 | 1.3451 | 0.5162 | 148.5964 |
| 0.31 | 0.0633 | 4.6082 | 0.3745 | 1.4285 | 0.5137 | 148.6084 |
| 0.33 | 0.0639 | 4.5784 | 0.3887 | 1.5109 | 0.5112 | 148.6240 |

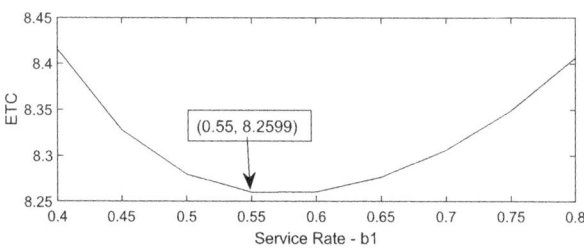

**Fig. 1.** $b_1$ vs ETC, $a = .35, b_2 = .15, c = .2, d = .15, s = 3, S = 10, N = 5$

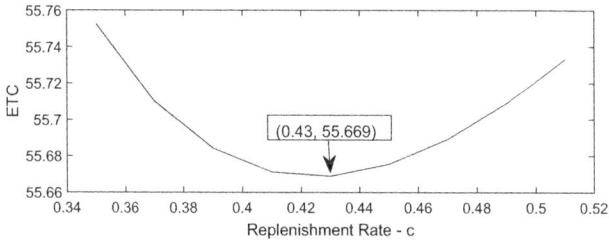

**Fig. 2.** c vs ETC, $a = .4, b_1 = .7, b_2 = .25, d = .15, s = 3, S = 10, N = 5$

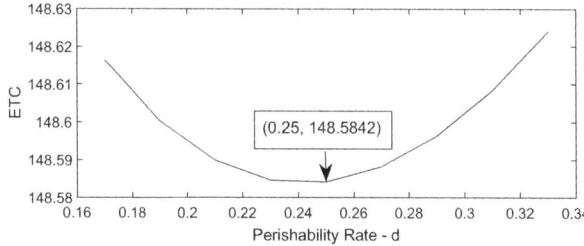

**Fig. 3.** d vs ETC, $a = .4, b_1 = .7, b_2 = .25, c = .2, s = 3, S = 10, N = 5$

**Table 4.** $(s, S)$ pair of the model for parameter values $N = 9, a = 0.25, b_1 = 0.42, b_2 = 0.15, c = 0.25, d = 0.15, c_0 = 1, c_1 = 1, c_2 = 1, c_3 = 1, c_4 = 13, c_5 = 8, c_6 = 1, c_7 = 1$

| S \ s | 1 | 2 | 3 | 4 | 5 | 6 |
|---|---|---|---|---|---|---|
| 14 | 49.15 | 48.60 | **48.40** | 48.42 | 48.60 | 48.87 |
| 15 | 49.43 | 48.99 | 48.87 | 48.94 | 49.14 | 49.43 |
| 16 | 49.77 | 49.42 | 49.36 | 49.47 | 49.70 | 50.01 |
| 17 | 50.15 | 49.89 | 49.88 | 50.03 | 50.28 | 50.60 |
| 18 | 50.56 | 50.37 | 50.41 | 50.59 | 50.86 | 51.19 |
| 18 | 51.00 | 50.88 | 50.96 | 51.17 | 51.46 | 51.80 |

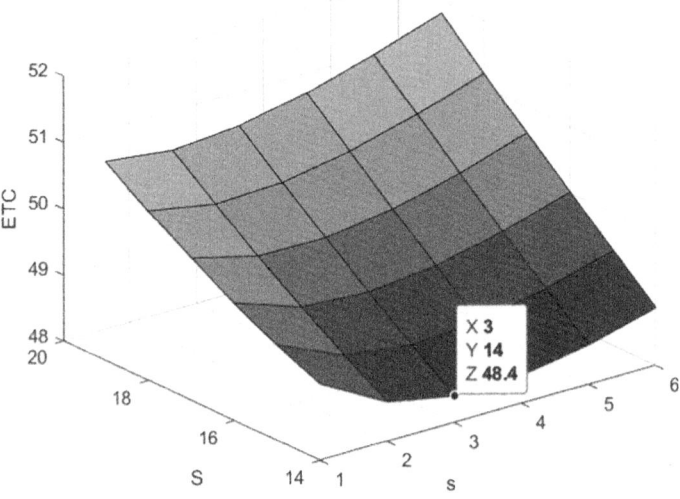

**Fig. 4.** $(s, S)$ Pair, $b = 0.55, c = 0.4, s = 5, S = 10, m = 15$

## 9 Conclusion

The paper analysed a discrete-time perishable inventory model with positive service and lead times under a modified N-policy framework. The model employed geometric distributions for inter-arrival times, service times, lead times, and perishability, offering a realistic representation of inventory dynamics. The modified N-policy effectively managed inventory and service processes, optimizing performance and minimizing the cost function. The Matrix Analytic Method was utilized to analyse system behaviour in the steady state. Numerical experiments were conducted to assess the impact of various system parameters, and an optimal (s, S) pair was derived for fixed parameter values based on the cost function.

For future work, one could extend the current study by considering arrivals to follow a Markov Arrival Process (MAP) and service times to follow a Phase-type (PH) distribution.

**Acknowledgments.** The authors acknowledge the financial support provided by FIST Program, Dept. of Science and Technology, Govt. of India, to the PG and Research Dept. of Mathematics through SR/FST/College-2018-XA 276(C).

# References

1. Krishnamoorthy, A., Deepak, T.G.: Modified N-Policy for M/G/1 queues. Comput. Oper. Res. **29**(12), 1611–1620 (2002)
2. Anbazhagan, N., Krishnamoorthy, A.: Perishable inventory system at service facility with N-Policy. Stoch. Anal. Appl. **26**(1), 120–135 (2008)
3. Alfa, A.S.: Applied Discrete-Time Queues. Springer, Heidelberg (2016)
4. Deepthi, C., Krishnamoorthy, A.: Discrete time inventory models with/without positive service time. Ph. D. Thesis. CUSAT (2013)
5. Lee, D.H., Yang, W.S.: The N-policy of a discrete time Geo/G/1 queue with disasters and its application to wireless sensor networks. Appl. Math. Model. **37**(6), 483–492 (2013)
6. Shajin, D., Narayanan, V.C., Krishnamoorthy, A.: Inventory with positive service time : a survey. In: Advanced Trends in Queueing Theory. ISTE & Wiley, London (2019)
7. Joy, J., Jose, K.P.: Perishable inventory system under (s, S) policy and age-dependent-requests. In: CCIS, vol. 1803, pp. 189–199 (2023)
8. Joy, J., Jose, K.P.: Stochastic analysis of a discrete-time production inventory model incorporating perishable items and positive servie time. Commun. Comput. Inf. Sci. **2163**, 63–75 (2024)
9. Anilkumar, M.P., Jose, K.P.: A discrete time Geo/Geo/1 inventory system with modified N-Policy. Malaya J. Matematik **8**(3), 868–876 (2020)
10. Anilkumar, M.P., Jose, K.P.: Analysis of a discrete time queueing-inventory model with back-order of items. In: 3c Empresa: investigacion y pensamiento critico, vol. 11, no. 2, pp. 50–62 (2022)
11. Raju, N., Krishnamoorthy, A.: Inventory with lead time the n-policy. Int. J. Inf. Manag. Sci. **9**(4), 45–52 (1998)
12. Moreno, P.: A discrete - time single-server queue with a modified n-policy. Int. J. Syst. Sci. **38**(6), 483–492 (2007)
13. Manuel, P., Sivakumar, B., Arivarignan, G.: Perishable inventory system with postponed demands and negative customers. J. Appl. Math. Decis. Sci. **2007**(94850), 1–12 (2007)
14. Reshmi, P.S., Jose, K.P.: A perishable system with inventory dependent arrival of customers. In: AIP Conference Proceedings, vol. 2261, p. 030041 (2020)
15. Varghese, R., Lakshmy, B., Krishnamoorthy, A.: An $(s, Q)$ inventory system with positive lead time and service time under n-policy. Culcutta Stat. Assoc. Bull. **66**(34), 241–260 (2014)
16. Lan, S., Tang, Y.: An N-policy discrete-time Geo/G/1 queue with modified multiple server vacations and Bernoulli Feedbacks. RAIRO Oper. Res. **53**(2), 367–387 (2019)
17. Zhang, Z., Xu, X.: Analysis for the M/M/1 queue with multiple working vacations and n-policy. Int. J. Inf. Manag. Sci. **19**(3), 495–506 (2008)

# Algorithm for Calculating the Stationary Probability Distribution of a System $M_2|1|(N_1, N_2)$ with Priorities

Natalia Haustova[1], Svetlana Moiseeva[1](✉), Ekaterina Pakulova[2], and Oybek Khurramov[3]

[1] National Research Tomsk State University, Tomsk, Russian Federation
smoiseeva@mail.ru
[2] Southern Federal University, Taganrog, Russian Federation
epakulova@sfedu.ru
[3] Karshi State University, Karshi, Uzbekistan
khurramov.os@qarshidu.uz
https://tsu.ru/, https://sfedu.ru, https://qarshidu.uz/

**Abstract.** In this paper we consider a model of information processing with two types of flows of customers—priority and non-priority. The service time of customers is random with an exponential probability distribution. The distribution parameters correspond to the customer type. Each type of customer has its own buffer, limited in waiting places. Information in the buffer is restricted by a time-to-live parameter, after which transmission of a customer may become irrelevant. Time-to-live parameter is a random variable that also has an exponential distribution. The contribution of the paper is to develop an algorithm for automatically calculating the stationary probability distribution. It allows to construct of coefficient matrices, simplify the one of the most intensive parts of working with priority systems manually. This algorithm provides the possibility of studying systems with priorities on large buffer volumes, which is extremely difficult to implement without automation.

**Keywords:** priority queuing systems · construction of coefficient matrices · calculation of probability distribution of the system · information with time-to-live restrictions

## 1 Introduction

Queuing systems have become an integral part of human life. Their continuous improvement and increase in functionality allows them to satisfy more and more needs. Queuing systems are widely applicable across various fields and industries such as telecommunications [4], computer science [7], manufacturing [6,13], transportation [12], supply chain management [8], healthcare [10] and others.

---
Supported by the Russian Science Foundation, research project No.24-21-00454, https://rscf.ru/project/24-21-00454/.

One of the specialized areas of application of the queuing system is the theory of teletraffic [1]. It considers the issues of data transmission in communication systems. One of the most relevant areas in teletraffic theory is the study of systems with priorities [2,3,5,11], which makes it possible to competently reassign network resources.

However, often we face with issues that require the use of optimization and algorithmization methods during the consideration of a particular system. The examples of such systems are the systems with priorities for an arbitrary number of places in the queue. This problem was identified when considering priority systems on small buffer volumes and constructing a coefficient matrix and a stationary probability distribution manually, which is physically impossible to do on large buffer volumes. In addition, customers received for servicing are impatient (unlike those considered in [9]). It means that the information has a lifetime, after which its transmission becomes irrelevant.

This paper presents a detailed algorithm for calculating a stationary probability distribution, which makes the study of priority systems with impatient customers and an arbitrary number of places in the buffer not only possible, but also the most convenient for the user.

## 2 Problem Statement

Let us consider single-server queuing system with Poisson input flows of impatient customers with parameters $\lambda_1$ and $\lambda_2$ and exponentially distributed service times with parameters $\mu_1$ and $\mu_2$ for each flow accordingly (Fig. 1).

The impatient means that a customer is in a buffer only for some period of time called "time to live". When this period is over the customer leaves the system. The buffers in a system are non-infinite.

The input flows differ in customers priority. The service time of customers depends on type of customer and given by an exponential probability distribution function with parameters $\alpha_i, i = 1, 2$. The intensities of customers leaving the system without servicing are expressed through parameters $i\alpha_1$ and $j\alpha_2$.

Let $i(t)$ be the number of priority customers in the buffer at time $t$, $j(t)$ be the number of non-priority customers in the buffer, $k(t)$ be the state of the server at time $t$: $k(t) = 0$—the server is free, $k(t) = 1$ - the server is occupied by a priority customer, $k(t) = 2$ - the server is occupied by a non-priority customer.

The contribution of this paper is to find the stationary probability distribution of a three-dimensional Markov process

$$\{k(t), i(t), j(t)\} : P_k(i,j) = \lim_{t \to \infty} P\{k(t) = k, i(t) = i, j(t) = j\}$$

with the following system parameters: $\lambda_p$—input flows intensities, $\mu_p$—service flows intensities, $\alpha_p$—intensities of customers leaving buffers, $N_p$—number of waiting places in the buffers, $p = 1, 2$.

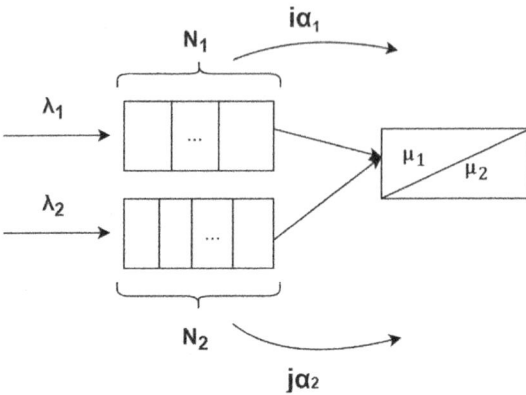

**Fig. 1.** Mathematical model of information processing with priorities and impatient customers

## 3 Formalization of Events in the System

This section is dedicated to events in the system. **System events for the priority input flow:**

- **A priority flow customer has arrived ($\lambda_1$)**
  1. If the server is free, then the state of the server is 1.
  2. If the server is busy, and there is space in the 1st buffer, then the number of customers in the 1st buffer increases by 1
- **A priority flow customer timeout expires ($i\alpha_1$)**
  3. If there are customers in the 1st buffer, then the number of customers in the 1st buffer is reduced by 1.
- **Servicing of priority flow customer is completed ($\mu_1$)**
  4. If there is a customer in the 1st buffer, then the number of customers in the 1st buffer is reduced by 1.
  5. If there are no customers in the 1st buffer, and there is a customer in the 2nd buffer, then the state of the server is 2, and the number of customers in the 2nd buffer is reduced by 1.
  6. If there are no customers in the 1st buffer and there are no customers in the 2nd buffer, then the state of the server is 0.

**System events for the non-priority input flow:**

- **A non-priority flow customer has arrived ($\lambda_2$)**
  7. If the server is free, then the state of the server is 2.
  8. If the server is busy and there is space in the 2nd buffer, then the number of customers in the 2nd buffer increases by 1.
- **A non-priority flow customer timeout expires ($j\alpha_2$)**
  9. If there are customers in the 2nd buffer, then the number of customers in the 2nd buffer is reduced by 1

- **Servicing of non-priority flow customer is completed ($\mu_2$)**
  10. If there are no customers in the 1st buffer, and there is a customer in the 2nd buffer, then the number of customers in the 2nd buffer is reduced by 1.
  11. If there is a customer in the 1st buffer, then the state of the server is 1, and the number of customers in the 1st buffer is reduced by 1.
  12. If there are no customers in the 1st buffer, and there are no customers in the 2nd buffer, then the state of the server is 0.

Figures 2 and 3 show the corresponding state transition graphs of the considering three-dimensional Markov process.

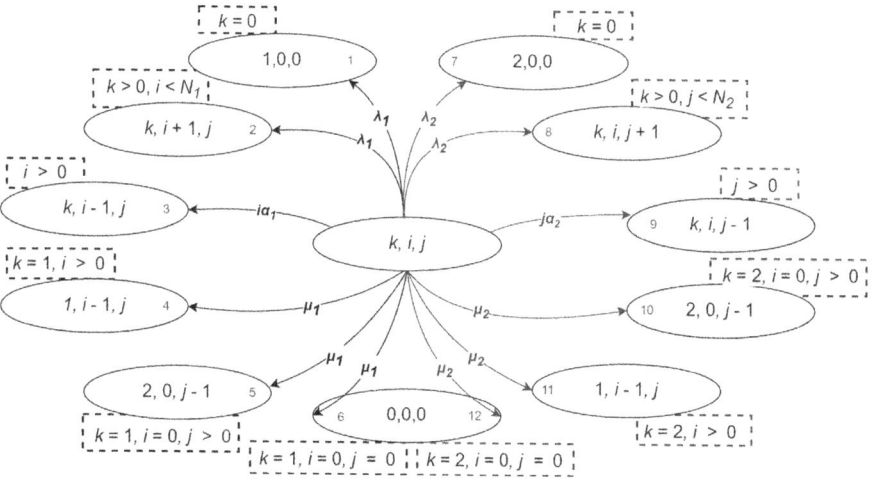

**Fig. 2.** Transition graph for diagonal elements of the process matrix $\{k(t), i(t), j(t)\}$

## 4 Mathematical Model of a System with Priorities for Impatient Requests and an Arbitrary Number of Places in the Queue

Denote $\pi_k(i,j)$ as the stationary probabilities of the three-dimensional process $P\{k(t) = k, i(t) = i, j(t) = j\} = P_k(i,j,t)$. Based on the transition graph, we compose a system of equations for stationary probabilities. This system of equations is obtained according to the mnemonic rule using the Queueing System (QS) state transition diagram, supplemented by the normalization condition. For an arbitrary length of queues, the system of linear equations has the following form.

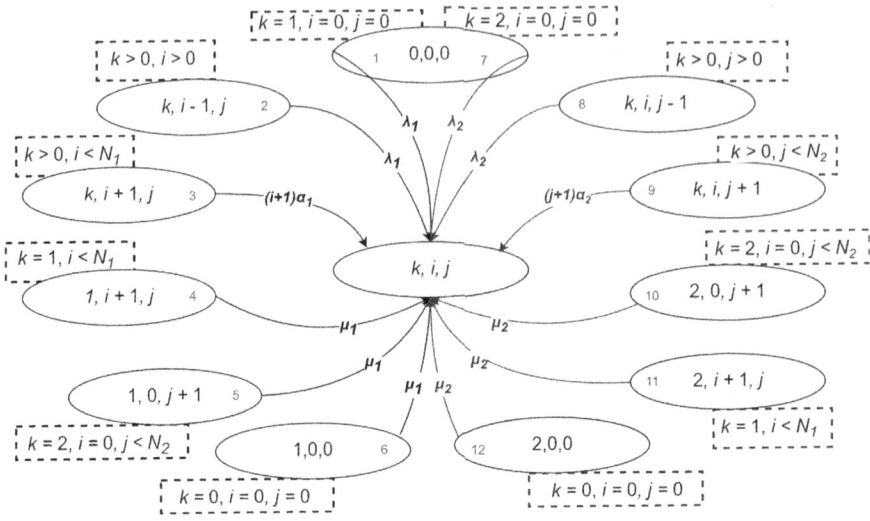

**Fig. 3.** Transition graph for system states $\{k(t), i(t), j(t)\}$

For $i = 0, j = 0$

$$\begin{cases} -(\lambda_1 + \lambda_2)\pi_0(0,0) + \mu_1\pi_1(0,0) + \mu_2\pi_2(0,0) = 0 \\ -(\lambda_1 + \lambda_2 + \mu_1)\pi_1(0,0) + \lambda_1\pi_0(0,0) + (\mu_1 + \alpha_1)\pi_1(1,0) + \alpha_2\pi_1(0,1) \\ +\mu_2\pi_2(1,0) = 0 \\ -(\lambda_1 + \lambda_2 + \mu_2)\pi_2(0,0) + \lambda_2\pi_0(0,0) + \mu_1\pi_1(0,1) + +\alpha_1\pi_2(1,0) \\ +(\mu_2 + \alpha_2)\pi_2(0,1) = 0. \end{cases}$$

For $k = 1, 2; i \neq 0, j = 0$

$$\begin{cases} -(\lambda_1 + \lambda_2 + \mu_1)\pi_1(i,0) + \lambda_1\pi_1(i-1,0) + \\ \quad +\mu_1\pi_1(i+1,0) + \mu_2\pi_2(i+1,0) = 0 \\ -(\lambda_1 + \lambda_2 + \mu_2)\pi_2(i,0) + \lambda_1\pi_2(i-1,0) = 0. \end{cases}$$

For $k = 1, 2; i = 0, j \neq 0$

$$\begin{cases} -(\lambda_1 + \lambda_2 + \mu_1)\pi_1(0,j) + \lambda_2\pi_1(0,j-1) + \\ \quad +\mu_1\pi_1(1,j) + \mu_2\pi_2(1,j) = 0 \\ -(\lambda_1 + \lambda_2 + \mu_2)\pi_2(o,j) + \lambda_2\pi_2(0,j-1) = 0. \end{cases}$$

For $k = 1, 2; i = 1 \ldots N_1 - 1, j = 1 \ldots N_2 - 1$

$$\begin{cases} -(\lambda_1 + \lambda_2 + \mu_1)\pi_1(i,j) + \lambda_1\pi_1(i-1,j) + \\ \quad +\lambda_2\pi_1(i,j-1) + \mu_1\pi_1(i+1,j) + \mu_2\pi_2(i+1,j) = 0 \\ -(\lambda_1 + \lambda_2 + \mu_2)\pi_2(i,j) + \lambda_1\pi_2(i-1,j) + \\ \quad +\lambda_2\pi_2(i,j-1) = 0. \end{cases}$$

For $k = 1, 2; i = N_1, j \neq 0$

$$\begin{cases} -(\lambda_1 + \mu_1)\pi_1(N_1, j) + \lambda_1\pi_1(N_1 - 1, j) + \\ \quad +\lambda_2\pi_1(N_1, j - 1) = 0 \\ -(\lambda_2 + \mu_2)\pi_2(N_1, j) + \lambda_1\pi_2(N_2, j) + \lambda_2\pi_2(N_1, j - 1) = 0. \end{cases}$$

For $k = 1, 2; i \neq 0, j = N_2$

$$\begin{cases} -(\lambda_1 + \mu_1)\pi_1(i, N_2) + \lambda_1\pi_1(i - 1, N_2) + \lambda_2\pi_1(i, N_2 - 1) + \\ \quad +\mu\pi_1(i + 1, N_2) + \mu_2\pi_2(i + 1, N_2) = 0 \\ -(\lambda_1 + \mu_2)\pi_2(i, N_2) + \lambda_1\pi_2(i - 1, N_2) + \\ \quad +\lambda_2\pi_2(i, N_2 - 1) = 0. \end{cases}$$

For $k = 1, 2; i = N_1, j = N_2$

$$\begin{cases} -(\mu_1)\pi_1(N_1, N_2) + \lambda_1\pi_1(N_1 - 1, N_2) + \\ \quad +\lambda_2\pi_2(N_1, N_2 - 1) = 0 \\ -(\mu_2)\pi_2(N_1, N_2) + \lambda_1\pi_2(N_1 - 1, N_2) + \\ \quad +\lambda_2\pi_2(N_1, N_2 - 1) = 0. \end{cases}$$

Considering the normalization condition $\sum_{k=0}^{2}\sum_{i=0}^{N_1}\sum_{j=0}^{N_2} \pi_k(i,j) = 1$, the system of linear equations can be solved by numerical methods.

## 5 Algorithm for Calculating Stationary Probability Distribution

The most intensive part of manually studying a system is compiling a coefficient matrix, because with an increase in the number of places in the waiting buffers by just one, there is a sharp increase in the number of equations, and therefore the dimension of the matrix. Moreover, it is important to associate the state index with the this state $\{k(t), i(t), j(t)\}$. For solving these tasks we developed the algorithm which was implemented as a Python program.

The algorithm for calculating the stationary probability distribution can be formulated as follows:

1. Establish a one-to-one correspondence between the state index $n \in [0, \ldots, 2*(N_1 + 1)*(N_2 + 1)]$ and the state $(k, i, j)$
2. Determine the zero matrix $A$ of dimension $|U|*|U|$
3. Fill matrix A with values according to Table 1. Here
4. Fill the last row of matrix A with ones.
5. Fill the vector of free elements b (one-dimensional array), in which all elements are equal to zero, except for the last one, which is equal to one.
6. Solve the system of linear algebraic equations $Ax = b$

The key result of the algorithm is the output of the probability distribution over all possible states of the system.

**Table 1.** Values for filling the coefficient matrix

| Condition | Action |
|---|---|
| $k=0, i=0, j=0$ | $A[n,n]-=\lambda_1+\lambda_2$ ; $A[n,(1,0,0)]+=\mu_1$ ; $A[n,(2,0,0)]+=\mu_2$ |
| $k=1, i>0$ | $A[n,n]-=\mu_1$ |
| $k=2, i>0$ | $A[n,n]-=\mu_2$ |
| $k=1, i=0, j=0$ | $A[n,n]-=\mu_1$ ; $A[n,(0,0,0)]+=\lambda_1$ |
| $k=2, i=0, j=0$ | $A[n,n]-=\mu_2$ ; $A[n,(0,0,0)]+=\lambda_2$ |
| $k=1, i=0, j>0$ | $A[n,n]-=\mu_1$ |
| $k=2, i=0, j>0$ | $A[n,n]-=\mu_2$ |
| $i>0$ | $A[n,n]-=i\alpha_1$ |
| $j>0$ | $A[n,n]-=j\alpha_2$ |
| $k>0, i>0$ | $A[n,(k,i-1,j)]+=\lambda_1$ |
| $k>0, j>0$ | $A[n,(k,i,j-1)]+=\lambda_2$ |
| $k>0, i<N_1$ | $A[n,n]-=\lambda_1$ ; $A[n,(k,i+1,j)]+=(i+1)\alpha_1$ |
| $k>0, j<N_2$ | $A[n,n]-=\lambda_2$ ; $A[n,(k,i,j+1)]+=(j+1)\alpha_2$ |
| $k=1, i<N_1$ | $A[n,(1,i+1,j)]+=\mu_1$ ; $A[n,(2,i+1,j)]+=\mu_2$ |
| $k=2, i=0, j<N_2$ | $A[n,(1,0,j+1)]+=\mu_1$ ; $A[n,(2,0,j+1)]+=\mu_2$ |

## 6 Numerical Experiment

The result of the algorithm is the output of the probability distribution $\pi_k(i_1, i_2)$ over all possible states of the system.

Then, summing up the corresponding probabilities, we obtain the following marginal probability distributions of the studied processes:

– stationary probability distribution of server states

$$\pi_k = P\{k(t)=k\} = \sum_{i_1=0}^{N_1}\sum_{i_2=0}^{N_2}(i_1, i_2)$$

– stationary probability distribution of the number of priority customers in the buffer

$$\pi(i_1) = P\{i_1(t)=i_1\} = \sum_{k=0}^{2}\sum_{i_2=0}^{N_2}\pi_k(i_1, i_2)$$

– stationary probability distribution of the number of non-priority customers in the buffer

$$\pi(i_2) = P\{i_2(t)=i_2\}i = \sum_{k=0}^{2}\sum_{i_1=0}^{N_1}\pi_k(i_1, i_2)$$

Let us give an example of a numerical implementation of the algorithm for the following **quality of service parameters**:

- the intensity of customers loss (priority and non-priority);
- the probability of loss and the proportion of customers that have left the buffer;
- the probability of being the customers in the buffer;
- the average number of customers of each type in the buffer;
- the average time a customer left in the buffer and in the system as a whole.

Let us assume the following **input parameters** to the system:

- $\alpha_1 = 0.1, \alpha_2 = 0.03$ is the average "time-to-live" of priority customers (10 s) and non-priority packets (33.3 s);
- $\lambda_1 = 0.2, \lambda_2 = 0.5$ Mbit/sec is the coming rate of priority and non-priority customers correspondingly;
- $\mu_1 = 0.2, \mu_2 = 0.3$ are the parameters of the exponential probability distribution of service duration for each flow;
- $N_1 = 5, N_2 = 20$ are the numbers of waiting places in buffers.

The Figs. 4 and 5 show the probability distributions for the number of customers in the corresponding buffers.

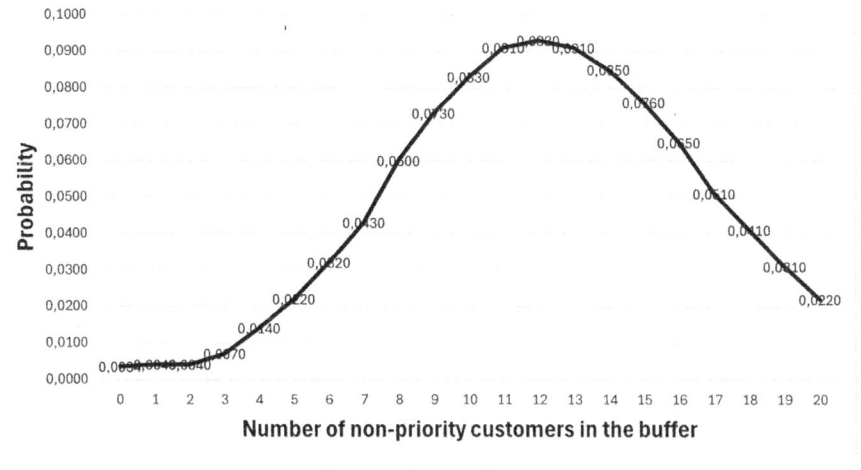

**Fig. 4.** Probability distribution of the number of non-priority customers in the buffer

Let us define the **technical characteristics of the quality of service**:

- The average number of customers in the buffer

$$E\{i_k\} = \sum_{i=0}^{N_k} i_k \pi(i_k) k = 0.1$$

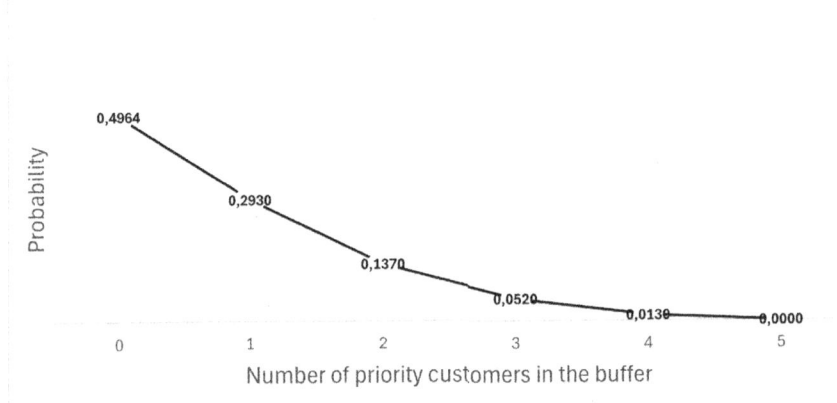

**Fig. 5.** Distribution of probabilities of the number of priority customers in the buffer

- The probability of denial of service $P = P_k(N_k)$, there are no customers in the corresponding buffer.
- The intensity of customers loss

$$\lambda_k^p = \alpha_k \sum_{i=0}^{N_k} i_k \pi(i_k) k = 0.1$$

- The fraction of occupation time of the server with priority customers $\pi_1$
- The fraction of occupation time of the server with non-priority customers $\pi_2$
- The fraction of the customers that have left the system without service after the expiration of the "time-to-live" $\frac{\lambda_k^p}{\lambda_k}$
- The average time of staying a customer in the queuing system

$$w_k = \frac{E\{i_k\}}{\lambda_k}$$

The results of a numerical experiment are given in Table 2. We define that the system load is very high $\rho_1 = 1, \rho_2 = \frac{5}{3}$. The channel occupancy rate is 99.98%.

From the results in Table 2 it is evident that despite the small values of the probabilities of denial of service due to lack of waiting places in the buffer, the fraction of leaving customers the system due to expired "time-to-live" parameter are very high.

Assuming that the parameters of incoming flows are fixed, then to ensure higher quality transmission it is necessary to increase the service speed.

Therefore, let's conduct another experiment by increasing the service intensity, for example, $\mu_1 = 0.4, \mu_2 = 1$. Numerical analysis showed that in this case it is possible to reduce the number of places in the buffers to $N_1 = 3, N_2 = 10$. The results of the experiments are presented in Table 3.

**Table 2.** Characteristics of quality of service for parameters $\mu_1 = 0.2, \mu_2 = 0.3$, $N_1 = 5, N_2 = 20$

| Parameters of quality of service | Input parameters | |
|---|---|---|
| | priority | non-priority |
| The average number of customers in the buffer | 0.78 | 11.96 |
| Dispersion of the number of customers in the buffer | 0.91 | 17.2 |
| The fraction of occupation time of the server | 0.58 | 0.41 |
| The probability of denial of service | 0 | 0.02 |
| The intensity of customers loss | 0.08 | 0.36 |
| The fraction of the customers that have left the system without service after the expiration of the "time-to-live" | 0.39 | 0.72 |
| The average time of staying a customer in the QS | 3.88 | 23.92 |

It is obvious from the table, the quality characteristics have significantly improved. At the same time, the channel occupancy factor for this example is 0.88, which is quite acceptable for practice.

**Table 3.** Characteristics of system quality of service for parameters $\mu_1 = 0.4, \mu_2 = 1$, $N_1 = 3, N_2 = 10$

| Parameters of quality of service | Input parameters | |
|---|---|---|
| | priority | non-priority |
| The average number of customers in the buffer | 0.31 | 2.55 |
| Dispersion of the number of customers in the buffer | 0.39 | 7.78 |
| The fraction of occupation time of the server | 0.41 | 0.41 |
| The probability of denial of service | 0.01 | 0.02 |
| The intensity of customers loss | 0.03 | 0.08 |
| The fraction of the customers that have left the system without service after the expiration of the "time-to-live" | 0.15 | 0.15 |
| The average time of staying a customer in the QS | 1.55 | 5.1 |

# 7 Conclusion

This paper presents a mathematical model of a multi-flow data transmission system with relative priority, one channel and different service rates for priority and non-priority impatient customers and buffers of arbitrary size. We develop numerical algorithm for calculating the probability distributions of the number of customers in each buffer and the state of the server, which allows to get technical characteristics of the system that have practical significance for the design of real information and telecommunication systems. In future work, we plan to generalize the results to a model with absolute priority and/or an unreliable channel.

## References

1. Akimaru, H., Kawashima, K.: Teletraffic: Theory and Applications. Springer, London (1999). https://doi.org/10.1007/978-1-4471-0871-9
2. Dudin, A., Dudin, S., Manzo, R., Rarità, L.: Analysis of multi-server priority queueing system with hysteresis strategy of server reservation and retrials. Mathematics **10**(20), 3747 (2022). https://doi.org/10.3390/math10203747
3. Dudin, A., Lee, M., Dudina, O., Lee, S.: Analysis of priority retrial queue with many types of customers and servers reservation as a model of cognitive radio system. IEEE Trans. Commun. (2016). https://doi.org/10.1109/tcomm.2016.2606379
4. Giambene, G.: Queuing Theory and Telecommunications: Networks and Applications. Springer, Heidelberg (2021). https://doi.org/10.1007/978-3-030-75973-5
5. He, Q., Xie, J., Zhao, X.: Priority queue with customer upgrades. Naval Res. Logist. (NRL) **59**(5), 362–375 (2012). https://doi.org/10.1002/nav.21494
6. Marsudi, M.: The application of mathematical model to analyze production system in a manufacturing industry (2023)
7. Mas, L., Vilaplana, J., Mateo, J., Solsona, F.: A queuing theory model for fog computing. J. Supercomput. **78**(8), 11138–11155 (2022). https://doi.org/10.1007/s11227-022-04328-3
8. Mohtashami, Z., Aghsami, A., Jolai, F.: A green closed loop supply chain design using queuing system for reducing environmental impact and energy consumption. J. Clean. Prod. **242**, 118452 (2020). https://doi.org/10.1016/j.jclepro.2019.118452
9. Moiseeva, S., Pakulova, E., Haustova, N.: Mathematical model of information processing with relative priority and different service intensity. In: 2023 19th International Asian School-Seminar on Optimization Problems of Complex Systems (OPCS), pp. 65–69. IEEE (2023). https://doi.org/10.1109/opcs59592.2023.10275769
10. Peter, P.O., Sivasamy, R.: Queueing theory techniques and its real applications to health care systems - outpatient visits. Int. J. Healthc. Manag. **14**(1), 114–122 (2019). https://doi.org/10.1080/20479700.2019.1616890
11. Sun, B., Lee, M.H., Dudin, A.N., Dudin, S.A.: Queueing system with absolute priority and reservation of servers. Math. Probl. Eng. **2014**, 1–15 (2014). https://doi.org/10.1155/2014/813150
12. Varghese, V., Verghese, V., Chandran, A.: Application of queuing theory in transportation. Int. J. Eng. Res. Techn **9**, 55–58 (2021)
13. Zhernovyi, K., Zhernovyi, Y.: Optimization of the part manufacturing process using analytical and simulation models of a closed queueing system. SCIREA J. Math. (2019). https://doi.org/10.54647/mathematics110502

# Profit Optimization of a Perishable Inventory System with Retrial of Customers and Unreliable Server

Bobina J. Mattam and K. P. Jose

Post Graduate and Research Department of Mathematics, St. Peter's College,
Kolenchery 682311, Kerala, India
kpjspc@gmail.com

**Abstract.** This paper considers a perishable inventory system with a single unreliable server where the customers arrive in an exponentially distributed time interval and demand a single item. The inventory is purchased from an outside source and the control policy is (s, Q). On arrival, a customer leads to service if the server is available and the inventory is non-empty. The server may break down while at service and it follows Poisson distribution. The arriving customer on finding the server busy or breakdown goes to a waiting place of infinite capacity called orbit, with pre-allotted probability or exits the system with complementary probability. Each customer in the orbit retries to enter the service facility following a Poisson distribution. After every unsuccessful retrial, the customer returns to the orbit with a pre-determined probability or is lost forever with a probability equal to its complement. An algorithmic solution to the problem is obtained using the Matrix Analytic Method. The mean number of customers lost before and after entering the system, the rate of successful retrials among overall retrials and some other performance measures of the system are derived. The impacts of system parameters on different measures are numerically studied. A suitable profit function is constructed and the optimum control policy is numerically obtained.

**Keywords:** Unreliable Server · Matrix Analytic Method · Perishable Inventory · Retrials

## 1 Introduction

Most of the existing perishable inventory models in the literature do not consider the inconveniences caused by the server. Due to this, industries/firms would lose

---

The authors acknowledge the financial support provided by FIST Program, Dept.of Science and Technology, Govt.of India, to the P. G. & Research Dept. of Mathematics, through SR/FST/College-2018-XA 276(C).

The authors would like to express sincere gratitude to the referees for their insightful comments and valuable suggestions, which greatly improved the quality of this paper.

their business. Normally, a firm must be prepared to tackle the unexpected constraints caused by the server to deliver the items on time. By performing the required repairing and maintenance of the server, a firm can suitably avoid lost sales or backlogged cases. This work proposes an inventory model of deteriorating items with unreliable server. The review articles [4,7,10] give an extensive summary of modeling of perishable inventory. Initially, an inventory system with positive service time and retrial of customers has been received limited attention in the literature. All the stochastic inventory models prior to Sigman and Simchi-Levi [13], assumed that the service time is negligible and does not follow any particular distribution. Later, Berman et al. [2] considered an inventory model with deterministic service time. Artalejo et.al. [1] introduced an alternative to classical approaches based either on backlogged or on lost sale cases. Ravichandran [11] investigated a continuous review perishable inventory system of $(s, S)$ type with positive lead time. The demands arrive according to a Poisson process. The usable age of items is distributed as Erlangian. Krishnamoorthy and Islam [5] introduced perishability in the retrial inventory model with a production unit. When the inventory level reaches zero, arriving demands are sent to the orbit which has finite capacity and try for their luck.

Sivakumar [14] considered a perishable inventory system under continuous review $(s, Q)$ policy with a finite number of demands. The lifetime of each item and lead time are assumed to be exponential. Also, assume that customers arrive during the stock-out period and enter into an orbit and these customers send out signals to access the server. Periyasamy [9] analysed a continuous review perishable inventory system with a single server and zero lead time. Customers arrive from a finite population according to a Poisson process. If the demand occurrs during busy period, it is directed to an obit and may retry from there. Also server searches for customers with a pre-assigned probability. Some important joint probability distributions are obtained in the steady state. Kumar and Elango [6] considered a single server queueing system of perishable items with finite waiting space. All the underlying processes are assumed as exponential. They modelled the problem as a Markov decision problem by using the value iteration algorithm to obtain the minimal average cost of the service. Reshmi and Jose [12] studied a queueing inventory system with perishable items and all underlying process are assumed as exponential. Items in the inventory perish at a linear rate. Later, Bobina and Jose [3] incorporated the concepts of production and unreliability to a deteriorating inventory with the same assumptions on distributions.

## 2 Mathematical Model of the System

Consider an $(s, Q)$ inventory system with a single server, retrial facility and perishable items. The lifetime of an item in the inventory is exponentially distributed with parameter $j\omega$, when there are $j$ items in the inventory. Customers arrive at the system following the Poisson distribution with parameter $\lambda$ and each customer demands a single item. If the server is idle at the arrival epoch of a customer, then that customer is immediately taken for service, provided the inventory is non-empty. The service time duration follows a negative exponential distribution with parameter $\mu$.

When the on-hand inventory level drops to $s$, an order of quantity $Q$ is placed and is replenished after a random lead time and is exponentially distributed with parameter $\beta$. The server may be subject to breakdown while at service and it follows the Poisson distribution with parameter $\delta_1$ and its repair times are exponentially distributed with parameter $\delta_2$. Any arriving customer, on finding the inventory level zero or the server busy or breakdown, either proceeds to join a waiting space of infinite capacity called orbit with probability $\gamma$ or exits the system with probability $1 - \gamma$. All customers who enter the orbit, generate requests for service at exponentially distributed time intervals with mean $\frac{1}{\theta}$ independently. The retrial customers who find the inventory out of stock or the server busy or breakdown, return to the orbit with probability $\delta$ or exit the system with probability $1 - \delta$.

Let $N(t)$ and $I(t)$ denote the number of customers in the orbit at time $t$ and the inventory level at time $t$ respectively. Further, let

$$C(t): \begin{cases} 0, & \text{if the server idle at time } t \\ 1, & \text{if the server busy at time } t \\ 2, & \text{if the server is breakdown at time } t. \end{cases}$$

Now, $\mathbf{X} = \{(N(t), C(t), I(t)) | t \geq 0\}$ constitutes a continuous time Markov chain with state space $F_0 \cup F_1 \cup F_2$, where

$$F_0 = \{(i, 0, j) | i \geq 0; 0 \leq j \leq S\},$$
$$F_1 = \{(i, 1, j) | i \geq 0; 1 \leq j \leq S\} \text{ and}$$
$$F_2 = \{(i, 2, j) | i \geq 0; 1 \leq j \leq S\}.$$

The generator matrix of the process is

$$Q_1 = \begin{bmatrix} A_{10} & A_0 & & & \\ A_{21} & A_{11} & A_0 & & \\ & A_{22} & A_{12} & A_0 & \\ & & A_{23} & A_{13} & A_0 \\ & & & \ddots & \ddots & \ddots \end{bmatrix}$$

where each element in $Q$ has size $(3S+1) \times (3S+1)$.

**Transitions of $A_0$**

- $(i, 0, 0) \xrightarrow{\lambda\gamma} (i+1, 0, 0); i \geq 0$
- $(i, 1, j) \xrightarrow{\lambda\gamma} (i+1, 1, j); i \geq 0, 1 \leq j \leq S$
- $(i, 2, j) \xrightarrow{\lambda\gamma} (i+1, 2, j); i \geq 0, 1 \leq j \leq S$

**Transitions of $A_{1i}$**

- $(i,0,j) \xrightarrow{\lambda} (i,1,j); i \geq 0, 1 \leq j \leq S$
- $(i,1,j) \xrightarrow{\mu} (i,0,j-1); i \geq 0, 1 \leq j \leq S$
- $(i,0,j) \xrightarrow{\beta} (i,0,Q+j); i \geq 0, 1 \leq j \leq s$
- $(i,1,j) \xrightarrow{\beta} (i,1,Q+j); i \geq 0, 1 \leq j \leq s$
- $(i,2,j) \xrightarrow{\beta} (i,2,Q+j); i \geq 0, 1 \leq j \leq s$
- $(i,0,j) \xrightarrow{j\omega} (i,0,j-1); i \geq 0, 1 \leq j \leq S$
- $(i,k,j) \xrightarrow{j\omega} (i,k,j-1); i \geq 0, 2 \leq j \leq S, k=1,2$
- $(i,1,j) \xrightarrow{\delta_1} (i,2,j); i \geq 0, 1 \leq j \leq S$
- $(i,2,j) \xrightarrow{\delta_2} (i,1,j); i \geq 0, 1 \leq j \leq S$
- $(i,0,j) \xrightarrow{\alpha_j} (i,0,j);$

$$\alpha_j = \begin{cases} -\lambda\gamma - \beta - i\theta(1-\delta); j=0 \\ -\lambda - \beta - j\omega - i\theta; 1 \leq j \leq s \\ -\lambda - j\omega - i\theta; s+1 \leq j \leq S \end{cases}$$

- $(i,1,j) \xrightarrow{\tau_j} (i,1,j);$

$$\tau_j = \begin{cases} -\lambda\gamma - \mu - \beta - i\theta(1-\delta); j=1 \\ -\lambda\gamma - \mu - \beta - i\theta(1-\delta) - j\omega - \delta_1; 2 \leq j \leq s \\ -\lambda\gamma - \mu - i\theta(1-\delta) - j\omega - \delta_1; s+1 \leq j \leq S \end{cases}$$

- $(i,2,j) \xrightarrow{\phi_j} (i,2,j);$

$$\phi_j = \begin{cases} -\lambda\gamma - \beta - \delta_2 - i\theta(1-\delta); j=1 \\ -\lambda\gamma - \beta - \delta_2 - j\omega - i\theta(1-\delta); 2 \leq j \leq s \\ -\lambda\gamma - j\omega - \delta_2 - i\theta(1-\delta); s+1 \leq j \leq S \end{cases}$$

**Transitions of $A_{2i}$**

- $(i,0,0) \xrightarrow{i\theta(1-\delta)} (i-1,0,0); i \geq 1$
- $(i,0,j) \xrightarrow{i\theta} (i-1,1,j); i \geq 0, 1 \leq j \leq S$
- $(i,1,j) \xrightarrow{i\theta(1-\delta)} (i-1,1,j); i \geq 0, 1 \leq j \leq S$
- $(i,2,j) \xrightarrow{i\theta(1-\delta)} (i-1,2,j); i \geq 1, 1 \leq j \leq S$

## 3 System Stability

To obtain the stability condition for the system study, we apply the Neuts-Rao truncation [8] by assuming $A_{1i} = A_{1N}$ and $A_{2i} = A_{2N}$ for all $i \geq N$. Then the generator matrix of the truncated system will be

$$Q = \begin{bmatrix} A_{10} & A_0 & & & & & \\ A_{21} & A_{11} & A_0 & & & & \\ & A_{22} & A_{12} & A_0 & & & \\ & & \ddots & \ddots & \ddots & & \\ & & & A_{2N} & A_{1N} & A_0 & \\ & & & & A_{2N} & A_{1N} & A_0 \\ & & & & & \ddots & \ddots & \ddots \end{bmatrix}.$$

Define $A_N = A_0 + A_{1N} + A_{2N}$; then $A_N = \begin{bmatrix} H_{11} & H_{12} & 0 \\ H_{21} & H_{22} & H_{23} \\ 0 & H_{32} & H_{33} \end{bmatrix}$.

We introduce some notations to describe the terms in the above matrix. $I_m$ denote an identity matrix of order $m$. $e$ is a column matrix of appropriate order where all entries are 1. $r_m(i)$ denote a $1 \times m$ row matrix whose $i$th entry is 1 and all other entries are zeros. $c_m(i)$ denotes the transpose of $r_m(i)$. $\otimes$ denotes Kronecker product of matrices. Thus each of the sub-matrices given in the matrix $A_N$ are as follows.

$$H_{11} = \beta(c_{S+1}(1) \otimes r_{S+1}(1)) + \sum_{i=2}^{s+1}(-\lambda - \beta - (i-1)\omega - N\theta)c_{S+1}(i) \otimes r_{S+1}(i)$$

$$+ \sum_{i=s+2}^{S+1}(-\lambda - (i-1)\omega - N\theta)c_{S+1}(i) \otimes r_{S+1}(i) + \sum_{i=2}^{S+1} i\omega(c_{S+1}(i) \otimes r_{S+1}(i-1))$$

$$+ \sum_{i=1}^{s+1} \beta(c_{S+1}(i) \otimes r_{S+1}(Q+i)), H_{12} = (\lambda + N\theta)\begin{bmatrix}0_{1 \times s} I_S\end{bmatrix}, H_{21} = \mu \begin{bmatrix}I_S & 0_{S \times 1}\end{bmatrix},$$

$$H_{22} = (-\beta - \mu - \delta_1)(c_S(1) \otimes r_S(1)) + \sum_{i=2}^{s}(-\beta - \mu - (i)\omega - \delta_1)c_S(i) \otimes r_S(i)$$

$$+ \sum_{i=s+1}^{S}(-\mu - (i)\omega - \delta_1)c_S(i) \otimes r_S(i) + \sum_{i=2}^{S}((i)\omega)c_S(i) \otimes r_S(i-1)$$

$$+ \sum_{i=1}^{s} \beta c_S(i) \otimes r_S(Q+i), H_{23} = \delta_1 I_S, H_{32} = \delta_2 I_S,$$

$$H_{33} = (-\beta - \delta_2)(c_S(1) \otimes r_S(1)) + \sum_{i=2}^{s}(-\beta - (i)\omega - \delta_2)c_S(i) \otimes r_S(i)$$

$$+ \sum_{i=s+1}^{S}(-(i)\omega - \delta_2)c_S(i) \otimes r_S(i) + \sum_{i=2}^{S}((i)\omega)c_S(i) \otimes r_S(i-1) +$$

$$\sum_{i=1}^{s} \beta c_S(i) \otimes r_S(Q+i).$$

The stationary probability vector $\pi_N$ of $A_N$ be partitioned as $\pi_N = (\pi_0^N, \pi_1^N, \pi_2^N)$, where

$$\pi_0^N = (\pi_{N,0,0}, \pi_{N,0,1} \ldots, \pi_{N,0,S}),$$
$$\pi_1^N = (\pi_{N,1,1}, \pi_{N,1,2}, \ldots, \pi_{N,1,S}),$$
$$\pi_2^N = (\pi_{N,2,1}, \pi_{N,2,2}, \ldots, \pi_{N,2,S}).$$

Then the relation $\pi_N A_N = 0$ along with the normalizing condition $\pi_N e = 1$ gives rise to the following equations:

$$\pi_0^N H_{11} + \pi_1^N H_{21} = 0, \Longrightarrow \quad \pi_0^N = -(\pi_1^N H_{21}) H_{11}^{-1},$$
$$\pi_0^N H_{12} + \pi_1^N H_{22} + \pi_2^N H_{32} = 0, \Longrightarrow \quad \pi_1^N = -(\pi_0^N H_{12} + \pi_2^N H_{32}) H_{22}^{-1},$$
$$\pi_1^N H_{23} + \pi_2^N H_{33} = 0 \Longrightarrow \quad \pi_2^N = -(\pi_1^N H_{23}) H_{33}^{-1}.$$

The matrices $H_{ii}; i = 1, 2, 3$ in the above set of equations are all diagonally dominant and hence invertible. Thus, all the terms in the previous set of equations exist and hence by Block Gauss-Seidel iteration, we can find the vector $\pi_N$.

Thus, the stability condition can be asserted as, the truncated system is stable if and only if $\pi_N A_0 e < \pi_N A_{2N} e$, where

$$\pi_N A_0 e = \pi_0^N (\lambda\gamma c_{S+1}(1) \otimes r_{S+1}(1))e + (\lambda\gamma(\pi_1^N + \pi_2^N))e,$$
$$\pi_N A_{2N} e = (\pi_0^N (N\theta(1-\delta))(c_{S+1}(1) \otimes r_{S+1}(1))e$$
$$+ ((\pi_0^N)N\theta \sum_{i=1}^{S} c_{S+1}(i+1) \otimes r_{S+1}(i))e + ((\pi_1^N + \pi_2^N)N\theta(1-\delta))e.$$

## 4 Steady State Distribution

Since **X** is a level dependent quasi-birth-death process, to calculate the steady state probability vector, we use the method described by Neuts-Rao [8]. Consider the steady state probability vector $\mathbf{x} = (x_0, x_1, x_2, \ldots)$ of $Q$, where

$$x_i = (z_{i,0,0}, z_{i,0,1} \ldots, z_{i,0,S}, z_{i,1,1}, z_{i,1,2}, \ldots, z_{i,1,S}, z_{i,2,1}, z_{i,2,2}, \ldots, z_{i,2,S}), (i \geq 0)$$

satisfies the relation

$$x_{N+k-1} = x_{N-1} R^k, \ k \geq 1$$

where the matrix $R$ is the unique non-negative solution of the matrix quadratic equation

$$R^2 A_2 + R A_1 + A_0 = \mathbf{0}$$

with $A_1 = A_{1N}, A_2 = A_{2N}$ and $R = \lim_{n \to \infty} R_n$, where $\{R_n\}$ is defined such that $R_{n+1} = -A_0 A_1^{-1} - R_n A_2 A_1^{-1}; n \geq 0$ and $R_0 = \mathbf{0}$.

The vectors $x_0, x_1, \ldots, x_{N-1}$ corresponding to boundary portion of $Q$ are obtained using Gauss-Siedel method subject to normalizing condition $\sum_{i=0}^{\infty} x_i \mathbf{e} = 1$.

## 5 Performance Measures

Using the above probability vectors, some important performance measures are given below,

1. Expected inventory level in the system,
$$E_{inv} = \sum_{i=0}^{\infty} \sum_{k=0}^{2} \sum_{j=1}^{S} j z_{i,k,j} + \sum_{i=0}^{\infty} \sum_{k=0}^{2} \sum_{j=1}^{S} j z_{i,k,j}$$

2. Mean number of customers in the orbit, $E_{orbit} = \Sigma_i \Sigma_k \Sigma_j i z_{i,k,j}$
3. Expected reorder rate
$$E_{ro} = \mu \sum_{i=0}^{\infty} z_{i,1,s+1} + (s+1)\omega \sum_{i=0}^{\infty} \sum_{k=0}^{2} z_{i,k,s+1}$$

4. Expected perishability rate,
$$E_p = \omega \left( \sum_{i=0}^{\infty} z_{i,0,1} + \sum_{i=0}^{\infty} \sum_{k=0}^{2} \sum_{j=2}^{S} j z_{i,k,j} \right)$$

5. Average number of departures after service completion,
$$E_{ds} = \mu \sum_{i=0}^{\infty} \sum_{j=1}^{S} z_{i,1,j}$$

6. Average number of customers lost prior to entering the orbit,
$$E_{la} = \lambda(1-\gamma) \sum_{i=0}^{\infty} \left( z_{i,0,0} + \sum_{j=1}^{S} z_{i,1,j} + \sum_{j=1}^{S} z_{i,2,j} \right)$$

7. Average number of customers lost during retrials,
$$E_{lr} = \theta(1-\delta) \sum_{i=0}^{\infty} i \left( z_{i,0,0} + \sum_{j=1}^{S} z_{i,1,j} + \sum_{j=1}^{S} z_{i,2,j} \right)$$

8. Overall rate of retrials, $\theta_1^* = \theta \left( \sum_{i=1}^{\infty} i x_i \right) e$
9. Successful rate of retrials, $\theta_2^* = \theta \sum_{i=0}^{\infty} i \left( \sum_{j=1}^{S} z_{i,0,j} \right)$
10. Ratio of successful rate of retrials, $R_{sr} = \frac{\theta_2^*}{\theta_1^*}$
11. Average rate of breakdown $A_{br} = \delta_1 \sum_{i=0}^{\infty} \left( \sum_{j=1}^{S} z_{i,1,j} \right)$
12. Average rate of repair $A_{rr} = \delta_2 \sum_{i=0}^{\infty} \left( \sum_{j=1}^{S} z_{i,2,j} \right)$

## 6 Profit Analysis

The objective is to obtain an adaptive $(s, Q)$ policy subject to some cost criteria. Since the objective profit function is not known explicitly, we define it as a combination of relevant system characteristics. One can determine the optimum values of i) s, the point at which the reorder is made and ii) $Q = S - s$, where S is the amount of inventory to be stored by maximizing the total profit. For this, the long-run profit function for this model, defined as the total profit per unit time is given by

$$E_{profit} = c_0 E_{ds} - c_1 E_{inv} - c_2 E_{orbit} - c_3 E_{ro} - c_4 E_p - c_5 (E_{la} + E_{lr}) - c_6 A_{br} - c_7 A_{rr},$$

where $c_0$=revenue per unit purchase, $c_1$=holding cost of inventory/unit/unit time, $c_2$=holding cost of customers/unit/unit time, $c_3$=reorder cost, $c_4$=cost due to decay of items/unit/unit time, $c_5$=cost due to loss of customers/unit/unit time, $c_6$=penalty due to breakdown of server/unit time, $c_7$=repair cost/unit time.

## 7 Numerical Experiments

In this section, we provide results of numerical illustration that has been carried out for studying the effects of variation of different parameters on various performance measures. Numerical experiments are conducted by considering some artificial data. Assume that the reorder level, $s = 5$ and the maximum permissible inventory level, $S = 20$. To study the variation of each parameter on system performances, we consider the following cases 7.1 to 7.7 with table representations.

### 7.1 Impact of Variation of Arrival Rate $\lambda$ on Various Performance Measures

As the arrival rate $\lambda$ increases, the number of customers that have to lead to the orbit $E_{orbit}$ also increases which in turn leads to the loss of arriving customers as well as retrying customers. The increase in $E_{orbit}$ results in the increase of $\theta_1^*$ and $\theta_2^*$ (see Table 1). Also, we can see that the increase in $\lambda$ ascends the breakdown rate visibly and the repair rate minutely.

Table 1. Effect of arrival rate $\lambda$ on various performance measures

| $\lambda$ | $E_{inv}$ | $E_{orbit}$ | $E_{ro}$ | $E_p$ | $E_{ds}$ | $E_{la}$ | $E_{lr}$ | $\theta_1^*$ | $\theta_2^*$ | $A_{br}$ | $Arr$ |
|---|---|---|---|---|---|---|---|---|---|---|---|
| 3.1 | 5.5647 | 1.5611 | 0.3697 | 3.1692 | 1.4617 | 0.4256 | 0.7349 | 3.1223 | 1.9392 | 0.1827 | 0.00039 |
| 3.2 | 5.5415 | 1.6366 | 0.3694 | 3.0092 | 1.4821 | 0.4449 | 0.7744 | 3.2732 | 2.0169 | 0.1852 | 0.00040 |
| 3.3 | 5.5193 | 1.7127 | 0.3692 | 2.8558 | 1.5016 | 0.4642 | 0.8145 | 3.4254 | 2.0944 | 0.1877 | 0.00040 |
| 3.4 | 5.4981 | 1.7894 | 0.3690 | 2.7090 | 1.5203 | 0.4836 | 0.8552 | 3.5789 | 2.1717 | 0.1900 | 0.00041 |
| 3.5 | 5.4779 | 1.8667 | 0.3687 | 2.5688 | 1.5381 | 0.5031 | 0.8963 | 3.7335 | 2.2489 | 0.1922 | 0.00041 |
| 3.6 | 5.4584 | 1.9446 | 0.3685 | 2.4351 | 1.5552 | 0.5226 | 0.9380 | 3.8892 | 2.3259 | 0.1944 | 0.00042 |
| 3.7 | 5.4398 | 2.0229 | 0.3683 | 2.3078 | 1.5715 | 0.5422 | 0.9801 | 4.0459 | 2.4027 | 0.1964 | 0.00042 |
| 3.8 | 5.4220 | 2.1017 | 0.3681 | 2.1868 | 1.5872 | 0.5619 | 1.0226 | 4.2035 | 2.4794 | 0.1984 | 0.00043 |

$S = 20; s = 5; \mu = 4; \omega = 0.8; \beta = 0.9; \theta = 2; \gamma = 0.8; \delta = 0.7; \delta_1 = 0.5; \delta_2 = 1.6$

### 7.2 Impact of Variation of Service Rate $\mu$ on Various Performance Measures

Intuitively, the service rate increases leads to a greater number of service completions. Therefore, $E_{ds}$ also increases and the number of customers in the orbit $E_{orbit}$ decreases. The overall and successful rate of retrials decreases since the $E_{orbit}$ is decreasing. Expected inventory level $E_{inv}$ get decreased when more and more customers get served. The number of unsatisfied customers decreases, that is $E_{la}$ and $E_{lr}$ decreases. It is also clear that, though overall retrials decrease, the successful retrials are increasing and the breakdown rate is also decreasing. Table 2 supports the intuition.

Profit Optimization of a Perishable Inventory System with Retrial 335

Table 2. Effect of service rate $\mu$ on various performance measures

| $\mu$ | $E_{inv}$ | $E_{orbit}$ | $E_{ro}$ | $E_p$ | $E_{ds}$ | $E_{la}$ | $E_{lr}$ | $\theta_1^*$ | $\theta_2^*$ | $A_{br}$ | $Arr$ |
|---|---|---|---|---|---|---|---|---|---|---|---|
| 3.6 | 5.4866 | 1.9216 | 0.3617 | 2.4693 | 1.4452 | 0.5156 | 0.9358 | 3.8433 | 2.2286 | 0.2007 | 0.0004 |
| 3.7 | 5.4842 | 1.9075 | 0.3635 | 2.4947 | 1.4693 | 0.5124 | 0.9255 | 3.8151 | 2.2344 | 0.1985 | 0.0004 |
| 3.8 | 5.4820 | 1.8937 | 0.3653 | 2.5197 | 1.4928 | 0.5093 | 0.9155 | 3.7874 | 2.2397 | 0.1964 | 0.0004 |
| 3.9 | 5.4799 | 1.8801 | 0.3670 | 2.5444 | 1.5157 | 0.5061 | 0.9058 | 3.7602 | 2.2445 | 0.1943 | 0.0004 |
| 4.0 | 5.4779 | 1.8667 | 0.3687 | 2.5688 | 1.5381 | 0.5031 | 0.8963 | 3.7335 | 2.2489 | 0.1922 | 0.0004 |
| 4.1 | 5.4760 | 1.8536 | 0.3704 | 2.5928 | 1.5599 | 0.5000 | 0.8871 | 3.7073 | 2.2529 | 0.1902 | 0.0004 |
| 4.2 | 5.4742 | 1.8408 | 0.3720 | 2.6165 | 1.5812 | 0.4971 | 0.8781 | 3.6816 | 2.2565 | 0.1882 | 0.0004 |
| 4.3 | 5.4725 | 1.8282 | 0.3736 | 2.6399 | 1.6020 | 0.4941 | 0.8693 | 3.6565 | 2.2598 | 0.1862 | 0.0004 |

$S = 20; s = 5; \lambda = 3.5; \omega = 0.8; \beta = 0.9; \theta = 2; \gamma = 0.8; \delta = 0.7; \delta_1 = 0.5; \delta_2 = 1.6$

### 7.3 Impact of Variation of Perishable Rate $\omega$ on Various Performance Measures

When the decay rate increases, obviously $E_p$ increases, which leads to a decrease in expected inventory level $E_{inv}$ as well as in expected departure from service $E_{ds}$. As $E_{inv}$ decreases, more customers join the orbit i.e. $E_{orbit}$ increases. When $E_{orbit}$ increases, we expect increase in measures like $E_{la}, E_{lr}, \theta_1^*$. But, due to the increase in $\omega$, $\theta_2^*$ ideally decreases. Table 3 supports these intuitions.

Table 3. Effect of perishable rate $\omega$ on various performance measures

| $\omega$ | $E_{inv}$ | $E_{orbit}$ | $E_{ro}$ | $E_p$ | $E_{ds}$ | $E_{la}$ | $E_{lr}$ | $\theta_1^*$ | $\theta_2^*$ | $A_{br}$ | $Arr$ |
|---|---|---|---|---|---|---|---|---|---|---|---|
| 0.3 | 7.2384 | 0.5461 | 0.2130 | 2.8276 | 1.4714 | 1.5011 | 0.2378 | 1.0923 | 0.6674 | 0.1839 | 0.00039 |
| 0.4 | 6.8499 | 0.5580 | 0.2512 | 3.6219 | 1.4513 | 1.5168 | 0.2450 | 1.1161 | 0.6924 | 0.1814 | 0.00039 |
| 0.5 | 6.5077 | 0.5704 | 0.2856 | 4.3553 | 1.4303 | 1.5332 | 0.2525 | 1.1410 | 0.7185 | 0.1787 | 0.00038 |
| 0.6 | 6.2020 | 0.5832 | 0.3168 | 5.0336 | 1.4088 | 1.5500 | 0.2603 | 1.1666 | 0.7454 | 0.1761 | 0.00038 |
| 0.7 | 5.9262 | 0.5962 | 0.3452 | 5.6617 | 1.3871 | 1.5669 | 0.2681 | 1.1924 | 0.7725 | 0.1733 | 0.00037 |
| 0.8 | 5.6754 | 0.6091 | 0.3711 | 6.2440 | 1.3655 | 1.5837 | 0.2760 | 1.2183 | 0.7998 | 0.1706 | 0.00037 |
| 0.9 | 5.4460 | 0.6220 | 0.3948 | 6.7848 | 1.3440 | 1.6004 | 0.2838 | 1.2440 | 0.8268 | 0.1680 | 0.00036 |
| 1 | 5.2353 | 0.6347 | 0.4166 | 7.2876 | 1.3229 | 1.6167 | 0.2915 | 1.2694 | 0.8536 | 0.1653 | 0.00035 |

$S = 20; s = 5; \lambda = 3.5; \mu = 4; \beta = 0.9; \theta = 2; \gamma = 0.8; \delta = 0.7; \delta_1 = 0.5; \delta_2 = 1.6$

### 7.4 Impact of Variation of the Replenishment Rate $\beta$ on Various Performance Measures

As the replenishment rate $\beta$ increases, the expected inventory $E_{inv}$ increases and hence the expected perishable rate $E_p$ increases. The production switch on rate also increases with an increase in $\beta$. When the inventory available to customers increases the service completion becomes faster, so $E_{ds}$. Accordingly, the expected number of customers in the orbit $E_{orbit}$ decreases, due to this, the measures $E_{la}, E_{lr}, \theta_1^*$ and $\theta_2^*$ decreases (Table 4).

**Table 4.** Effect of replenishment rate $\beta$ on various performance measures

| $\beta$ | $E_{inv}$ | $E_{orbit}$ | $E_{ro}$ | $E_p$ | $E_{ds}$ | $E_{la}$ | $E_{lr}$ | $\theta_1^*$ | $\theta_2^*$ | $A_{br}$ | $Arr$ |
|---|---|---|---|---|---|---|---|---|---|---|---|
| 0.3 | 2.6709 | 1.0566 | 0.1819 | 2.6936 | 0.7551 | 2.0071 | 0.5689 | 2.1132 | 1.8143 | 0.0943 | 0.00020 |
| 0.4 | 3.3404 | 0.9312 | 0.2255 | 3.4644 | 0.9126 | 1.9038 | 0.4848 | 1.8625 | 1.5224 | 0.1140 | 0.00024 |
| 0.5 | 3.9272 | 0.8343 | 0.2630 | 4.1557 | 1.0405 | 1.8170 | 0.4208 | 1.6687 | 1.3004 | 0.1300 | 0.00028 |
| 0.6 | 4.4443 | 0.7582 | 0.2956 | 4.7722 | 1.1454 | 1.7438 | 0.3711 | 1.5166 | 1.1287 | 0.14317 | 0.00031 |
| 0.7 | 4.9024 | 0.6977 | 0.3240 | 5.3208 | 1.2322 | 1.6818 | 0.3321 | 1.3955 | 0.9938 | 0.1540 | 0.00033 |
| 0.8 | 5.3104 | 0.6489 | 0.3490 | 5.8090 | 1.3046 | 1.6289 | 0.3010 | 1.2979 | 0.8863 | 0.1630 | 0.00035 |
| 0.9 | 5.6754 | 0.6091 | 0.3711 | 6.2440 | 1.3655 | 1.5837 | 0.2760 | 1.2183 | 0.7998 | 0.1706 | 0.00037 |
| 1 | 6.0035 | 0.5763 | 0.3907 | 6.6326 | 1.4171 | 1.5448 | 0.2556 | 1.1528 | 0.7292 | 0.1771 | 0.00038 |

$S = 20; s = 5; \lambda = 3.5; \mu = 4; \omega = 0.8; \theta = 2; \gamma = 0.8; \delta = 0.7; \delta_1 = 0.5; \delta_2 = 1.6$

### 7.5 Impact of Variation of the Retrial Rate $\theta$ on Various Performance Measures

As the retrial rate $\theta$ increases, one would expect a decrease in expected number of customers in the orbit $E_{orbit}$. As the production switch on rate increases, expected inventory level $E_{inv}$ and $E_p$ increases. When $\theta$ increases, the expected number of service completion increases, that is $E_{ds}$. The decrease in $E_{la}$ is very negligible because $E_{inv}$ is increasing. From Table 5, as $\theta$ increases, most of the retrying customers fail to access a free server so $E_{lr}$ increases.

**Table 5.** Effect of retrial rate $\theta$ on various performance measures

| $\theta$ | $E_{inv}$ | $E_{orbit}$ | $E_{ro}$ | $E_p$ | $E_{ds}$ | $E_{la}$ | $E_{lr}$ | $\theta_1^*$ | $\theta_2^*$ | $A_{br}$ | $Arr$ |
|---|---|---|---|---|---|---|---|---|---|---|---|
| 1.5 | 5.6760 | 0.7522 | 0.3712 | 5.4872 | 1.3659 | 1.5844 | 0.2486 | 1.1284 | 0.7351 | 0.1707 | 0.00049 |
| 1.6 | 5.6756 | 0.7181 | 0.3711 | 5.6592 | 1.3660 | 1.5845 | 0.2547 | 1.1490 | 0.7499 | 0.1707 | 0.00046 |
| 1.7 | 5.6754 | 0.6871 | 0.3711 | 5.8200 | 1.3660 | 1.5844 | 0.2604 | 1.1681 | 0.7636 | 0.17075 | 0.00043 |
| 1.8 | 5.6753 | 0.6588 | 0.3711 | 5.9705 | 1.3659 | 1.5843 | 0.2659 | 1.1859 | 0.7764 | 0.1707 | 0.00041 |
| 1.9 | 5.6753 | 0.6329 | 0.3711 | 6.1116 | 1.3657 | 1.5840 | 0.2710 | 1.2026 | 0.7885 | 0.1707 | 0.00039 |
| 2 | 5.6754 | 0.6091 | 0.3711 | 6.2440 | 1.3655 | 1.5837 | 0.2760 | 1.2183 | 0.7998 | 0.1706 | 0.00037 |
| 2.1 | 5.6756 | 0.5871 | 0.3711 | 6.3685 | 1.3651 | 1.5834 | 0.2807 | 1.2331 | 0.8104 | 0.1706 | 0.00035 |

$S = 20; s = 5; \lambda = 3.5; \mu = 4; \omega = 0.8; \beta = 0.9; \gamma = 0.8; \delta = 0.7; \delta_1 = 0.5; \delta_2 = 1.6$

### 7.6 Impact of Variation of the Probability $\gamma$ on Various Performance Measures

As the probability $\gamma$ increases, more unsatisfied customers are pushed into the orbit, causing $E_{orbit}$ to rise. This, in turn leads to a reduction in the loss of customers upon arrival. However, as the customers in the orbit increases retrials become more unsuccessful. As predicted, the production switch-on rate decreases, and inventory levels also drop, resulting in a reduction of $E_p$. These trends are confirmed by the data in Table 6.

Table 6. Effect of probability $\gamma$ on various performance measures

| $\gamma$ | $E_{inv}$ | $E_{orbit}$ | $E_{ro}$ | $E_p$ | $E_{ds}$ | $E_{la}$ | $E_{lr}$ | $\theta_1^*$ | $\theta_2^*$ | $A_{br}$ | $Arr$ |
|---|---|---|---|---|---|---|---|---|---|---|---|
| 0.1 | 5.7628 | 0.1903 | 0.3722 | 8.7201 | 1.2890 | 1.9358 | 0.0839 | 0.3807 | 0.2594 | 0.1611 | 0.0003 |
| 0.2 | 5.7186 | 0.3934 | 0.3716 | 7.3964 | 1.3276 | 1.7657 | 0.1759 | 0.7868 | 0.5262 | 0.1659 | 0.0003 |
| 0.3 | 5.6753 | 0.6091 | 0.3711 | 6.2440 | 1.3654 | 1.5837 | 0.2760 | 1.2182 | 0.7998 | 0.1706 | 0.0003 |
| 0.4 | 5.6331 | 0.8373 | 0.3706 | 5.2505 | 1.4024 | 1.3899 | 0.3842 | 1.6746 | 1.0796 | 0.1753 | 0.0003 |
| 0.5 | 5.5921 | 1.0776 | 0.3701 | 4.4013 | 1.4382 | 1.1845 | 0.5005 | 2.1553 | 1.3652 | 0.1797 | 0.0003 |
| 0.6 | 5.5524 | 1.3297 | 0.3696 | 3.6814 | 1.4729 | 0.9679 | 0.6248 | 2.6594 | 1.6557 | 0.1841 | 0.0004 |
| 0.7 | 5.5143 | 1.5929 | 0.3691 | 3.0755 | 1.5062 | 0.7406 | 0.7569 | 3.1859 | 1.9505 | 0.1882 | 0.0004 |
| 0.8 | 5.4779 | 1.8667 | 0.3687 | 2.5688 | 1.5381 | 0.5031 | 0.8963 | 3.7335 | 2.2489 | 0.1922 | 0.0004 |

$S = 20; s = 5; \lambda = 3.5; \mu = 4; \omega = 0.8; \beta = 0.9; \theta = 2; \delta = 0.7; \delta_1 = 0.5; \delta_2 = 1.6$

### 7.7 Impact of Variation of the Probability $\delta$ on Various Performance Measures

The fall in the average loss during retrial is the direct impact of the rise in $\delta$ values. Also, as $\delta$ rises, customers who fail to access the server return to the orbit more quickly which leads to the higher orbit population. However, with more customers in orbit, the server experiences higher demand, leading to an increase in the expected loss upon arrival, $E_{la}$. The observations are given in Table 7,

Table 7. Effect of probability $\delta$ on various performance measures

| $\delta$ | $E_{inv}$ | $E_{orbit}$ | $E_{ro}$ | $E_p$ | $E_{ds}$ | $E_{la}$ | $E_{lr}$ | $\theta_1^*$ | $\theta_2^*$ | $A_{br}$ | $Arr$ |
|---|---|---|---|---|---|---|---|---|---|---|---|
| 0.1 | 5.7301 | 0.3248 | 0.3717 | 7.6854 | 1.3166 | 1.5342 | 0.4292 | 0.6496 | 0.4273 | 0.1645 | 0.00011 |
| 0.2 | 5.7245 | 0.3515 | 0.3716 | 7.5252 | 1.3215 | 1.5392 | 0.4142 | 0.7031 | 0.4629 | 0.1651 | 0.00013 |
| 0.3 | 5.7180 | 0.3833 | 0.3716 | 7.3422 | 1.3273 | 1.5450 | 0.3965 | 0.7667 | 0.5051 | 0.1659 | 0.00015 |
| 0.4 | 5.7103 | 0.4218 | 0.3715 | 7.1310 | 1.3341 | 1.5520 | 0.3754 | 0.8436 | 0.5560 | 0.1667 | 0.00018 |
| 0.5 | 5.7011 | 0.4692 | 0.3714 | 6.8845 | 1.3424 | 1.5603 | 0.3497 | 0.9385 | 0.6185 | 0.1678 | 0.00021 |
| 0.6 | 5.6897 | 0.5295 | 0.3712 | 6.5931 | 1.3525 | 1.5706 | 0.3176 | 1.0592 | 0.6972 | 0.1690 | 0.00027 |
| 0.7 | 5.6754 | 0.6091 | 0.3711 | 6.2440 | 1.3655 | 1.5837 | 0.2760 | 1.2183 | 0.7998 | 0.1706 | 0.00037 |
| 0.8 | 5.6569 | 0.7197 | 0.3709 | 5.8207 | 1.3825 | 1.6009 | 0.2195 | 1.4395 | 0.9393 | 0.1728 | 0.00056 |

$S = 20; s = 5; \lambda = 3.5; \mu = 4; \omega = 0.8; \beta = 0.9; \theta = 2; \gamma = 0.8; \delta_1 = 0.5; \delta_2 = 1.6$

## 7.8 Impact of Variation of the Breakdown Rate $\delta_1$ on Various Performance Measures

Along with the increase in the breakdown rate, $\delta_1$, more unsatisfied customers tend to leave without entering the orbit. This, in turn leads to an increased loss of customers upon arrival, i.e. $E_{la}$ decreases. As $E_{orbit}$ increases retrials become unsuccessful that force to an increase in $E_{lr}$. As $E_{orbit}$ increases, we expect increase in $E_{ds}, \theta_1^*$ and $\theta_2^*$. Table 8 supports these intuitions. As expected the reorder rate, $E_{ro}$ decreases, and hence the inventory level and thus the expected perishability, $E_p$.

Table 8. Effect of probability $\delta_1$ on various performance measures

| $\delta_1$ | $E_{inv}$ | $E_{orbit}$ | $E_{ro}$ | $E_p$ | $E_{ds}$ | $E_{la}$ | $E_{lr}$ | $\theta_1^*$ | $\theta_2^*$ | $A_{br}$ | $Arr$ |
|---|---|---|---|---|---|---|---|---|---|---|---|
| 0.4 | 5.7819 | 0.6150 | 0.3847 | 6.3898 | 1.4240 | 1.5688 | 0.2768 | 1.2300 | 0.7882 | 0.1424 | 0.00030 |
| 0.5 | 5.6754 | 0.6091 | 0.3711 | 6.2440 | 1.3655 | 1.5837 | 0.2760 | 1.2183 | 0.7998 | 0.1706 | 0.00037 |
| 0.6 | 5.5756 | 0.6043 | 0.3582 | 6.1047 | 1.3113 | 1.5978 | 0.2755 | 1.2087 | 0.8112 | 0.1966 | 0.00042 |
| 0.7 | 5.4819 | 0.6004 | 0.3460 | 5.9715 | 1.2610 | 1.6111 | 0.2754 | 1.2009 | 0.8225 | 0.2206 | 0.00047 |
| 0.8 | 5.3937 | 0.5973 | 0.3344 | 5.8444 | 1.2143 | 1.6237 | 0.2755 | 1.1947 | 0.8336 | 0.2428 | 0.00052 |
| 0.9 | 5.3106 | 0.5948 | 0.3234 | 5.7229 | 1.1707 | 1.6356 | 0.2758 | 1.1896 | 0.8444 | 0.2634 | 0.00057 |
| 1 | 5.2320 | 0.5928 | 0.3129 | 5.6069 | 1.1299 | 1.6469 | 0.2763 | 1.1856 | 0.8550 | 0.2824 | 0.00061 |
| 1.1 | 5.1576 | 0.5912 | 0.3029 | 5.4961 | 1.0918 | 1.6577 | 0.2768 | 1.1825 | 0.8653 | 0.3002 | 0.00065 |

$S = 20; s = 5; \lambda = 3.5; \mu = 4; \omega = 0.8; \beta = 0.9; \theta = 2; \gamma = 0.8; \delta = 0.7; \delta_2 = 1.6$

## 7.9 Impact of Variation of the Repair Rate $\delta_2$ on Various Performance Measures

When the repair rate $\delta_2$ increases, it is obvious that the expected rate of repair increases. i.e. $A_{rr}$ increases. Also, the average services completed, $E_{ds}$ also increases. Table 9 supports these intuitions.

Table 9. Effect of probability $\delta_2$ on various performance measures

| $\delta_2$ | $E_{inv}$ | $E_{orbit}$ | $E_{ro}$ | $E_p$ | $E_{ds}$ | $E_{la}$ | $E_{lr}$ | $\theta_1^*$ | $\theta_2^*$ | $A_{br}$ | $Arr$ |
|---|---|---|---|---|---|---|---|---|---|---|---|
| 1.4 | 5.6752 | 0.6091 | 0.3711 | 6.2439 | 1.3654 | 1.5837 | 0.2760 | 1.2183 | 0.7998 | 0.1706 | 0.00032 |
| 1.5 | 5.6753 | 0.6091 | 0.3711 | 6.2440 | 1.3654 | 1.5837 | 0.2760 | 1.2183 | 0.7998 | 0.1706 | 0.00034 |
| 1.6 | 5.6754 | 0.6091 | 0.3711 | 6.2440 | 1.3655 | 1.5837 | 0.2760 | 1.2183 | 0.7998 | 0.1706 | 0.00037 |
| 1.7 | 5.6754 | 0.6091 | 0.3711 | 6.2441 | 1.3655 | 1.5837 | 0.2760 | 1.2183 | 0.7998 | 0.1706 | 0.00039 |
| 1.8 | 5.6755 | 0.6091 | 0.3711 | 6.2442 | 1.3655 | 1.5837 | 0.2760 | 1.2183 | 0.7997 | 0.1706 | 0.00041 |
| 1.9 | 5.6756 | 0.6091 | 0.3711 | 6.2443 | 1.3656 | 1.5837 | 0.2760 | 1.2183 | 0.7997 | 0.1707 | 0.00044 |
| 2 | 5.6757 | 0.6091 | 0.3711 | 6.2444 | 1.3656 | 1.5837 | 0.2760 | 1.2183 | 0.7997 | 0.1707 | 0.00046 |

$S = 20; s = 5; \lambda = 3.5; \mu = 4; \omega = 0.8; \beta = 0.9; \theta = 2; \gamma = 0.8; \delta = 0.7; \delta_1 = 0.5$

## 7.10 Graphical Illustrations

**Effect of Variation of Different Parameters on Profit.** Fix all the parameters except the varying one as $\lambda = 3.5$, $\mu = 4$, $\gamma = 0.8$, $\omega = 0.8$, $\theta = 2$, $\beta = 0.9$, $\delta = 0.7$. $\delta_1 = 0.5$, $\delta_2 = 1.6$ (See Figs. 1, 2 and 3). The variation caused by $\delta_1$ and $\delta_2$ are obvious decreasing and increasing graphs respectively.

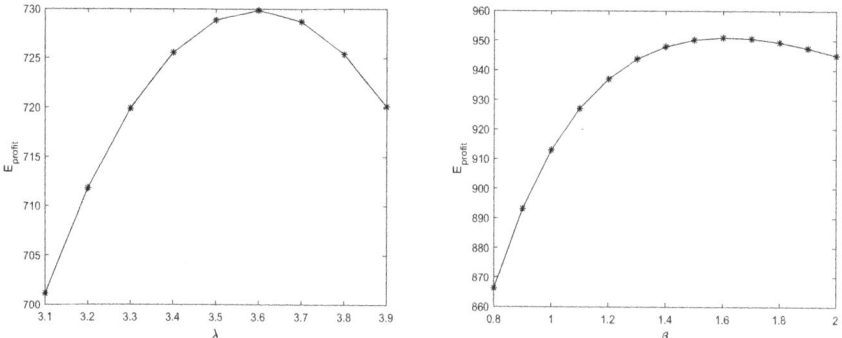

**Fig. 1.** Effect on profit due to the variation of $\lambda$ and $\beta$

For $c_0 = 1400$, $c_1 = 5$, $c_2 = 520$, $c_3 = 100$, $c_4 = 140.c_5 = 20$, $c_6 = 10$ and $c_7 = 40$, we plot values of total profit by varying $\lambda$ and get the maximum $E_{profit} = 729.85$ at $\lambda = 3.6$. Also, for $c_0 = 900$, $c_1 = 5$, $c_2 = 20$, $c_3 = 100$, $c_4 = 140.c_5 = 20$, $c_6 = 10$ and $c_7 = 40$, we plot values of total profit by varying $\beta$ and obtain the maximum $E_{profit} = 950.99$ at $\beta = 1.6$ (see Fig. 1).

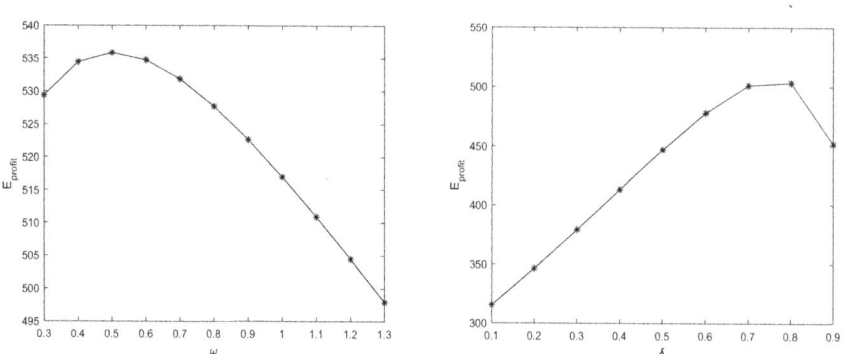

**Fig. 2.** Effect on profit due to the variation of $\omega$ and $\delta$

For $c_0 = 800$, $c_1 = 100$, $c_2 = 10$, $c_3 = 10$, $c_4 = 40.c_5 = 20$, $c_6 = 10$ and $c_7 = 40$, we plot values of total profit by varying $\omega$ and get the maximum $E_{profit} = 535.79$ at $\omega = 0.5$. Also, for $c_0 = 900$, $c_1 = 5$, $c_2 = 230$, $c_3 = 100$,

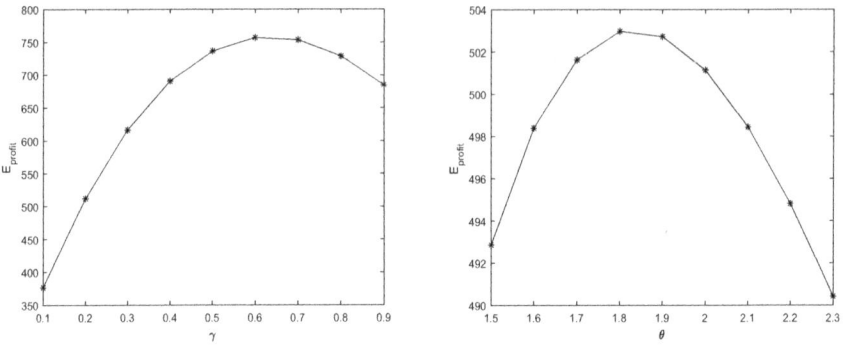

**Fig. 3.** Effect on profit due to the variation of $\gamma$ and $\theta$

$c_4 = 140. c_5 = 20$, $c_6 = 10$ and $c_7 = 40$, we plot values of total profit by varying $\beta$ and obtain the maximum $E_{profit} = 950.99$ at $\beta = 1.6$ (see Fig. 2).

For $c_0 = 1400$, $c_1 = 5$, $c_2 = 520$, $c_3 = 100$, $c_4 = 140. c_5 = 20$, $c_6 = 10$ and $c_7 = 40$, we plot values of total profit by varying $\gamma$ and get the maximum $E_{profit} = 756.81$ at $\gamma = 0.6$. Also, for $c_0 = 900$, $c_1 = 5$, $c_2 = 230$, $c_3 = 100$, $c_4 = 140$. $c_5 = 20$, $c_6 = 10$, $c_7 = 40$ we plot values of total profit by varying the probability $\delta$. For given parameter values, the maximum $E_{profit} = 503.35$ is obtained at $\delta = 0.8$ (see Fig. 3).

**Optimal $(s, Q)$ Pair.** This section explores the behaviour of the total profit by varying reorder level $s$ and maximum inventory level $S$, keeping other parameters fixed. Using the above defined total profit, the total profit is tabulated for some set of $(s, S)$ pair. From Table 10, the optimal $(s, S)$ pair is $(3, 16)$ and hence the optimal $(s, Q)$ pair is $(3,13)$ and the corresponding optimal profit is 1707.5. In Table 10, underlined value denotes the column maximum and bold face value denotes the row maximum (Fig. 4).

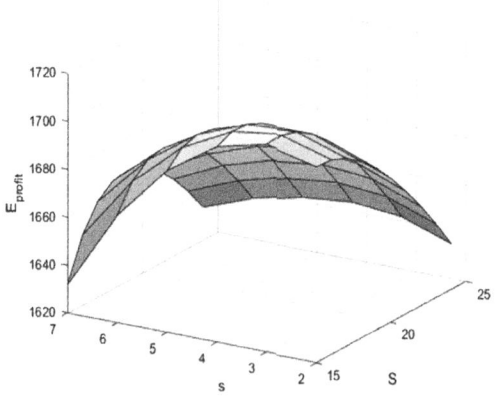

**Fig. 4.** (s,S) Pair Vs Profit

Table 10. Effect of $s$ and $S$ on total profit

| $s\backslash S$ | 15 | 16 | 17 | 18 | 19 | 20 | 21 | 22 | 23 | 24 |
|---|---|---|---|---|---|---|---|---|---|---|
| 2 | **1701.8** | 1701.3 | 1698.5 | 1693.8 | 1687.6 | 1679.9 | 1671.1 | 1661.3 | 1650.6 | 1639.1 |
| 3 | 1706.7 | **1707.5** | 1705.5 | 1701.3 | 1695.3 | 1687.8 | 1678.9 | 1669.0 | 1658.1 | 1646.3 |
| 4 | 1701.6 | **1704.5** | 1704.1 | 1701.1 | 1695.9 | 1688.9 | 1680.5 | 1670.8 | 1660.0 | 1648.3 |
| 5 | 1687.6 | 1693.8 | **1695.8** | 1694.6 | 1690.8 | 1684.9 | 1677.3 | 1668.1 | 1657.7 | 1646.3 |
| 6 | 1664.7 | 1675.5 | 1681.1 | **1682.6** | 1680.8 | 1676.4 | 1670.0 | 1661.8 | 1652.1 | 1641.3 |
| 7 | 1631.7 | 1649.1 | 1659.6 | 1664.8 | **1665.9** | 1663.7 | 1659.0 | 1652.1 | 1643.5 | 1633.5 |

$\lambda = 3.5; \mu = 4; \theta = 2; \beta = 0.9; \omega = 0.8; \gamma = 0.8; \delta = 0.7; \delta_1 = 0.5; \delta_2 = 1.6; c_0 = 900; c_1 = 5; c_2 = 10; c_3 = 100; c_4 = 140; c_5 = 20; c_6 = 10; c_7 = 40$

## 8 Concluding Remarks

This paper examined a perishable inventory system with an infinite orbit to accommodate retrial customers. Both inter-arrival and service times followed an exponential distribution, with inventory replenishment governed by an (s, Q) policy. Customers were directed to join the orbit if the inventory level reached zero, or if the server was busy or experiencing a breakdown. The matrix analytic method was employed to determine the stationary probability vector, enabling the calculation of key performance measures. A suitable profit function was developed, and the optimal (s, Q) pair was identified. Numerical illustrations demonstrated the impact of parameter variations. The future extensions of this model could include the incorporation of a finite buffer, server vacations, or a production mechanism.

## References

1. Artalejo, J.R., Krishnamoorthy, A., Lopez-Herrero, M.J.: Numerical analysis of (s, S) inventory systems with repeated attempts. Ann. Oper. Res. **141**(1), 67–83 (2006)
2. Berman, O., Kaplan, E.H., Shevishak, D.G.: Deterministic approximations for inventory management at service facilities. IIE Trans. **25**(5), 98–104 (1993)
3. Mattam, B. J.,Jose, K. P.: Stochastic analysis of a production inventory system with deteriorating items, unreliable server and retrial of customers. In: Information Technologies and Mathematical Modelling. Queueing Theory and Applications, vol. 2163, pp. 91–105. Springer, Cham (2024). doi: https://doi.org/10.1007/978-3-031-65385-8_7
4. Goyal, S., Giri, B.C.: Recent trends in modeling of deteriorating inventory. Eur. J. Oper. Res. **134**(1), 1–16 (2001)
5. Krishnamoorthy, A., Islam, M.E.: Production inventory with random life time and retrial of customers. In Proceedings of the Second National Conference on Mathematical and Computational Models, pp. 89–110 (2003)
6. Kumar, R.S., Elango, C.: Markov decision processes in service facilities holding perishable inventory. Opsearch **49**(4), 348–365 (2012)

7. Nahmias, S.: Perishable inventory theory: a review. Oper. Res. **30**(4), 680–708 (1982)
8. Neuts, M.F., Rao, B.: Numerical investigation of a multiserver retrial model. Queue. Syst. **7**(2), 169–189 (1990)
9. Periyasamy, C.: A finite population perishable inventory system with customers search from the orbit. Int. J. Comput. Appl. Math. **12**(1) (2017)
10. Raafat, F.: Survey of literature on continuously deteriorating inventory models. J. Oper. Res. Soc. **42**(1), 27–37 (1991)
11. Ravichandran, N.: Probabilistic analysis of a continuous review perishable inventory system with markovian demand, erlangian life and non-instantaneous lead time. OR Spect. **10**(1), 23–27 (1988)
12. Reshmi, P.S., Jose, K.P.: A queueing-inventory system with perishable items and retrial of customers. Malaya J. Matematik **7**(2), 165–170 (2019)
13. Sigman, K., Simchi-Levi, D.: Light traffic heuristic for anm/g/1 queue with limited inventory. Ann. Oper. Res. **40**(1), 371–380 (1992)
14. Sivakumar, B.: A perishable inventory system with retrial demands and a finite population. J. Comput. Appl. Math. **224**(1), 29–38 (2009)

# Author Index

**A**
Anilkumar, M. P.   303
Astafiev, Sergey   193

**C**
Cao, Yu   29

**D**
Dudin, Alexander   54, 230
Dudin, Sergei   230
Dudina, Olga   54, 230
Dushatov, Nurlan T.   219

**E**
Efrosinin, Dmitry   85

**F**
Fedorova, Ekaterina   98, 171

**H**
Haustova, Natalia   316

**I**
Ibrohimova, Yorqinoy   17
Igolkin, V.   260
Imomov, Azam A.   17, 158

**J**
Jose, K. P.   70, 303, 327
Joseph, Binumon   70
Joy, Jijo   303

**K**
Kazakov, Alexander   1
Khurramov, Oybek   316
Kitaeva, Anna   29

**L**
Lempert, Anna   1
Lizyura, Olga   119

**M**
Markovich, Natalia   44
Mattam, Bobina J.   327
Melikov, Agassi   288
Meloshnikova, Natalya   171
Moiseeva, Svetlana   316
Muradov, Rustamjon S.   219

**N**
Nazarin, Artem   108
Nazarov, Anatoly   98, 119
Nazarov, Zuhriddin A.   158
Nikolaeva, Daria   98
Nurullaev, Mirkhon Muhammadovich   273

**O**
Ozkar, Serife   288

**P**
Pakulova, Ekaterina   316
Phung-Duc, Tuan   181

**R**
Rudić, Branislav   85
Rumyantsev, A.   260
Rusilko, Tatiana   146
Ryzhov, Maksim   44

**S**
Sakai, Yuta   181
Salnikov, Dmitry   146
Sopin, Eduard   108

Sturm, Valentin 85
Sztrik, János 134

**T**
Tóth, Ádám 134
Tsodikov, V. L. 244

**V**
Vu, Giang 1

**Z**
Zharkov, Maxim 1
Zorine, Andrei V. 208, 244

Made in the USA
Monee, IL
03 May 2026